Knowledge Studies in Higher Education

Volume 3

Scope of the Series

Even though knowledge is the main content of teaching and universities are key knowledge producers, scholars have only recently begun to actively explore research on knowledge studies in higher education. As this field of study has grown, it has increasingly overlapped with the research focus of other fields, namely research and science policy, and information studies. However, these three fields have developed independently with little interaction between them, causing our understanding of knowledge to be limited, compartmented, and lacking a multidimensional perspective. This book series is designed to improve knowledge studies in higher education by stimulating interactions between these different approaches.

Coverage in this series includes:

- University and knowledge production
- R & D funding systems
- Education reforms
- Innovation systems for emerging regions
- School curriculum and knowledge
- Social utility of knowledge production
- University research and in-house research
- Research collaborations.

With its comprehensive overview and multidisciplinary perspective, this series provides scholars and policymakers with the theory and data they need to make more informed decisions regarding knowledge research in higher education.

More information about this series at http://www.springer.com/series/11778

Hugo Horta • Manuel Heitor • Jamil Salmi
Editors

Trends and Challenges in Science and Higher Education

Building Capacity in Latin America

Springer

Editors
Hugo Horta
Faculty of Education
University of Hong Kong
Hong Kong, Hong Kong

Manuel Heitor
Instituto Superior Técnico
University of Lisbon
Lisbon, Portugal

Jamil Salmi
University Diego Portales
Santiago, Chile

Knowledge Studies in Higher Education
ISBN 978-3-319-20963-0 ISBN 978-3-319-20964-7 (eBook)
DOI 10.1007/978-3-319-20964-7

Library of Congress Control Number: 2015957582

Springer Cham Heidelberg New York Dordrecht London

Printed on acid-free paper

Springer International Publishing AG Switzerland is part of Springer Science+Business Media (www.springer.com)

Prologue I. Engaged Development: Paying Homage to Albert Hirschman and Oscar Niemeyer

Why Pay Homage?

Albert Hirschman (April 7, 1915 – December 10, 2012) and Oscar Niemeyer (December 15, 1907 – December 5, 2012) were men of their time and of the future.

Albert Hirschman was always uncomfortable with attempts made by others to confine him to a particular discipline or to a specific group of thinkers. He ventured into as many disciplines as necessary to unfold a phenomenon that caught his attention. One of his most well-known statements is that "the idea of trespassing" was a constant in his thinking. He permanently called us to search for and propose (new) categories that could lead to a better understanding of development processes.

Oscar Niemeyer clearly contributed to changing the aesthetics of buildings and cities worldwide. While his designs are landmarks, Niemeyer was equally an outstanding personage in the political domain to the extent that he would consider architecture less important than a political engagement to change society and, above all, to celebrate life and friends. He constantly called for us to dream, to make things happen.

The lives and works of Hirschman and Niemeyer inspired me to question what development is (or should be). To what extent has the breadth and depth of the very concept of development been explored and expanded since the end of WWII when it first entered the agendas of intellectuals and policymakers? Is not development time- and place-specific in the sense that it may have different meanings and dynamics for each and every society? Is not every society permanently challenging and pursuing changes to its status quo? If history shows that there is no end in the search of development, does not this "time- and place-specific" concept has to be elastic enough to constantly incorporate new and, to a great extent, unforeseen dimensions?

Substantive theoretical, historical, and empirical efforts are urgently required to investigate and propose frameworks that can be analytically and policy-wise useful. Different approaches must be tried to experiment with those meanings, determinants, and policy implications surrounding the concept of development, understood as changes (incremental or radical) on a specified state of advancement over time. It is necessary to go far beyond generic—and relatively indisputable—statements, for example, that development is associated with transformation to and determined by a complex interaction of different realms of life (e.g., cultural, environmental, economic, social, technological, political).

This article is but a small contribution to the debate but with very specific purposes: to propose the need for a political standpoint for the concept of development. Departing from the Schumpeterian concept of paradigm change, this article suggests the world is going through a transition period with a prevalence of uncertainties (in the sense of unpredictability) in different dimensions. Such prevalence of uncertainty imposes a series of questions concerning what is considered "development." Such questioning then opens the way for the proposition that the time has come to bring into the limelight, explicitly, a political dimension to the development processes. Thus, this article introduces not an analytical but a normative concept—"engaged development"—as a concept to be taken up by those political actors involved with the promotion of processes of change in different societies.

Where to, World? The Trends and Uncertainties Around Us

History—and Hirschman—teaches that processes of change occur as a result of a conjunction of extraordinary circumstances. It is quite undeniable that the world is going through a time of intense, difficult changes. Such intensity still requires a strong narration, as provided by John Steinbeck in *The Grapes of Wrath*, exposing the shocks between the "old" and the "new" during the great depression in the United States.

In Schumpeterian terms, the world may very well be going through a period of paradigm change, not only in the technological domain but far beyond, reaching into, for example, political, environmental, and competitive domains. As the transition unfolds, old values, assets, and competencies become obsolete but resist disappearing while the determined search for innovative alternatives continues. However, even among innovators, mortality rates are very high. Financial crises become "functional" as they accelerate the process of asset liquidation. Uncertainty prevails until an emerging paradigm (in Schumpeterian terms) and their associated norms and institutions become dominant.

Can such generic statements be translated into factual and credible narrative? Is it possible to "organize" how to observe intense changes? Again, Niemeyer and Hirschman are valuable. Niemeyer defended that it is not important to look at forests or trees but at the spaces among trees (and forests!) when observing processes of social change. Hirschman always took a stand that continuous and unexpected changes require the constant need to search for new explanatory ideas that could, indeed, contradict one's old and long standing concepts.

The concepts of diversity and heterogeneity, derived from the structuralist school of the United Nations Economic Commission for Latin America and the Caribbean,

Table 1 World trends and uncertainties

Trends ⇒ diversities?
More voices in world decisions: old protagonists remain; newcomers (trying to) joining in
Indignados ("the outraged") worldwide… Nontraditional mass protests
Crisis of long duration
Inclusion of many into markets
Pressure on resources—of all kinds
Fierce competition for the generation, appropriation, and distribution of wealth
Strong, unstoppable rhythm of technical progress
Activism by states: protection against crisis; defense of national interests
Extreme availability of and access to information
Uncertainties ⇒ heterogeneities?
Negotiated or conflictive multipolarity?
Indignados: common or different outrage? Democracy? Democracies?
Out of the crisis: When? Who?
Emerging countries or emerging middle classes? Aspirations? Aesthetic ideals?
Resources and the environment: are we doomed?
Competitive practices… which ones will prevail?
Is mass-customized production achievable?
The state: which public goods to provide? What are public goods, nowadays?
Information abound; but Eliot's knowledge and wisdom?

will be used to narrate trends and uncertainties in political, economic, social, environmental, and technological domains. The consequence of applying the concept of diversity for the description of trends is that there are strong indications that the "one size fits all" solution is doomed to disappear for each and every domain considered here. Concurrently, when and where uncertainty prevails is very difficult, if not impossible, to foresee the directions of change that open the possibility that heterogeneity may prevail. That is, the capacity to act and react upon events may be very different among different actors, leading to the maintenance of strong divisions among and within societies, even if these differences are expressed in new forms (Table 1).

The table below presents, in a stylized form, world trends and uncertainties for political, economic, social, environmental, cultural, and technological dimensions.

Regarding the geopolitical dimension, while the established post-WW II order is long gone and substituted by others, such as the G20, it is still not clear whether the upcoming multipolarity will be negotiated one or whether a conflictive ambiance will remain for a long period.

As for the political dimension, the world is seeing the eruption of mass protests in different regions, usually organized by the young, deeply dissatisfied with dictators, autocrats, or traditional forms of political representation. The rage and demands are directed at different targets in different countries, e.g., for political freedom; for more, better, and more efficient states; against the economic and financial establishments; and against the economic policies being implemented in reaction to the 2008 crisis. What is not known is, first, the extent to which these claims will be taken up by decision makers, and second, in places where political change has occurred, what type of new order will come about, and how long it will take for a new set of norms to prevail.

The economic dimension has been very much in the limelight recently. After 6 or more years of deep economic crisis, it is now possible to state that this is a crisis of long duration. What has been seen is the unfolding of a succession of economic problems of different natures, hitting different countries. Regardless of the very welcome willingness of policymakers to apply heterodox formulas to face specific challenges, uncertainty still prevails as the different engines of growth (understood as most relevant economies) have not and are not running at the same revolution and speed.

Remaining still with the economic dimension, the most relevant phenomenon the twenty-first century has seen is the process of the social and economic inclusion of dozens, even hundreds of millions of people into the mass markets of different countries, the so-called emerging markets. They are benefiting from the growth process these countries are experiencing as well as by the deliberate economic policies to increase minimum wages above inflation rates. As a result, these "emerging middle classes" are becoming consumers of goods and services that are helping to spur economic activities for local and foreign producers. However, this process of inclusion remains unconsolidated, at least in respect to the following three dimensions. First and more importantly, their consolidation as middle classes is still to be achieved; regression is still a plausible scenario. Second, the level of quality and sophistication of goods and services being consumed still has a long way to go to reach the quality of that being consumed by the established middle classes or by the middle classes in the United States, Europe, and Japan. Third, these emerging middle classes are experimenting with new ways of life that are not yet considered cultural or aesthetic norms.

Regarding the environmental dimension, for some time now, there has been an increasing amount of clear demonstrations that the pressure on resources—of all kinds—has been steadily mounting. The process of the economic inclusion of millions applies further stress to the sustainable reproduction capacity of economic systems. Is the world doomed to face the consequences of the overuse of its resources? Are decision makers and citizens willing to engage in investments and to change behaviors towards a different approach to resources?

In the competitive domain, the world has seen an increasing transnationalization of production: production systems, enterprises, and suppliers are increasingly involved in a deep and extensive web of international transactions. What is really at stake here is an environment of very fierce competition by economic agents for the generation, appropriation, and distribution of wealth. Established patterns of competition—cost or differentiation, for example—are just simplistic manifestations of modes of competition. It is relatively unknown which code of competitive practices will be dominant in the years to come.

Looking at the technological domain, regardless of the high intensity of the recent crisis, the rhythm of technical progress has not abated. Scientific advances are a constant and are being translated into new devices. However, uncertainty is very much embedded in technical progress. First, general-purpose technologies, like information and communications technologies and new materials, are being applied to any type of investment for increased efficiency and quality of services. Second, contemporary innovations potentially leading to disruptive changes are increasingly dependent on advanced scientific knowledge and on the convergence of different technologies. Innovation is nowadays more complex and interdisciplinary. That is, innovations depend on cooperation among firms, scientific institutions, and technology labs encompassing deep and wide capabilities, the so-called systems of innovation. Stand-alone research laboratories within a particular firm simply do not have the means to accumulate all the necessary competencies for a given innovation challenge. Thus, uncertainties arising from innovation are not only associated with pursuing something "inexistent" but also with the extent to which innovators—firms or research institutions—are able to bring together partners that have complementary technological capabilities for a "convergent process of building up innovations."

Contrary to the prevailing norms of the 1990s and early 2000s, the period after the economic crisis brought about, to Western countries, an activism by national states that were thought to be long buried and never to return. This was not the case for a country where high and steady growth rates have been a constant for the past 30 years: China. However, even though not explicitly supporting their economies, Western countries, in different formats, have never eluded the task of supporting their productive systems. What the crisis clearly highlighted was that in different modes, national states have always had a relevant role in economic development. In short, national states have played relevant roles in the many dimensions of development. However, what is unknown is through which mechanisms and how effective these institutions will be to deal with the related development uncertainties.

The political, economic, social, and technological trends and uncertainties stylized above could not come about if an increasing mass of information was not made available. Big data is useful to increase the efficiency of urban transport in urban centers, to decode the DNA of the human body, and to control mass protests. Information abounds but, paraphrasing T.S. Eliot, one can question whether information is leading to knowledge and, above all, to wisdom.

The Time Has Come for Engaged Development

Can development, understood as changes (incremental or radical) on a specified state of advancement over time, be foreseen in any of the above dimensions? The trends and uncertainties depicted above, if convincing, open up the need to reconsider the perspective within which to place the very concept of development. The point of departure is history. First, development is time-, place-, and dimension-specific. Second, the search for progress does not lessen, despite periods of uncertainty. In short, even with different meanings and dynamics for each and every

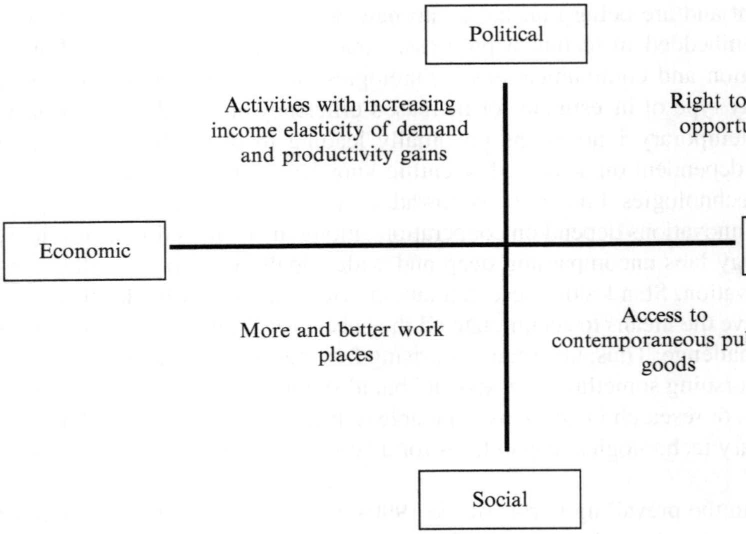

Fig. 1 Dimensions and directions of development: a normative vision

society, and in a context of uncertainty, it is very unlikely that societies will abandon the search for changes in their current status quo. As a consequence, the concept of development must be elastic enough to constantly incorporate new dimensions.

Even if theorizing development may be a difficult, almost impossible, endeavor, it is possible to postulate that development implies and demands societal choices; it is a political process by its very nature. Political choices and processes, largely by definition, may be hard to be conceptualize, but this does not impede the proposal of appropriate frameworks to the fitting of policy directives as a contribution to the political debate.

Four dimensions—political, economical, social, and ethical—are proposed in this article and organized in two axes, as shown in Fig. 1. For each quadrant a development directive is suggested. As development is time- and place-specific, this diagram must be specified to the stage of advance of each society.

Regarding the combination of political and economic dimensions, the normative directive is to pursue and foster dynamic economic activities, in two senses: the expansion of (i) markets and (ii) productivity. This directive is on a political–economic quadrant because if successful, it will generate income and taxes that will induce productive and policy sustainability.

For the economic–social quadrant, the development directive concerns more and better workplaces. Such a directive encapsulates most of the concerns of workers as it leads to their concrete well-being. Such a development directive is powerful not only for the concrete contribution to the well-being of workers but also because it includes a very relevant component closely associated to the previous quadrant. Not only does the quantitative expansion of workplaces matter, but these must qualitatively improve in time.

Regarding the social–ethical quadrant, the development directive is access to contemporaneous public goods. Societies today are urgently calling for fundamental public goods—education, transport, safety, and health—to be made readily available. However, there are other public goods that must be made widely available as demonstrated by a number of societies already: culture, leisure, and a clean environment.

The last quadrant, a combination of the ethical and political dimensions, is to a great extent relatively intangible but is possibly the most relevant in ideologically influencing the other quadrants: the right to equal opportunities. Development, in the last instance, should imply an environment in which rights of any kind are respected and enforced. It is a directive towards a transparent, efficient, and accountable rule of law, but is also more than that: it is a directive for a set of rights to directly imply the right to equal opportunities, in which all citizens have the similar chance to acquire the necessary skills and competencies for access to progress. Individuals then will be differentiated according to competencies, having had similar opportunities to access and acquire them.

The combination of these development directives implies one specific postulate: development is a political process. Thus, which development form is appropriate? I advocate engaged development, the ethical commitment to processes leading to inclusiveness, competitiveness, and sustainability. However, the question arises: is this a utopian ideal?

João Carlos Ferraz Professor (on leave) at the Instituto de Economia, Universidade federal do Rio de Janeiro. This article is based on two conferences I was invited to attend: The Sir Patrick Gillam Lecture, at the London School of Economics in January 2013, and ALTEC 2013, Porto, in October 2013. The views expressed here are my own. They do not reflect those of the Brazilian Development Bank, the institution I was affiliated to (as Executive Director) at the time of these conferences and of writing this article (October 2014).

Instituto de Economia, Universidade federal do Rio de Janeiro,
Rio de Janeiro, RJ, Brazil
jcferraz@bndes.gov.br

João Carlos Ferraz

Prologue II. *Todos por la Educación*: Advocacy for Education in Colombia. Everyone for Education

The social movement *Todos por la Educacion* (Everyone for Education) is one of the main social innovations promoting education in Colombia. The movement was launched on January 30, 2014, by a group of young community leaders. Our objective is to ensure that education is a key priority in Colombia. The movement aims to address inequality and promote peace via high-quality education for all Colombians.

Country Context

Colombia has high levels of poverty and inequality. Although poverty has decreased by 9.7 percentage points in the last 5 years, Colombia is still one of the most inequitable countries in the world. Colombia's Gini coefficient is 0.593, unaltered between 2012 and 2013 (DANE 2014). Inequality is acute between urban and rural areas, regions, and municipalities. For example, 49.9 % of the population of Quibdó (194 miles from Bogota) is considered to be living in poverty, compared with 10.2 % of the population in Bucaramanga (185 miles from Bogota) and 10.3 % in Bogota.

In Colombia, African Colombians, indigenous populations, children, and women represent the populations with the highest poverty levels, vulnerability, and exclusion. According to the 2011 National Report on Human Development, 63 % of Colombia's indigenous populations live under the national poverty line, 47.6 % in extreme poverty, and 28.6 % of those under age 15 are illiterate. Additionally, in the regions with high concentrations of African Colombians, poverty is higher still: numbers exceed 30 % of the population, and the proportion of poor people based on Unmet Basic Necessities (Necesidades Báscias Insatisfechas) (43.1 %) is almost double the national average.

Other social statistics are also alarming. In Colombia, 19.5 % of teenagers between 15 and 19 years old have children or are pregnant. Teenage pregnancy affects education because most teenagers who are pregnant or parents do not attend school, dropout, or attend a lower grade than corresponds to their age. According to the Ministry of Education, in 2011 between 20 % and 45 % of the school dropout rate was attributed to teenage maternity or paternity.

Education in Colombia

The most important education challenge in the country is improving the quality of education, while also increasing enrollment rates in preschool and higher education. In terms of quality, Colombia's poor results in the 2012 OECD PISA exams ignited a strong national debate. Compared with Latin American results, Colombia was second to last, only above Peru. In terms of enrollment, in 2013, 1,706,807 children were enrolled in pre-kindergarten programs. Regarding high schools, the country needs to increase access and reduce the number of grade repetitions and late enrollments, which affect the difference between gross and net enrollment (Table 1).

Regarding quality, and linked to the results of the PISA tests, challenges such as those identified by Vasco (2006) should be noted. Vasco, as a member of the *Misión de Sabios* (Sabios 1996) claims Colombia should "reconcile the need for high levels of education in mathematics, natural science and technology with the growing apathy of young people regarding these areas." Furthermore, he states that we need to "reconcile pluralism and postmodern amorality through the teachings of coexistence, ethics, morality, democracy and citizenship" and to "move from teaching and assessment based on achievements and specific goals to competency assessment."

In 2012, the coverage rate for higher education reached 37 %, a rather modest percentage made worse by the fact that approximately 45 % of incoming students at this level do not graduate (MEN 2013). In addition to the limited access are various quality issues: the educational gap regarding the basic skills of students when they enter university, the lack of relevance of the curricula, and restrictions of both access and quality faced by students with fewer financial resources.

Another important factor is the inequality among regions regarding both coverage and quality of education. This disparity can be clearly observed by comparing coverage figures and the SABER State exams for fifth, ninth, and eleventh grade students (eleventh grade is the senior year in Colombia). For example, in 2012, the gross coverage rate for secondary school education ranged among regions from 33.19 % in Guainía to 91.74 % in Boyacá (MEN 2013). Regarding SABER exams, Barrera-Osorio (2012) analyzed the results of these tests by region and socioeconomic status, finding that "the knowledge and skills of students are quite unevenly distributed. Students in rural areas have a worse performance than in urban areas

Table 1 Education coverage rate for Colombia 2012

Enrollment	Preschool	1st to 5th grades	6th to 9th grades	10th and 11th grades
Gross	97.10 %	110.99 %	101.89 %	75.40 %
Net	63.39 %	87.10 %	71.48 %	40.98 %

Source: Ministry of Education

and students from lower socioeconomic strata have worse performance than those of higher socioeconomic strata. All these differences are statistically significant. A similar message is obtained by calculating the average results in these tests by region: the gap between regions is also important" (CEDE 2012).

It is also essential to consider the level of skills of teachers in Colombia. In this regard, Osorio (2012) analyzed a number of variables such as educational characteristics of new teachers and the Teachers' Statute and concluded that Colombian teachers have low skill levels compared with those of other professionals. Thus, Osorio (2012) advised that a policy should be introduced whereby appropriate incentives are provided to attract better-educated graduates to teaching, the provision of more and better undergraduate programs, and the strengthening of the status of teaching in terms of regulation, evaluation, and career incentives.

A recent study shows the possible benefits of increased funding to transform Colombia's education policy. The Compartir Foundation study, *After teaching excellence: Improving the quality of education for all Colombians* (2014), proposes a systemic policy focused on teaching excellence. Research shows that it is financially feasible to achieve the suggested management changes and adjustment of funding sources: "the initial annual cost is approximately 1.8 billion pesos, which would amount to about 3.4 billion dollars per annum.... The cost would not exceed in any year 0.3 % of Gross Domestic Product (GDP), 1.7 % of the central government or 9 % of the Ministry of National Education's budget."

A greater investment in education is one of the main determinants when addressing the improvements required for Colombia's educational policy and its implementation. It is clear that we need to achieve increased and sustainable public investment where allocations to education are sustainable, given that the education investment in Colombia (4.4 % of GDP) is lower by more than one point of GDP than Mexico, Brazil, Uruguay, Venezuela, Costa Rica, and Argentina.

Birth of a Movement

Everyone for Education is a citizen-organized movement seeking to include education as a national priority. Its young founders view Colombia's poor-quality education system with great concern, as it increasingly marks inequalities in the country, particularly in rural areas and vulnerable populations. Recognizing that education is the main tool to build socially inclusive development and therefore peace, the movement has focused its efforts on proposals to structurally transform education.

Inspired by models like those of Brazil, Chile, and Mexico, Colombia's Everyone for Education is built upon permanent citizen mobilization and an awareness of the importance of education as a national priority, as a matter that pertains all: families, young people, teachers, social, political, and economic sectors, and the media. In a diverse country marked by armed conflict for over 50 years, Colombia requires an inclusive educational process that takes into account local realities and draws on the strengths and initiatives that have emerged, even amidst conflict.

Colombia has undertaken various construction processes for proposals that seek to improve the quality of education in different areas; for many years, there has been collective construction and documents containing specific proposals to implement improved quality and access to education.

In its initial phase, the Everyone for Education movement raised an agenda to transform the vision of education in the country, and this has been endorsed by over 9,000 members who joined via the website www.todosporlaeducacion.co. Using the movement's 10-point proposal, "Pact for Education," members have gained the support of various institutions, other citizens, and politicians towards the national mobilization around education. Furthermore, 26 elected members of congress and five 2014 presidential candidates have joined and expressed a commitment to install education as a national priority of the next government.

The 10 Points of the Pact for Education

Education Is a National Priority

The fate of the country depends on education. We believe that investing in education is a pathway to economic and social development for Colombia. Therefore, education should always be a priority of government at all levels, political parties, labor organizations, and civil society in general. In particular, the priority should be reflected by the state in national plans and regional development, financial projections, and resource allocation. This requires coordination and work beyond what is already established.

The Objective of Education Is to End Inequality and Build a Democratic and Peaceful Society

Education should be directed to form analytical citizens, independent, respectful of diversity, innovative, peacemakers, and responsive to local and global needs. Global needs demand innovative and creative solutions, people who are able to solve problems, work in groups, speak other languages, and understand technologies. Education must be accessible to any person born in Colombia—regardless of their ethnicity, gender, social class, religious creed, visible or invisible disability, political affiliation, or sexual orientation—to ensure that all persons have access to the same opportunities to enter the labor market and have the necessary skills to develop as engaged citizenry. This concept of education will build a democratic, egalitarian, prosperous, and competitive population, able to meet the challenges peace-building brings society. To achieve this, it is essential that education is a priority to close the existing gaps between urban and rural areas, public and private schools, and socioeconomic differences. Instead of a space of segregation, a school should be a place for social integration.

Quality in the Education Policy

Recognizing the significant advances achieved in coverage in primary and secondary education, Colombia's education policy should focus on quality improvement via the following measures:

(i) The effective implementation of the standards defined in law and education plans with particular emphasis on the development of quantitative skills and literacy.
(ii) Increasing and constant maintenance of school equipment and infrastructure, and municipal schools should be supported by national, local government, and civil society.
(iii) The implementation of extended daily attendance at school with a view to the gradual implementation of a full day at school at all levels of preschool, elementary, and secondary education (currently, many students study part-time to enable other students to study in the afternoons) via extracurricular activities in spaces designed to enhance their learning, such as libraries and sports complexes.
(iv) The continued implementation of universal and controlled tests that will be corrected and improved with a defined periodicity, and to inform students, schools, and teachers about their progress and areas for improvement. Tests should be focused on promoting students' critical and analytical skills.
(v) Respecting the ethnic and linguistic diversity of our country, bilingualism should be a priority on education and teachers and students should be able to improve their skills in a second language that is internationally competitive and provide them with access to knowledge developed in other countries. Student exchanges and salary incentives for teachers and an adequate mastery of a second language must be taken into account when designing a new public policy in education.

Excellent Teachers for Better Education

Teachers are a key element in improving the quality of education in Colombia. Thus, teaching excellence deserves an extra point in the present pact. We propose the following:

(i) Strengthen teaching degree programs and search tools to improve previous and in-service training with actions such as financing the costs of college prospects or developing special programs in teaching service.
(ii) Encourage the vocation of teaching in the best graduates and students through scholarships for their training as teachers.
(iii) Encourage the involvement of professionals who are not teachers, exploring options such as scholarships for a university degree in education.
(iv) Support those entering teaching with strategies such as mentor support with quality teachers to ensure proper development in the early years of service.

(v) Ensure the effective implementation of periodicals for the continuous improvement of teaching quality assessments.
(vi) Implement a competitive compensation for faculty that encourages a greater range of salary scales or performance bonuses to teachers and those better skilled and qualified according to assessments, to acquire a higher vocational training and those working in remote areas.
(vii) Undertake an awareness campaign for the dignity of the teaching profession and increase recognition for teaching.

Education Is the Point of Integration for Families, Communities, the Government, and the Corporate Sector

We need to increase the involvement of families in education and create awareness at the community level of the importance of promoting education and the quality of education received by children and young people in Colombia. Communities need to strengthen their participation in decision making for the development of schools, and the government needs to ensure effective options. Once done, all educational establishments must publicly post the annual results of institutional assessments so that the entire educational community is aware of the scope of the educational quality of the institution. Education and infrastructure should be seen as a cornerstone for building cities and creating community, offering open schools to families. In addition, the family will participate in the governing bodies of the schools and in the choice of governing bodies by education departments.

Improve Control, Monitoring, and Accountability

Mechanisms and tools that guarantee the proper implementation of education policy and more effectively monitor and regulate the education sector must be created. Follow-ups must be conducted on the quality of each campus and ensure that any complaints and grievances of parents and students are heard and incorporated without compromising the decentralized nature of the education sector in Colombia. Local success stories should be replicated while monitoring educational quality.

Increase Funding for Education

Education spending must respond to the fact that education is a top priority of Colombia. The distribution of resources among different categories of domestic spending should reflect this. The financing of the objectives in this pact, with a focus on the quality of education, should become viable with additional tax revenue or reallocations in the national budget.

Understanding Early Childhood as the Basis of the Educational Process

The educational component must be consolidated from a fundamental axis in the effective implementation of a public policy of comprehensive early childhood care. The early years of a child's life are crucial to their chances of future development and are the basis of the entire education system. Thus, the possibility of improving the quality of education and closing the existing achievement gap can come from the appropriate public provision of pre-kindergarten programs, sufficient food, and access to quality educational spaces for children during their early years of life. There must be an integration with preschool education and the strengthening of the technical capacities of national and regional institutions and actors (including the professionalization of community caregivers) who are responsible for the care and development of children in their early childhood.

Diversified and Inclusive Access to Technical, Technology, and Professional Education

Technical and technological education is an engine of economic progress and higher education, and a key to innovation that also enables mobility and social integration. We seek to promote the following:

(i) An increase in the resources invested in science and technology. In particular, resource royalties for innovation, which should be used in a relevant, timely, and efficient way along with assessments and monitoring of their destination.
(ii) Substantial systems of public and private scholarships to students with good academic performances who demonstrate financial need. Civil society support initiatives of universities and tax incentives for donations will be strengthened. The state will fund scholarships for the most vulnerable population.
(iii) The integration of technical and technological education and the business sector, using mechanisms such as dual training. SENA (Colombia's technical public school) must focus on increasing teacher quality and relevance. It is necessary that the state explore alternatives to increase coverage for scholarships at other institutions providing technical and technological training program accreditations.
(iv) Investing in improvements in the quality of education in public universities via a quality faculty. The number of PhDs in teaching must also increase.
(v) Increase coverage, especially in regions with reduced access and lower-income populations.
(vi) Enter a general integral approach to the internationalization of higher education.
(vii) Strengthen career exploration of students. Internships should be encouraged especially in the final grades of school, enabling better informed decisions.

We Will See that This Pact Is Fulfilled

The supervision and monitoring group that will accompany the development of the present pact will comprise a technical team as well as a Technical Advisory Committee. The committee will define indicators and targets to ensure compliance and accompany government actions so that the pact's goals are achieved. The group will continue month-to-month meeting commitments, with reports made through appropriate channels.

After the publication of the pact and the commitment of over 22,000 citizens throughout Colombia, the movement created a 30-year plan, the Grand National Education Agreement, outlining essential goals and indicators for the new period of government. For this, the movement has the support of organizations like Fundación Empresarios por la Educación, Fundación Compartir, Fundación Corona, the Chamber of Commerce of Bogotá, Ford Foundation, Federación Colombiana de Educadores, Naciones Unidas, and various universities and higher education institutions including CESA, EAFIT, CENTRAL, and UTP.

We have invited all sectors of society to engage, discuss, and contribute to the formation of the agreement, which is based on an initial seven-point agenda that will be made available to Colombia's citizens during a consultation process, which aims to bring together various stakeholders. We see the stakeholders as not only experts and decision makers but also those who are involved in the daily functioning of the education system. The methodology included virtual consultations, participatory workshops in different parts of the country, and expert panels on each of the points. The contributions were intended to feed the agenda and lead to the construction of policy guidelines that will be eventually accepted by the relevant authorities and the general public, to guide the management of education in the medium and long term.

Eight roundtable discussions included consultation with experts, focused on four themes: Early Childhood, Primary and Secondary Education, Higher Education, and Funding. Two discussions will include meetings with members of the Mission of the Wise for the validation of the 30-year vision for education in Colombia, and eight regional roundtable discussions will focus on regional themes in Caquetá, Boyacá, La Guajira, Santander, Valle del Cauca, Atlantic, Antioquia, and Bogotá. Additionally, the education agreement will include forums and debates at universities and other social settings on the importance of education and the objectives of the initiative. Furthermore, the initiative will include fun days, marathons for education, symbolic activities, social mobilization, and communication networks to increase awareness about the importance of ensuring that education is regarded as a national priority.

Juan David Aristizábal BA in Business, CESA Business School. Co-founder of Buena Nota, Ashoka Fellow, awarded Fulbright scholarship, one of the 30 Under 30 by Forbes Magazine, MTV Millennial Award, and a professor and member of the CESA board.

CESA Business School, Carleton University Juan David Aristizábal
Ottawa, ON, Canada

References

(s.f.). Obtenido de http://idbdocs.iadb.org/wsdocs/getdocument.aspx?docnum=37782672.
Barrera-Osorio (2012). *Calidad de la educación básica y media: diagnóstico y propuestas*. Bogotá: Universidad del Rosario
CEDE (2012). *Calidad de la educación y propuestas*. Bogota, Colombia.
DANE. https://www.dane.gov.co/files/investigaciones/condiciones_vida/pobreza/bol_pobreza_13.pdf. Accessed 15 Mar 2014.
Ministerio de Educación Nacional. (2013). *Programa para la Transformación de la Calidad Educativa*. Bogota: MEN.
Sabios, M.D. (1996). http://www.icesi.edu.co/investigaciones_publicaciones/images/pdf/colombia_filo_de_la_oportunidad.pdf. Accessed 20 Jan 2015.
Vasco, C. (2006). http://www.eduteka.org/pdfdir/RetosEducativos.pdf. Accessed 4 Dec 2014.

Contents

Chapter 1
Introduction. Democratizing Higher Education and Science in Latin America

Manuel Heitor and Hugo Horta

1.1 Introduction

This introductory chapter reflects the overall goal of this book to discuss the role that integrated science and higher education policies may play in further democratizing and promoting socioeconomic development in Latin America. The chapter suggests that this may be achieved via two complementary goals: (i) broadening access to knowledge via the formal learning processes of higher education, as well as fostering informal processes of science culture; and (ii) promoting the advanced qualification of people while strengthening research institutions. The rationale for our approach is related to the need for a sufficiently stable environment to train and supply talented people, including researchers, for knowledge-intensive, fast-paced, and uncertain labor markets. This gains relevance vis-à-vis the growing demand for higher education by populations seeing education as necessary for social mobility (Altbach et al. 2009). In response to the explosive social demand for higher education, competitive globalized markets, and the vast sociopolitical transformations already induced by new waves of educated youth, countries face the need and the opportunity for greater investment in science, technology, and higher education (Roberts and Hite 2007). This is a fundamental issue for Latin America. Social

M. Heitor (✉)
Center for Innovation, Technology and Policy Research, IN+, Instituto Superior Técnico, University of Lisbon, Lisbon, Portugal
e-mail: mheitor@ist.utl.pt

H. Horta
Division of Policy, Administration and Social Sciences Education, Faculty of Education, The University of Hong Kong, Hong Kong, Hong Kong

Center for Innovation, Technology and Policy Research, IN+, Instituto Superior Técnico, University of Lisbon, Lisbon, Portugal
e-mail: hugo.horta@ist.utl.pt; horta@hku.hk

H. Horta et al. (eds.), *Trends and Challenges in Science and Higher Education*, Knowledge Studies in Higher Education 3,
DOI 10.1007/978-3-319-20964-7_1

unbalance in education and the need to enlarge the human capital pool have been longstanding challenges for Latin American countries attempting to sustain socio-economic development (Carlson 2002).

For most Latin American countries, the main challenges are threefold: first, to broaden access to higher education; second, to make this access more socially balanced (see, for example, Kohli 2009); and third, to move ahead in terms of investment in knowledge production and diffusion. For example, Brazil's gross enrollment ratio in higher education was 29 % in 2012. Although this figure is above the level of gross enrollment ratio defined by Martin Trow (2007) for typical "elitist" higher education systems (i.e., approximately 15 %), the country's percentage of the population aged 25–64 years that has attained tertiary education was below 12 % in 2011 (Fig. 1.1). A similar trend in net enrollment rates is observed in other Latin American countries, including El Salvador (25 %) and Mexico (29 %), and to a lesser extent in Paraguay (35 %) and Peru (43 %). Puerto Rico (86 %), Argentina (79 %), and Chile (74 %)[1] are evolving towards "universal" higher education systems, with net enrollment rates well above 60 %. Still, the gap in qualifications in these countries is a reality. For example, the percentage of the population aged 25–64 years that has attained tertiary education in Argentina was below 14 % in 2011 (Fig. 1.1). With the notable exception of Chile (as it is characterized by relative qualification levels similar to France and Germany), Latin American countries have low-skilled active populations, sharing with Mediterranean basin countries the challenge to further qualify their population (Giordano and Pagano 2013).

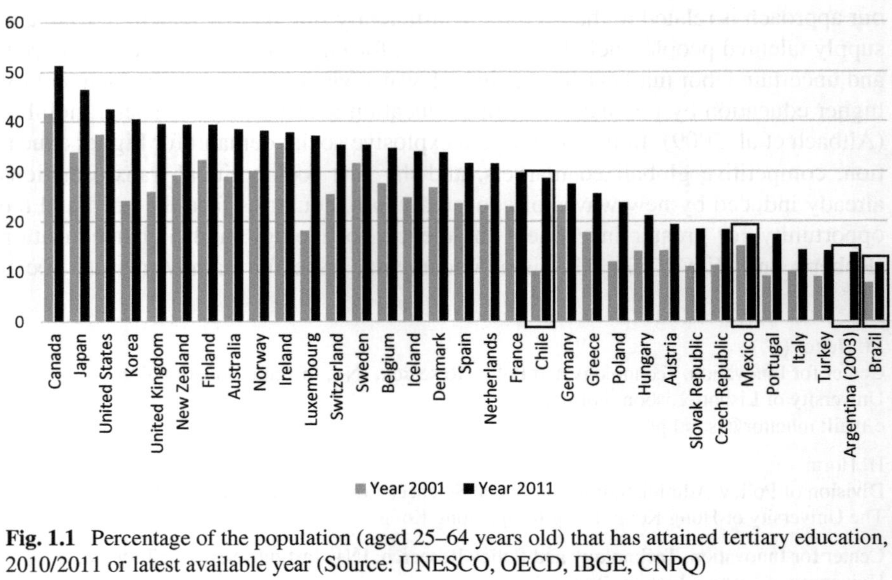

Fig. 1.1 Percentage of the population (aged 25–64 years old) that has attained tertiary education, 2010/2011 or latest available year (Source: UNESCO, OECD, IBGE, CNPQ)

[1] Data from UNESCO for 2012 or the last available year; data from Brazil refer to the National Higher Education Census of 2013.

Our analysis suggests that increasing the qualifications of the labor force must be achieved together with the challenge of making access to education more socially balanced. The current unbalance can be seen in Fig. 1.2, which shows that the majority of students enrolled in higher education in Latin American countries are still mostly from those families with the highest income. In Chile, a country with a universal higher education system, 62 % of the highest income quartile population of tertiary education age is enrolled in higher education compared with only 21 % of those from the lowest income quartile. In Brazil, approximately 47 % of the highest income quartile population of tertiary education age is enrolled in higher education, while only 5 % of the lowest income quartile population is enrolled in tertiary education. The social unbalance in higher education access is affecting Latin American countries in different stages of higher education development and maturity. This needs to be tackled to meet globally accepted equity standards towards sustainable societies, where broad access to education potentiates the formation of highly trained labor forces and their contribution to social and economic development.

A trend is, however, observable in Fig. 1.2: the relative participation of students from different income quartiles is always greater in countries with greater enrollment in higher education, underlining the need for policies that foster the enrollment in tertiary education and guarantee that education will be at the center of social mobility processes. To increase participation in higher education, there is the need for a qualified body of teachers, and not only for higher education institutions.

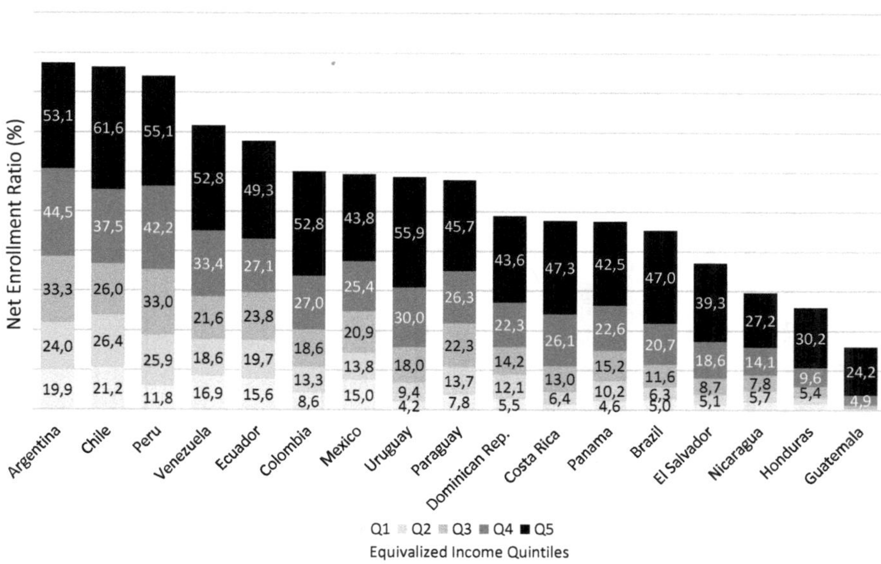

Fig. 1.2 Net enrollment rates for tertiary education by income quintiles, 2012 or latest available year (Source: SEDLAC (Socio-Economic Database for Latin America and the Caribbean) – Universidad Nacional de la Plata and the World Bank)

Students in primary and secondary education must also be prepared with solid learning bases. The analysis of Brazil's STEM education by Horta and Noronha Lisboa Filho (2014) highlights the importance of training qualified teachers for all educational levels to support learning at tertiary education and also the performance of research and development (R&D). However, investment in R&D, critical to facilitate conditions to nurture knowledge intensive environments where researchers, academics, and teachers can be trained and develop learning activities, remains relatively small (Battelle 2013). The gross expenditure on R&D in Brazil has not surpassed 1.3 % of GDP, and in Argentina it has been as low as 0.6 % (Fig. 1.3). Overall, the region lags in R&D capacity to support new knowledge creation and the adequate training of teachers for all levels of education.

Taking this context into account, the advancement of knowledge institutions together with higher education assumes critical relevance in Latin America. However, how can integrated science and higher education policies be appropriately framed to foster this process? This question has driven the research behind this chapter and the other chapters of this book. The chapters emphasize the way public policy can open new opportunities for modernizing higher education systems in Latin America in the years to come, as well as improve access to higher education, and to better qualify the labor force. In doing so, international references are used wherever appropriate to facilitate the discussion of lessons learned elsewhere, but we avoid the temptation of using them for emulation purposes. Thus, the approach of this book strengthens the message of Nowotny et al. (2003) that science is contextualized and needs to consider the social construction of knowledge-based systems (Bijker et al. 1987), together with the fostering of "inclusive learning" (Conceição and Heitor 2002).

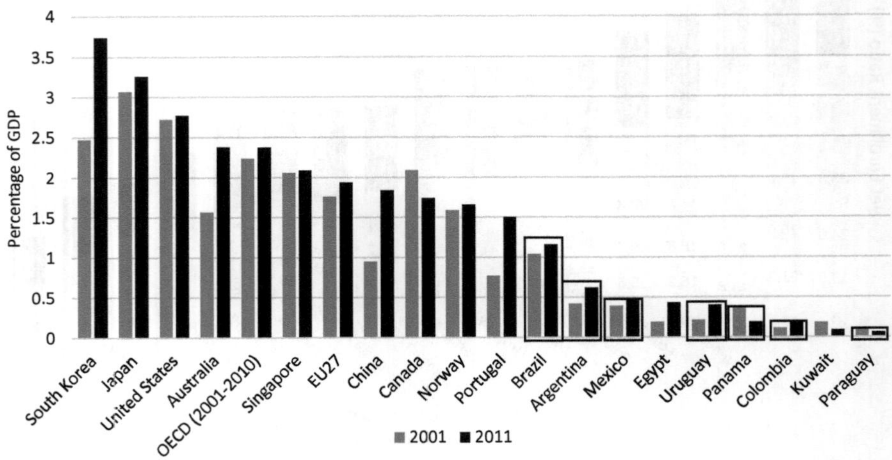

Fig. 1.3 Gross expenditure on R&D as a percentage of GDP, selected countries, 2001 and 2011

1.2 Research Framework and Analysis

In the following paragraphs, we present and discuss our main arguments about higher education systems and related public policies in Latin America. Our analysis draws from international comparative studies, fieldwork, and interviews conducted over the last 3 years, in addition to our own experience as researchers and policymakers in the field of science and higher education policy. On-site visits and many discussions with researchers and policymakers were carried out in Brazil (Rio de Janeiro, São Paulo, Brasilia, Rio Grande do Sul, Fortaleza), Colombia (Bogota, Medellin, Cartagena), Argentina (Buenos Aires), and Chile (Santiago), addressing challenges for higher education and science policies. The work involved participation in a policy research workshop organized by the World Bank, the OCDE and the Brazilian Development Bank in Rio de Janeiro (October 2011), a school of advanced studies at the Federal University of Rio de Janeiro (UFRJ, March 2013), two research workshops in Fortaleza, Ceará (September 2012 and December 2013), and an education and innovation summit in Bogota (May 2014). A major event and research workshop was organized in Porto, Portugal, in October 2013, bringing together over 800 experts in science, technology, and innovation (ST&I) policies based in Latin America.[2]

These debates, discussions, and exchange of ideas have underlined the need to consider the process of developing human capital, but also the role that higher education and scientific institutions have in facilitating it. Human capital is vital for the creation and dissemination of knowledge (Lall 1990), and striving towards greater human capital is of the utmost importance for both developed and developing countries. This ultimate goal requires, per se, policies and strategies towards effective institutional autonomy and integrity of knowledge institutions. For universities, this is particularly important in a context where alliances and partnerships among universities worldwide, and with corporations, gain significant relevance. Our analysis shows that universities need to be both *adaptable* and *resilient*, and this requires policies towards their institutional autonomy and integrity. It also highlights the need to give constant priority to people and knowledge in a way that provides networks of institutions with a critical mass capable of promoting the international standing of scientific and higher education institutions.

Two further issues should be noted. First, innovation must be considered together with competence building and advanced training in individual skills through the complex interactions between formal and informal qualifications (Helpman 2004). This requires broadening the social basis for knowledge activities, including higher education enrollment, and strengthening the upper levels of the research system leading to top-quality knowledge production. Figure 1.4 underlines the importance of having a critical mass of researchers to create and reshape knowledge. The most developed countries have high rates of researchers per 1000 labor force, and are striving to increase these rates further (see the case of the OECD countries). With

[2] ALTEC 2013, http://www.altec2013.org/

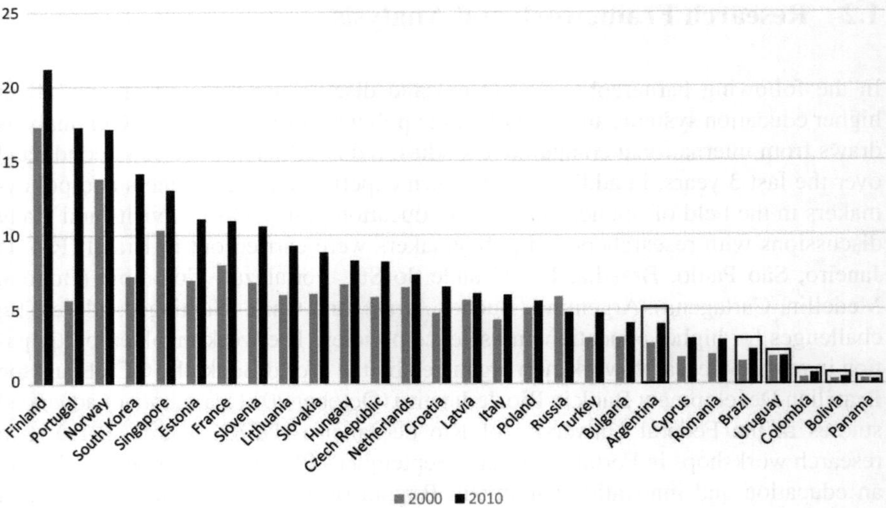

Fig. 1.4 Total number of researchers (head count) per 1000 labor force, 2010/2011 or latest available year (Source: UNESCO, OECD, IBGE, CNPQ)

the exception of Russia, these countries are endeavoring to attract and accumulate talented people. Latin American countries are still characterized by very low concentrations of researchers per 1000 labor force: the rate in Finland in 2011 was five times greater than that of Argentina, the Latin American country with the most researchers per 1000 labor force.

Second, strengthening the networks of institutions involves flows of people, independent of their socioeconomic background. It is the organized cooperation among networks of qualified workers, together with variations of users across the entire social fabric of our societies that help diffuse innovation and the design of products and services (Ernst and Kim 2002). However, establishing these innovation communities requires the systematic development of routines of collaboration on the basis of formal education programs, research projects, and a diversified and nonstructured array of informal processes of networking (see Saxenian 2006; Tung 2008). Figure 1.5 shows that Latin American countries still rely on comparatively low Education Equity Indexes (EEI) when compared internationally. Brazil's EEI in 2008 was below 12 % and the entire Latin American EEI was close to 23 %, corresponding to about half the average value for all countries considered in the statistical sample shown in Fig. 1.5.

The analysis drives us to consider, in detail, the key policy issues associated with the penetration of higher education in Latin American countries. First, the most important contextual change, shared by all countries in Latin America, is the rapid expansion of the higher education sector associated with social and economic aspirations. Second, along with an expanding demand, there is a need for greater diversity in the provision of education, to better address the diverse "populations" across

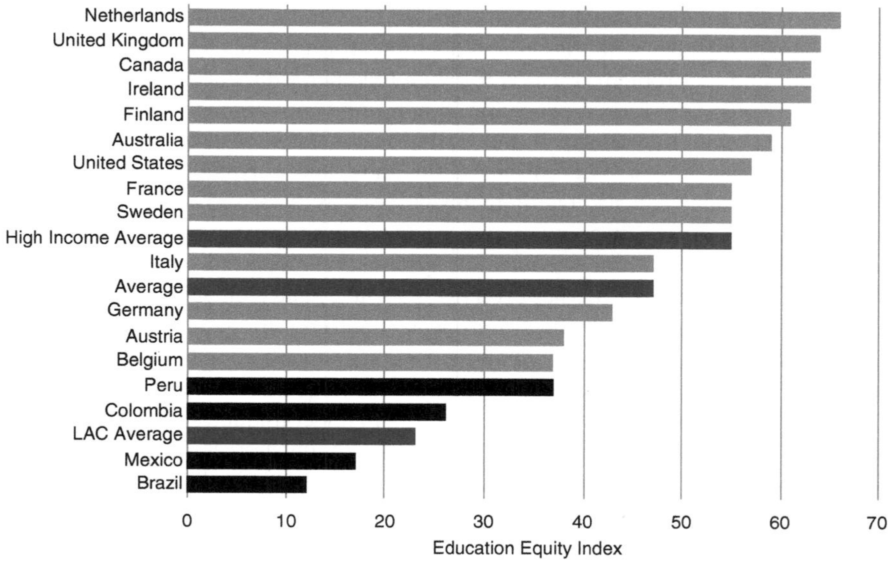

Fig. 1.5 Education Equity Index (EEI) for 2008 or latest available year (Source: SEDLAC (Socio-Economic Database for Latin America and the Caribbean) – Universidad Nacional de la Plata and the World Bank)

the relatively diversified socioeconomic societal background. Third, competition among the emerging regions of the world is becoming intense, with the realization that quality higher education and research, and its permeability in the economy, represents critical steps towards economic growth. Fourth, pressures for change are also coming from national and local governments towards more responsive institutions associated with new public management approaches, assessment drive, and governance models.

1.2.1 Research Hypothesis: Further Diversifying Higher Education in Latin America

Universities in Latin America (and elsewhere) are increasingly pressed to respond to the needs of society, including unforeseeable changes in employment markets. This includes expanding their operations and preparing their graduates with learning skills beyond merely technical competences (see Werfhorst 2014). However, universities cannot be expected to foresee the demands of the employment market in advance (not even firms are able to do so). Our hypothesis is that this can be addressed by further developing a diversified higher education system, encompassing various institutions with different vocations, together with policies and a regulatory framework towards their increasing diversification (Teichler 2008). In many parts of the world, this issue has been addressed by strengthening the role of scientific research, together with master and doctorate formal education, in universities, while focusing on problem-oriented research together with short cycles of professional and vocational formal education in non-university higher education institutions (i.e., community colleges in the United States, *fachhochschulen* in Germany and Switzerland, *hogeschools* in the Netherlands, and polytechnics in Portugal). "Distance learning" can also facilitate this process, as discussed below.

The literature on the diversity of higher education shows that fostering the structural diversification of higher education systems has occurred either by creating dual system institutions (internal diversity) or via a binary model (referring to the systemic diversity adopted by most European countries; Horta et al. 2008). The latter creates an institutional differentiation of higher education, with university institutional missions mostly associated with undertaking research and postgraduate education, while providing a general education to undergraduates. In contrast, non-university higher education institutions focus mostly on short-term learning cycles and vocational and professional training (Teixeira et al. 2012). The main challenge worldwide has been associated with the need to foster the social acceptability and technical credibility of both types of institutions.

From our previous research (Horta et al. 2008), we conclude that this requires adequate and diversified science policies, creating and promoting incentives for all institutions and, simultaneously, differentiating even the type of research performed at universities and non-university higher education institutions. Non-university higher education institutions are expected to be regionally focused and position themselves near to the labor markets, with fast responses to changes in the demand for qualified people. Thus, the learning that they offer and the research they engage in needs to be more applied, vocational and close to local market demands. This requires that they develop institutional abilities to identify the needs of local employment trends and promote regional clusters of innovation. In these activities, their students can play a pivotal role, as these institutions increasingly tend to accommodate adult students engaging in lifelong education and already possess a good knowledge of the market needs (Saar et al. 2014; Koljatic and Silva 2013).

This reinforces the need to open and strengthen diversified systems of higher education in Latin America, allowing for different learning and teaching frameworks in professionally oriented and academic-driven programs. Only a diversified higher education system can cope with the growing diversity in student characteristics, learning demands, and societal needs (Werfhorst 2014; Boliver 2011). Diversified higher education systems could also ensure sustained adaptation and flexibility, capable of providing society with the instruments it needs to deal with instability in employment and, more generally, the inevitable changes in technology trends, markets and needs. Moreover, this seems to be a way to meet the challenge of maintaining quality, together with mass education systems.

In fostering institutional diversification in higher education in Latin America, "distance learning" may play a key role. We prefer to use "distance learning" (DL) as opposed to "distance education" (DE) to emphasize the centrality of learners in the teaching and learning process. In addition, it is also important to clarify that the "distance" component of DL is a continuum that ranges from a largely face-to-face or "presential" mode of delivery to a largely distance delivery mode. The international experience with DL shows two dominant trends. First, DL has moved to virtual and interactive teaching and learning processes, which represent a shift from the dimension of "space" in distance learning to the dimension of "time". Second, the use of blended models by traditional universities, which use the new DL pedagogy in support of classical face-to-face instruction, has spread rapidly and potentially represents the wave of the future.

A leading example worldwide is that of Brazil, where DL has been successfully attained through the Open University of Brazil (Universidade Aberta do Brasil; UAB) and Centre for Distance Education for Rio de Janeiro State (Centro de Ensino a Distância do Estado do Rio de Janeiro; CEDERJ) in Rio de Janeiro. UAB has more than 100,000 students in undergraduate and graduate programs. It is a publicly funded joint program of Brazil's Ministry of Education with various municipalities, cooperating with 70 public higher education institutions. These institutions share some 550 learning centers distributed all over Brazil and various facilities, including materials production and delivery, internet facilities, training in DL for professors, tutors and technical staff, among others. The learning centers and infrastructure are partially provided by the state and municipal governments. CEDERJ is a consortium composed of six public universities, offering nine programs to more than 20,000 students, with 33 learning centers distributed in the state of Rio de Janeiro. To support these activities, a state foundation, the Centre for Science of Rio de Janeiro State, was created with a specific budget to produce the materials with the university professors and to administer the DL process, learning centers, the platform, and teaching and DL tutorial systems.

A major argument for DL as a tool to promote higher education access is its lower marginal cost per student compared with face-to-face instruction. Following this logic, the expansion of DL could be achieved with comparatively smaller investment (Larson and Murray 2008). Although setting up DL programs requires large initial investments, after several years, payoffs can be achieved via lower marginal costs per student and higher levels of student intake. DL could be, therefore, a way to not only promote access to tertiary education but also offer a distinctive form of education that can contribute to a diversified education offering, contributing to more diverse higher education systems.

1.2.2 A Main Challenge: Framing the Advanced Training of Human Resources and Teaching Staff

Goldin and Katz (2008) claim that it was by no accident the last century was both the "American century" and the "human capital century". The connection between the two concerns the role of education in economic growth and productivity. A greater level of education results in higher labor productivity, fostering a higher rate of aggregate growth. The authors claim that the nation that invested the most in education (and did much of that investment during the twentieth century when education would critically matter) was the nation that had the highest level of per capita income. Recently, Phelps (2013) argued that prosperity exploded in some nations between the 1820s and 1960s, creating not just unprecedented material wealth, but "flourishing" meaningful work, self-expression and personal growth for more people than ever before. Phelps makes the case that the wellspring of this growth was modern values such as the desire to create, explore and meet challenges. Interestingly, most of the world's innovation has not been driven by a few isolated visionaries such as Henry Ford or Mark Zuckerberg. Rather, it has been driven by millions of people empowered to think, develop, and market innumerable new products and processes and improve existing ones.

These authors, similarly to us, cannot state that economic growth is a simple matter of investing in education, but it is clear that significant and continuous investment in education over time tends to lead to higher levels of technology and productivity, facilitating economic growth and higher standards of living (Heitor 2008). These outcomes, however, do not affect everyone in the same way. Goldin and Katz (2008) also state that the benefits from economic growth might be unequally distributed and a high average standard of living might not translate into betterment for all. The implications of these remarks for Latin America are relevant, considering the overall challenges of promoting capacity building and the quality of provision in mass and diversified systems of higher education.

The higher education system in the United States, among many others, can provide some insights. According to the Carnegie Foundation for the Advancement of Teaching, a few hundred "research universities" operate within a system of about 6500 higher education institutions, including those providing vocational training. This institutional diversity shows an appropriate response to the rapid changes in the employment market, particularly by those institutions more oriented towards teaching and with shorter graduation times, without undue pressure on research universities (Locke and Wellhausen 2014). From the standpoint of research universities, the establishment of graduate schools has further contributed to institutional diversity in the United States and fostered the advanced training of human resources and teaching staff. Graduate schools have been developed worldwide over the past decade in various ways that range from interdisciplinary structures and based in a single university, to subject-specific inter-university structures. In general they provide a better link between research training and research environments and provide flexible postgraduate learning structures that traditional university departments cannot.

Graduate schools can also further develop current postgraduate activities contributing decisively to training programs for teachers and researchers, while also enhancing the research capacity and knowledge base of the relevant higher education institutions. This is of fundamental importance for Latin American countries, where the number of students enrolled in graduate education, and particularly in PhD programs, is limited when compared internationally. With the notable exception of the University of São Paulo (Universidade de São Paulo) and the University of Campinas (Universidade Estadual de Campinas) in Brazil, most leading universities across Latin America still rely on undergraduate education, with the

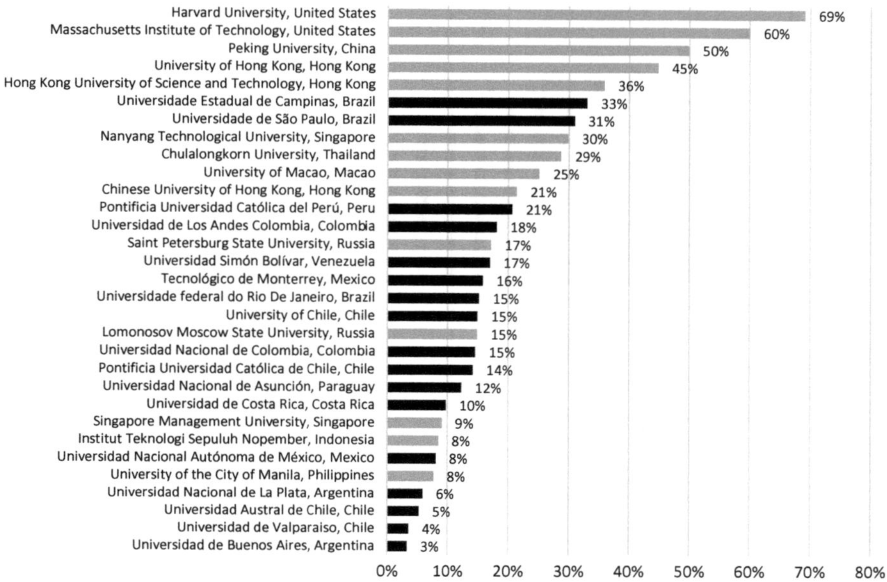

Fig. 1.6 Percentage of students enrolled in M.Sc. and Ph.D. programs in relation to the total number of enrollments in a sample of selected universities for 2011 or latest available year (Source: university websites

fraction of graduate students accounting for less than 20 % of the overall number of students (Fig. 1.6).

Doctoral education remains the most significant form to promote the advanced education of people and to build reliable universities capable of facing the challenges emerging from technical change. Doctoral education is also a fundamental step in developing national strategies for capacity building within the entire education system, with an emphasis on training teachers and researchers for modern societies. In this regard, Fig. 1.7 suggests a significant need for change in Latin American countries. Brazil, for example, has doubled its number of PhD graduates over the last decade, but still graduates only 0.6 new PhDs per 10,000 inhabitants. In contrast, the European countries shown in Fig. 1.7 graduate above 1.5 per 10,000 inhabitants, with Germany and the United Kingdom graduating more that 3.2 new PhDs per 10,000 inhabitants. Beyond its fundamental role to differentiate universities in a diversified education system, the doctoral system allows the strengthening of the top tier of the knowledge pyramid by challenging institutions to approach leading edges of common knowledge and to facilitate the production of new knowledge. It does not necessarily contribute to economic growth within a short-term framework, but instead represents a critical dimension of the fostering of "university–science" relationships, which is a long-term process (Freitas et al. 2013).

The implications of fostering doctoral education for university–science relationships are straightforward. Doctoral education is closely linked to the production of

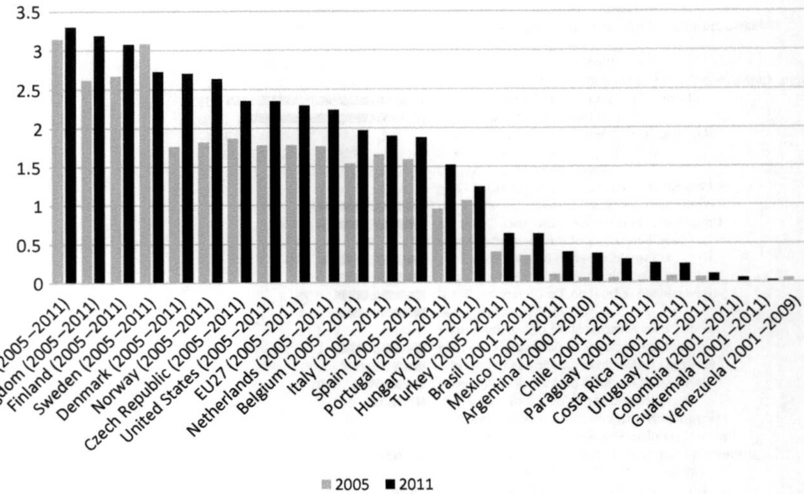

Fig. 1.7 New Ph.D.s per 10,000 inhabitants (2000/2001/2005 and 2009/2010/2011) (Source: RICYT, EuroStat, OECD)

new knowledge and requires the development of adequate learning environments that can only be provided by research-driven institutional frameworks. Academic research centers, developed together with graduate schools in universities, have shown to provide such environments across all disciplines in research-driven doctoral and post-doctoral education (Caree et al. 2014). The main issue is that they require a long-term vision to provide the necessary investment in science to facilitate the conditions to nurture knowledge intensive environments where researchers and academics can be trained (Zoller et al. 2014).

The long-term commitment to investment in R&D is necessary because new knowledge results from cumulative processes, together with learning routines able to contribute to the necessary continuous update of research and teaching practices required to keep up with fast changing global and knowledge markets (Zeigler 2012). Figure 1.8 quantifies the cumulative gross domestic expenditure on R&D for the first decade of this century (i.e., 2000–2010) for a sample of selected countries, showing very low cumulative levels for Latin American countries when compared internationally. The accumulation of investment in R&D in Brazil between 2000 and 2010 was half of that in Korea, three times smaller than that in Germany and near 3.6 times smaller than in China. These very low levels of investment in R&D do not foster adequate advanced training levels of skilled people and lag behind modern values, as suggested by Phelps (2013) and Goldin and Katz (2008), to foster the desire to create, explore, and meet emerging challenges.

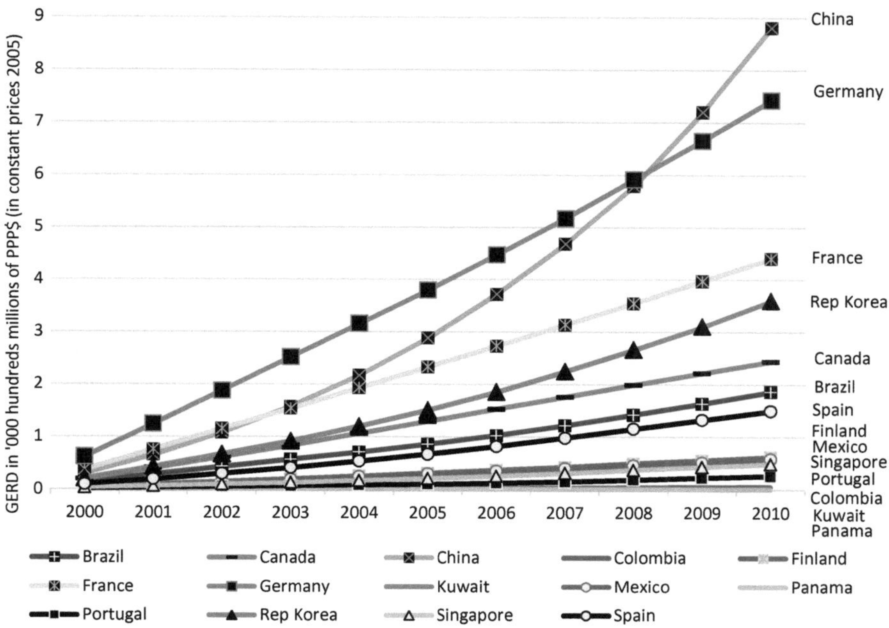

Fig. 1.8 Cumulative gross domestic expenditure on R&D between 2000 and 2010 (GERD in '000 PPP$' in constant prices 2005) in selected countries (Source: UNESCO, OECD)

1.3 Discussion and Policy Implications: Promoting Access and Qualifications in Latin America

Human capital is commonly considered a critical condition for the creation and dissemination of knowledge. Greater human capital is of the utmost importance for any country, but this ultimate goal requires policies towards more open and diverse higher education systems and institutions that consider the experiences of people and nonlinear life paths (Saar et al. 2014). This entails expanding the social basis for scientific and technological development and the appropriation of a scientific and technological culture (Majewski 2013). Our analysis suggests that the success of such policies imply the need for institutional autonomy and the integrity of modern universities in a context where innovation must be considered together with competence building and the advanced training of people via complex interactions between formal and informal qualifications (Helpman 2004). Figure 1.9 schematically illustrates our main argument: policies are required to broaden the social basis for knowledge activities and to strengthen the top tiers of the research system leading to knowledge production. This argument is discussed in the following paragraphs.

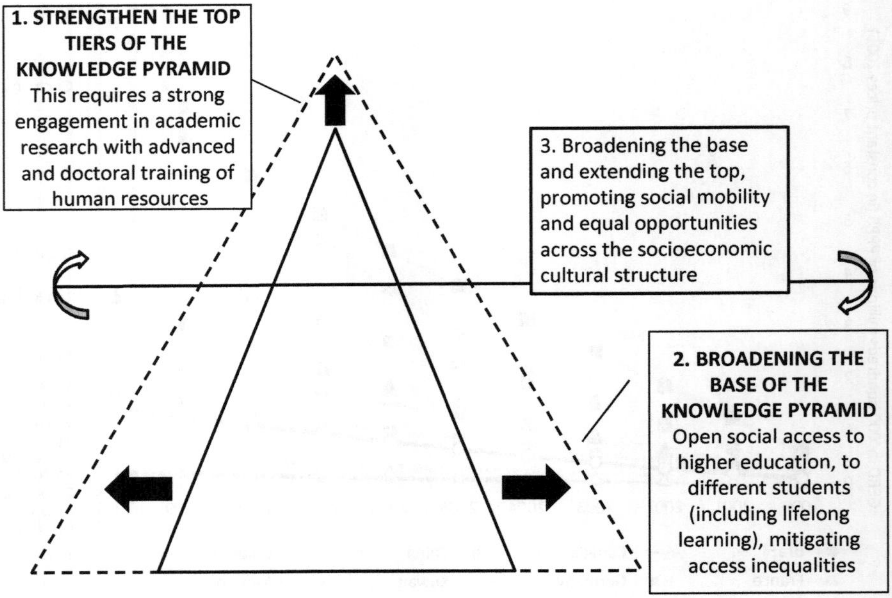

Fig. 1.9 Schematic representation of proposed "orthogonal" policies to foster access and qualifications, reducing scientific and knowledge gaps

1.3.1 Policies to Broaden the Basis for Knowledge, Promoting Access to Knowledge and Learning

The main reason for governments to increase funding for higher education is to increase participation rates and to extend the recruitment base (Barr 2004; Barr and Crawford 2005). Furthermore, new opportunities are required to give students more flexible pathways across different types and levels of educational qualifications. A key issue in Latin America is the need to open up higher education by strengthening the "bottom of the pyramid" shown in Fig. 1.9. We suggest this can be addressed on two different but related levels: (i) operationally, looking at the process of attracting and funding students; and (ii) strategically, looking at society, in general, and the process of gaining societal trust through a vigorous relationship between universities and the remaining education system.

1.3.2 Attracting and Funding Students

The underlined assumption is that "students matter", and as a result, increased diversified systems are required, as discussed in the previous section of this chapter. However, funding mechanisms to support students need to be modernized. The key issue is how to balance loans and grants for students, develop innovative loan systems and combine them with flexible legislation to accommodate reasonable student incomes through part-time work. Barr (2004) reminds us that the goal is to provide free education to all students, by guaranteeing graduates that the costs will be shared. Thus, the issue is how to ensure that the cost is shared fairly between taxpayers and graduates, as well as other private sources; as yet, the perfect solution has not been empirically proven (at least not based on scientific grounds). Although income-contingent loan systems are a typical reference worldwide, as acknowledged by the OECD, their applicability is dependent on the characteristics of the existing fiscal systems, which can be inadequate to establish such a system, as may be the case for many countries in Latin America.

Portugal faced a similar challenge. The solution was to create a system of student loans with a mutual guarantee underwritten by the state to complement the traditional system of public grants, thereby improving access to higher education for all students (Heitor et al. 2015). Michael Gallagher, in a personal comment stated that, "the Portuguese initiative satisfies the key policy criteria: it is a horizontally equitable scheme; it represents good value for students; it is financially sustainable at higher volumes of student take-up; it is low risk for government and financial institutions; it avoids the need for additional administrative infrastructure. The loan facility reduces disincentives to study by covering reasonable living costs while deferring repayment obligations until after graduation. The allowable repayment period (twice the period of study) is normally sufficient to permit students to make loan repayments without committing a disproportionate share of their income after graduation".

Still concerning the Portuguese Loan System, Nicholas Barr, also in a personal comment, recently stated that, "(1) the scheme is universal; (2) supplements existing grants rather than replacing them, hence extends students' options; (3) has no blanket interest subsidy; (4) has a very innovative mutuality element, which is the key that makes it possible for the scheme; (5) makes use of private finance". The loans scheme also has incidental benefits, by virtue of the progression requirements and the incentives for improving grade point averages. In particular, it has the potential to encourage students to complete their studies, and it may encourage students to undertake courses that are more likely to lead to positive employment outcomes.

1.3.3 Gaining Societal Trust: Promoting Relationships Between Universities and the Education System

Increased levels of public understanding of science have been clearly associated with modern societies worldwide. This issue has been discussed in the United States and Europe over the last five decades (e.g., Ziman 1978, 2000), either in terms of renewing science education or creating a science culture. Miller et al. (2002) recognizes the difficult climate in promoting a science (and knowledge) culture in society. The continued implementation of actions fostering "science for all" is a practice worth following, where the concept of "knowledge-integrated communities" appears particularly suitable to facilitate the meeting of universities and basic and secondary schools in specific projects driving society at large (Heitor and Bravo 2010). However, this requires new knowledge about social behaviors, as well as new methodological developments to help move emerging regions worldwide towards a knowledge society in a fast moving landscape. The objective is to integrate systems of knowledge and ways of practicing, where schools interact with universities in systematic ways, building routines of cooperative work.

In this context it is also important to note that it is crucial to understand that the relationship among higher education, social contexts and public policy dialogs also includes a modern rationale that considers the broad value of "research and learning". This debate requires universities to better understand "how people learn". It is clear that learning systems vary considerably across the full spectrum of disciplines, but if the ultimate goal is to enlarge participation rates in higher education, the debate will gain from current knowledge of basic and secondary education. Given the many changes in student populations, technology resources and society's demands, new pedagogical approaches that are more student-centered and more culturally sensitive are needed.

The potential revolution for discovering what the "networked world" provides is the ability to create scalable environments for learning that engage tacit and explicit dimensions of knowledge. The term used by Brown and Douglas (2010), borrowed from Polanyi, is "indwelling". Understanding this notion requires a connection among experience, embodiment and learning. First, the world of the twenty-first century is characterized by a sense of constant change, which requires a further rethink of the notions of interaction with new knowledge towards a deeper understanding of participation (*knowing*). Second, the notion of experience (and participation) within new media contexts has shifted from a traditional sense of experiencing content to using content as a context to construct a social world with others (*making*). Third, understanding the means by which social media supports a kind of play (*playing*) that allows people to navigate the complexities of a constantly shifting world. What may be most important to understand is that each of these dimensions of learning is in the process of evolving in response to new unpredictable demands. In our societies, *knowing, making* and *playing* emerge as critical components of "becoming". In relation to this, the training of a teaching body via university–science relationships can be highly beneficial.

1.3.4 Role of Science Policy in Enhancing the Qualification System

The reinforcement of the top tiers of education systems in Latin America via science policies able to attract and train a quality teaching body and foster the specialization of universities relies on the creation of conditions to strengthen institutions and to form the critical masses to facilitate high quality research. In this respect, and following some of the issues initially raised by Ziman (1968) and later Ernst (2003), one critically important institutional issue refers to the training of doctoral students and young scientists to provide them with core competencies. Such endowments can help them to become successful researchers, prepared with "transferable skills" for the job market outside research and academia. The issue can be further oriented in three different lines of discussion.

First, this requires adequate levels of public funding to train a teaching body and attract skilled people, making use of proper research environments (Heitor and Horta 2014). It must consider the concentration of funds to spur forms of international academic and scientific cooperation oriented towards the research training of young scientists and future teachers. Increasing the investment in science and technology in Latin America is critical to facilitate the conditions to nurture knowledge-intensive environments where researchers and academics can be trained and develop knowledge-based activities. Figure 1.10 complements the analysis of the previous sections and shows that the gross expenditure on R&D per capita in any Latin American country is still well below acceptable values at an international level. Furthermore, any increases over the last decade are still well below that of most industrialized countries. After correcting for parity and at constant prices of 2005, the annual per capita expenditure on R&D in Brazil is nearly eight times smaller than that in Germany and 1/10th of that in the United States. Although the rate in

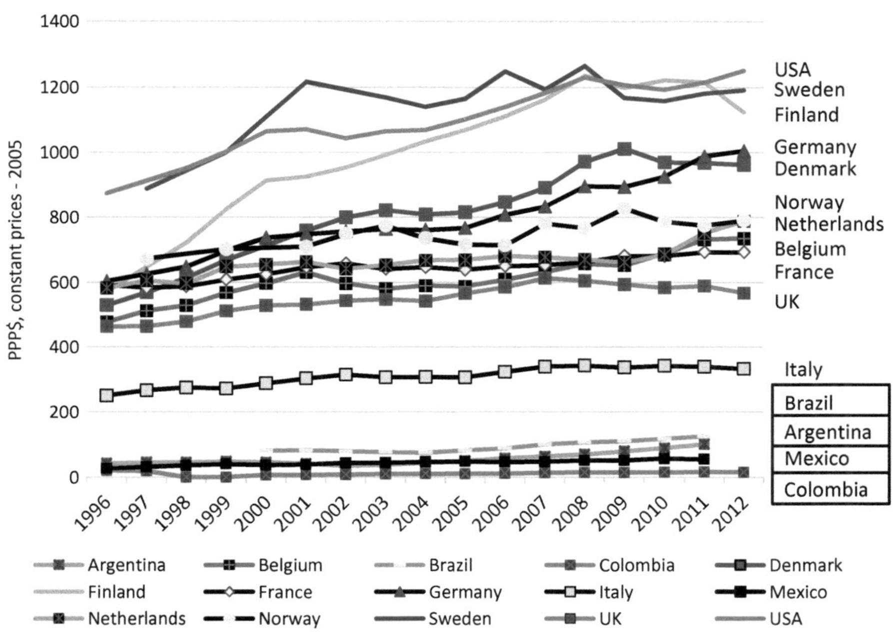

Fig. 1.10 GERD per capita (in PPP$, constant prices 2005) (Source: UNESCO Institute for Statistics)

Brazil has increased approximately 50 % between 2000 and 2011, the same increase in Germany was close to 67 % and 37 % in the United States. These figures clearly show a wide gap in the way Brazilian (and in general, Latin American) society considers the investment in R&D as a matter of priority.

Second, at an institutional level, attracting skilled people and teaching staff to modernize and strengthen research universities can be facilitated by establishing graduate schools, which have been progressively developed worldwide over the past decade, as discussed before in this chapter. Our research also shows that international academic and scientific cooperation seems to emerge as a major influencing factor for development at an unprecedented level to address these issues (Heitor 2015). It is well known that academic institutions from developed countries are now operating internationally, not only addressing potential students individually (following the traditional internationalization paradigm) but also increasingly developing new types of institutional arrangements such as international graduate schools. These could be considered a major goal for Latin American countries if developed in an "inclusive way", alongside the further development of current post-graduate activities in many existing universities.

Third, it should also be noted that the literature suggests that the progress of scientific and technological knowledge is a cumulative process, depending on the long-term process of the widespread disclosure of new findings. For example, David (2007) has systematically shown that "open science is properly regarded as uniquely well suited to the goal of maximizing the rate of growth of the stock of reliable knowledge". As a result, universities should behave as "open science" institutions and provide an alternative to the intellectual property approach when dealing with difficult problems in the allocation of resources for the production and distribution of information. Consequently, the main challenge for public policies is to keep a proper balance between open science and commercially oriented R&D based upon proprietary information.

In this discussion it should be made clear that in allocating public funding for higher education and the promotion of institutional diversification (Conceição et al. 2003), it is important to separate real resources from monetary resources and to explain their relationship with other factors relevant to the operation of universities. The traditional approach in many regions worldwide is that resources (faculty, administrative staff, students, R&D projects) and the university's facilities directly generate income. The latter, in turn, must be spent on maintaining resources. In fact, in many funding systems, resources are the only element considered in determining public core funding, the logic of determining income being limited to supporting resources.

An alternative model was developed in the framework of a "context-input-process-output-outcome" system where a university's financial flows should explicitly deal with activities and results, as well as resources and infrastructure (Conceição and Heitor 2005). This model of university operation presents the university as a "knowledge productive system", that is, resources and infrastructure (inputs) are used in activities, producing results (outputs). These results in turn generate income that finance activities, and as they unfold, entail expenditure to support resources. This second part of the cycle represents the financial flows associated with the operation of the university. The model itself is based on the conceptualization of the university's relationship with society and on the modeling of their operation, highlighting the existence of intangible flows as well as the tangible flows identified with teaching and research.

In this sense, the university ought to be acknowledged as an autonomous institution that creates knowledge and promotes creativity, supported by public funding policies. Horta et al. (2008) suggest the use of science policy instruments based on research funding-related mechanisms as a major policy instrument to diversify higher education. There are several benefits to using such competitive funding strategies. One relates to the competitive nature associated with a well-consolidated evaluation framework, especially concerning research activities.

Research evaluation practices tend to follow three important conditions, in addition to the requirement to be set apart from teaching (see Jauch 1976). The first is the use of international expertise in research evaluation processes, a lesson learned from the Portuguese experience since the mid-1990s, where it played an important driving mechanism for the evolution of the entire research and academic system (Horta 2010). The second is that research assessments should not be solely based on quantitative indicators. Although quantitative evaluations may complement peer-review practices in view of the recognized criticisms of "pure" peer-review assessments (Relman 1990), the "extreme" quantification of the academic activity may impose a culture of "bureaucratization of knowledge". The third is that the evaluation process should be articulated with a continuous and periodic opening of competitive calls for research grants and fellowships that ought to take into account the missions of higher education institutions to foster institutional differentiation (see Conceição and Heitor 2005).

The other benefit of a competitive funding strategy for research is that the allocation of such funding it is not strongly impacted by positional goods as it is with respect to education (particularly at the undergraduate level). Funding for the education component of higher education in Latin America should be mostly noncompetitive to support the development of learning infrastructures and activities. Even when competition exists in the education market, promoting diversity exclusively on that basis is relatively ineffective (Tavares and Cardoso 2013). This occurs because of a lack of relevant information for students and the powerful impact of positional goods, which model students' choices. In addition, where research evaluation is a well-consolidated assessment, it is still difficult to evaluate teaching performance (Dixit 1998). Although certain authors defend the robustness of teaching evaluation processes such as student ratings (Marsh 2007), academics and higher education administrators continue to express doubts about their meaningfulness and suitability (Gilliot 2001).

1.3.5 Challenging the Institutional Context: Strengthening and Promoting Institutional Autonomy

A special comment is warranted regarding the need to grant effective autonomy and independent legal status to universities in Latin America, to give them the capacity to self-govern and function in pursuit of work that is deemed essential to society (see Estermann et al. 2011). Scholars such as Hasan (2007) identified the relevance of strengthening the regulatory regime to facilitate more autonomous institutions in line with the requirements of public interest. At an operational level, the attribution of independent legal status to universities has become a common practice in Europe to promote public institutions governed by private law. Hasan (2007) identified four conditions for a successful implementation. First, accepting the new legal status should be voluntary. Second, because not all universities are either willing or capable of taking up the new legal status option, the process should be planned on a case-by-case basis. Third, the level of autonomy granted has to be meaningful and based on carefully decided strategic research and academic agendas. Fourth, the transition to a new legal status requires many support structures and arrangements, and a professional management structure.

New legal regimes of higher education, such as the one described above, can be a potential way for universities in Latin American countries to establish modern organizational principles, consolidate autonomy and accountability principles, establish governing boards with external participation, launch consortia, and recognize research centers as part of university management frameworks. This requires reforms addressing various aspects of higher education autonomy, ensuring that higher levels of autonomy are appropriately monitored so that they are put into the service of public interest.

1.4 Introducing the Book

This introductory chapter is followed by two prologs. João Carlos Ferraz uses the theories of Albert Hirschman and Oscar Niemeyer to emphasize the need for a political standpoint for the concept of development, thus helping Latin American countries to cope with an uncertain world in times of multifaceted transition. He advocates for engaged development as an ethical commitment to processes that underline inclusiveness, competitiveness and sustainability, asking in the end if attaining such ideal might not be a utopia. In the second prolog, Juan David Aristizábal addresses the critical issue of education for all. He advocates for education in Colombia and discusses the creation of a student movement, in terms of placing education as a political priority to lessen inequalities and sustain the creation of a peaceful and democratic society.

The remainder of the book is organized in three parts, comprising 13 chapters addressing critical challenges and opportunities for innovation and knowledge-based change in Latin America.

Part I considers emerging issues in science and higher education in Latin America and includes five chapters. Chapter 2, by Guillermo A. Lemarchand, considers the scientific productivity and dynamics of self-organizing networks within Ibero-American and Caribbean countries between 1966 and 2013 in all scientific fields. By generalizing a mathematical model developed by the author, it is shown that the evolution of self-organizing scientific networks depends on the mathematical function, which describes the number of full-time equivalent researchers against time.

Chapter 3, by Anne-Marie Maculan, considers the critical issues in the move towards a knowledge-based economy in Brazil. The author argues that the new orientation towards knowledge is characterized by several factors, including a steady increase in research investment, the expansion of higher education, greater financial support for innovative firms, the mission given to universities in technology transfer and by measures that favor the structuring of a knowledge market.

Chapter 4, by Marta Losada, discusses current challenges in higher education in Colombia and their implication for R&D based on key indicators benchmarked at the international level. From the significant gaps identified, the author proposes policy orientations that could sustain knowledge capacity in both science and technology and higher education systems.

Chapter 5, by Jae Park, sheds light on the establishment of scientific, cultural and educational linkages between China and Latin America. The author does so by presenting a historiography of power relations concerning Confucius institutes in Peru. His analysis shows that the power relations between Peru and China have evolved with distinctive forms and techniques, reaching the twenty-first century with language education and cultural exchange as their dominant currencies, and mediated by higher education as a state apparatus.

The final chapter of Part I, by Elizabeth Balbachevsky, discusses the current condition of the academic profession in Latin America. The author, although considering the differences between Latin American countries, describes the academic profession in Latin America as a layered profession, where old and new profiles co-exist with inherent tensions relating to their origin, place of work and identity and preference between teaching and research.

The second part of the book addresses policies and instruments to foster scientific capacity. It considers four chapters.

Chapter 7, by Leandro Lepratte, presents the results of an exploratory study of policy proposals for ST&I based on a convergent framework between social studies of science and technology and neo-Schumpeterian evolutionary economics oriented to complex systems. The ontological, epistemological and theoretical assumptions of the proposed convergent framework are discussed through a theorizing process based on various research lines on innovation, technological change and structural change. These research lines are related to a ST&I policy proposal aimed at solving the problems of development in Latin America, highlighting the important role of universities in such policies.

Chapter 8, by Salvador Estrada and Joao Aguirre, discusses research on the evaluation of public research centers (PRCs) in Mexico, arguing for public support. The authors show that not all PRCs are covered by this evaluation, and that PRCs are both mechanisms and recipients of public policy. The challenges of the evaluation stem from assumptions of direct impacts or outcomes from PRC services and related perceptions of certain causes and effects resulting from their activities on their products and services, goals and strategic purposes.

Chapter 9, by Arancibia Gutiérrez and Fabio Neto, analyzes initiatives, focusing on the programs established by Brazil and Chile, to stimulate the insertion of researchers into the firms, within their respective ST&I policies. The authors find some degree of convergence in the programs, suggesting a number of issues that could improve the effectiveness of both countries' programs.

Chapter 10, by Juan Sotuyo and Maria Marques, uses the creation and development of the Itaipu Technological Park (ITP) as an example of a multidimensional strategy oriented towards local development. The authors highlight the role of the ITP, particularly in terms of its contribution to capacity building and local critical mass formation, stating that it differs from conventional technology park development.

The third part of the book discusses forms of networking and industry–science relationships. It considers three chapters.

Chapter 11, by Eva Stal and Asa Fujino, analyses the evolution of university–industry relations in Brazil for innovation, from a survey of articles published in leading national journals or presented at related Brazilian and regional conferences between 1980 and 2012. The authors found that American and European papers assume a natural context of cooperation, while in Brazil there is still a debate on whether this collaboration should occur, and if this is a legitimate role for universities. This is related to cultural, organizational and institutional aspects of the Brazilian scientific and academic environment.

Chapter 12, by Eduardo Robles-Belmont presents a framework for the characterization and evaluation of the implementation and development of emerging technologies through an analysis of technical–economic networks and social networks in Mexico. The results find heterogeneous actors within the various collaboration network configurations in the different development poles of these emerging technologies.

The final chapter in Part III, by Fernanda Di Meglio, Luísa Mayoral and Maria Araya, focuses on cooperation networks involving universities and other societal institutions by analyzing the Faculty of Veterinary Sciences at the National University of the Center of Buenos Aires Province. The authors identify the main aspects of its academic and extra-academic networks, describing them in pursuit of a deeper reflection of their role for the building of research capabilities and institutional positioning for Argentina and Latin America.

The book concludes with a brief summarizing chapter addressing new horizons for scientific and higher education development in Latin America.

1.5 Summary

The growing demand for education and the prospects for its rapid evolution in many Latin American countries in the years to come requires the need to better integrate science and education policies to further democratize and promote socioeconomic development through two complementary goals: (i) open access to the knowledge base through higher education; and (ii) promote advanced qualification of skilled people and strengthen research institutions through adequate consideration of human resources and institutional issues in the process of technical change. We argue that this requires a better understanding of diversity in higher education and further democratic access in general, resulting in great social balance.

To cope with such a variety of demands and with a continuously changing environment, this introductory chapter argues that higher education institutions should continue to promote the necessary *institutional integrity* to facilitate students to experience environments of free knowledge production and diffusion (Conceição et al. 2003). In internal organizational terms, this requires *adaptable* and *resilient* institutions. In public policy terms, by directing governmental activity to human resources and strengthening institutional autonomy, we require political actions to concentrate on critical pillars of democracy.

In other words, by focusing public policies on the "external" dimensions of knowledge institutions through attracting people and fostering opportunities for knowledge production, governments and major public and private stakeholders ask that higher education institutions strengthen their capacity to make the critical internal changes. By doing so, these institutions will establish and modernize their systems of teaching and research on a path of *diversity* and *specialization*, without compromising quality. Furthermore, by strengthening their institutional integrity and enhancing their external links with society, higher education institutions are asked to carefully improve their relationships with economic, social and political actors, thereby creating "new" reinforced institutions that gain *societal trust*.

Our research across Latin America in the last 3 years suggests that this requires broadening the social basis for knowledge activities and strengthening the upper reaches of the research system leading to knowledge production at the highest level. Such objectives clearly require significant attention toward the advanced qualification of skilled people and teaching staff for the entire education system and its links with the economy and all society. This is a continuous process, evolving a long-term framework, and requiring a proper understanding of the effective role played by science–university relationships, beyond the most "common places". Furthermore, this process should dominate policies of science through short-term, demand-driven economic development issues. Our final observation is that the effective institutional autonomy of higher education institutions and the diversity of higher education systems must be promoted in a context where the building of human capital is a priority for Latin America.

References

Altbach, P., Reisberg, L., & Rumbley, L. E. (2009). *Trends in global higher education: Tracking an academic revolution*. Paris: UNESCO.

Barr, N. (2004). Higher education funding. *Oxford Review of Economic Policy, 20*(2), 264–283.

Barr, N., & Crawford, I. (2005). *Financing higher education: Answers from the UK*. London: Routledge.

Battelle. (2013). 2014 Global R&D Funding Forecast. http://www.battelle.org/site/home/global-2014-r-d-funding-forecast. Accessed 10 Dec 2013.

Bijker, W. E., Hughes, T. P., & Pinch, T. (1987). *The social construction of technological systems*. Cambridge: MIT Press.

Boliver, V. (2011). Expansion, differentiation, and the persistence of social class inequalities in British higher education. *Higher Education, 61*(3), 229–242.

Brown, J. S., & Douglas, T. (2010). *A new culture of learning: Cultivating the imagination for a world of constant change*. Lexington: Create Space.

Carlson, B. (2002). Education and the labor market in Latin America: Confronting globalization. *CEPAL Review, 77*, 117–134.

Carree, M., Malva, A. D., & Santarelli, E. (2014). The contribution of universities to growth: Empirical evidence from Italy. *The Journal of Technology Transfer, 39*(3), 393–414.

Conceição, P., & Heitor, M. V. (2002). Knowledge interaction towards inclusive learning – Promoting systems of innovation and competence building. *Technological Forecasting and Social Change, 69*(7), 641–651.

Conceição, P., & Heitor, M. V. (2005). *Innovation for all? Learning from the Portuguese path to technical change and the dynamics of innovation*. London: Praeger.

Conceição, P., Heitor, M. V., & Veloso, F. M. (2003). Infrastructures, incentives and institutions: Fostering distributed knowledge bases for the learning society. *Technological Forecasting and Social Change, 70*(7), 583–617.

David, P. (2007). *The historical origin of 'open science' – An essay on patronage, reputation and common agency contracting in the scientific revolution*. Stanford: Stanford Institute for Economic Policy Research.

Dixit, A. K. (1998). *The making of economic policy: A transition-cost politics perspective – Munich lectures in economics*. Cambridge: MIT Press.

Ernst, R. R. (2003). The responsibility of scientists, a European view. *Angewandte Chemie International Edition, 42*, 4434–4439.

Ernst, D., & Kim, L. (2002). Global production networks, knowledge diffusion, and local capability formation. *Research Policy, 31*(8–9), 1417–1429.

Estermann, T., Nokkala, T., & Steinel, M. (2011). *University autonomy in Europe II*. Brussels: European University Association.

Freitas, I. M. B., Marques, R. A., & Silva, E. M. P. (2013). University-industry collaboration and innovation in emergent and mature industries in new industrialized countries. *Research Policy, 42*(2), 443–453.

Gilliot, D. (2001). Incentives in academia. In M. Dewatripont, F. Thys-Clemenat, & L. Wilkin (Eds.), *The strategic analysis of universities: Microeconomic and management perspectives* (pp. 57–71). Brussels: Editions de L'Université de Bruxelles.

Giordano, A., & Pagano, A. (2013). Brazil in the transition towards knowledge economy: Between qualification and internationalization of human capital. *Transition Studies Review, 20*(1), 19–31.

Goldin, C., & Katz, L. F. (2008). *The race between education and technology*. Cambridge/London: The Belknap Press of Harvard University Press.

Hasan, A. (2007). *Independent legal status and universities as foundations*. Lisbon: Portuguese Ministry of Science, Technology and Higher Education.

Heitor, M. V. (2008). A system approach to tertiary education institutions: Towards knowledge networks and enhanced societal trust. *Science and Public Policy, 35*(8), 607–617.

Heitor, M. V., & Bravo, M. (2010). Portugal on the crosstalk of change, facing the shock of the new: People, knowledge and ideas fostering the social fabric to facilitate the concentration of knowledge integrated communities. *Technological Forecasting and Social Change, 77*, 218–247.

Heitor, M., & Horta, H. (2014). Further democratizing Latin America: Broadening access to higher education and promoting science policies focused on the advanced training of human resources. *Journal of Technology Management and Innovation, 9*(4a), 64–82.

Heitor, M., Horta, H., & Leocádio, M. (2015). Enlarging the social basis towards mass flourishing: Lessons learned from a risk-sharing loan scheme to foster access to higher education in Portugal. *Technological Forecasting and Social Change*, under review.

Heitor, M. V. (2015). How university global partnerships may facilitate a new era of international affairs and foster political and economic relations? *Technological Forecasting and Social Change, 95*, 276–293.

Helpman, E. (2004). *The mystery of economic growth*. Cambridge: Harvard University Press.

Horta, H. (2010). The role of the state in the internationalization of universities in catching-up countries: An analysis of the Portuguese higher education system. *Higher Education Policy, 23*, 63–81.

Horta, H., Huisman, J., & Heitor, M. V. (2008). Does a competitive research funding encourage diversity in higher education? *Science and Public Policy, 35*(3), 146–158.

Horta, H., & Noronha Lisboa Filho, P. (2014). A keystone to the future of Brazil. Fostering general and STEM education for an inclusive development. In B. Freeman, S. Marginson, & R. Tytler (Eds.), *The age of STEM: Educational policy and practices across the world in science, technology, engineering and mathematics* (pp. 266–277). New York: Routledge.

Jauch, L. R. (1976). Relationships of research and teaching: Implications for faculty evaluation. *Research in Higher Education, 5*(1), 1–13.

Kohli, A. (2009). Nationalist versus dependente capitalista development: Alternate pathways of Asia and Latin America in a globalized world. *Studies in Comparative International Development, 44*(4), 386–410.

Koljatic, M., & Silva, M. (2013). Opening a side-gate: Engaging the excluded in Chilean higher education through test-blind admission. *Studies in Higher Education, 38*(10), 1427–1441.

Lall, S. (1990). *Building industrial competitiveness in developing countries*. Paris: OECD.

Larson, R. C., & Murray, M. E. (2008). Distance learning as a tool for poverty reduction and economic development: A focus on china and Mexico. *Journal of Science Education and Technology, 17*(2), 175–196.

Locke, R. M., & Wellhausen, R. L. (2014). *Production in the innovation economy*. Cambridge: MIT Press.

Majewski, E. (2013). Higher education reform: Matching education to labour market needs. *European View, 12*(2), 179–188.

Marsh, H. W. (2007). Students' evaluations of university teaching: A multidimensional perspective. In P. P. Raymond & C. S. John (Eds.), *The scholarship of teaching and learning in higher education: An evidence based perspective* (pp. 319–384). New York: Springer.

Miller, S., Caro, P., Koulaidis, V., de Semir, V., Staveloz, W., & Vargas, R. (2002). Report from the expert group benchmarking the promotion of rtd culture and public understanding of science. http://www.jinnove.com/upload/documentaire/PP-fe-106.pdf.

Nowotny, H., Scott, P., & Gibbons, M. (2003). Mode 2' revisited: The new production of knowledge. *Minerva, 41*, 179–194.

Phelps, E. S. (2013). *Mass flourishing: How grassroots innovation created jobs, challenge, and change*. New York: Princeton University Press.

Relman, A. S. (1990). Peer review in scientific journals – What good is it? *Western Journal of Medicine, 153*, 520–522.

Roberts, J. T., & Hite, A. B. (2007). *The globalization and development reader: Perspectives on development and global change*. Malden: Blackwell.

Saar, E., That, K., & Roosalu, T. (2014). Institutional barriers for adults' participation in higher education in thirteen European countries. *Higher Education, 68*(5), 691–710.

Saxenian, A. (2006). *The new Argonauts: Regional advantage in a global economy*. Cambridge: Harvard University Press.

Tavares, O., & Cardoso, S. (2013). Enrolment choices in Portuguese higher education: Do students behave as rational consumers? *Higher Education, 66*(3), 297–309.

Teichler, U. (2008). Diversification? Trends and explanations of the shape and size of higher education. *Higher Education, 56*(3), 349–379.

Teixeira, P. N., Rocha, V., Biscaia, R., & Cardoso, M. F. (2012). Competition and diversity in higher education: An empirical approach to specialization patterns of Portuguese institutions. *Higher Education, 63*(3), 337–352.

Trow, M. (2007). Reflections on the transition from elite to mass to universal access: Forms and phases of higher education in modern societies since WWII. In J. J. F. Forest & P. G. Altbach (Eds.), *International handbook of higher education* (pp. 234–280). Dordrecht: Springer.

Tung, R. L. (2008). Brain circulation, diaspora, and international competitiveness. *European Management Journal, 26*, 298–304.

Van de Wefhorst, H. G. (2014). Changing societies and four tasks of schooling: Challenges for strongly differentiated educational systems. *International Review of Education, 60*(1), 123–144.

Zeigler, D. (2012). Evolution and the cumulative nature of science. *Evolution Education and Outreach, 5*(4), 585–588.

Ziman, J. (1968). *Public knowledge: The social dimension of science*. Cambridge: Cambridge University Press.

Ziman, J. M. (1978). *Reliable knowledge: An exploration of the grounds for belief in science*. Cambridge: Cambridge University Press.

Ziman, J. M. (2000). *Real science: What it is, and what it means*. Cambridge: Cambridge University Press.

Zoller, F. A., Zimmerling, E., & Boutellier, R. (2014). Assessing the impact of the funding environment on researchers' risk aversion: The use of citation statistics. *Higher Education, 68*(3), 333–345.

Part I
Emerging Issues in Science and Higher Education in Latin America

Chapter 2
The Scientific Productivity and the Dynamics of Self-organizing Networks: Ibero-American and Caribbean Countries (1966–2013)

Guillermo A. Lemarchand

2.1 Production Knowledge in Mainstream Journals (1966–2013)

To understand the behavior of scientific productivity among various Ibero-American and Caribbean countries, this study analyzes the evolution of the production of scientific articles published in mainstream journals in all fields of knowledge. The analysis covers the period from 1966 to 2013.

Within the study period, more than 99 % of the publications listed in mainstream journals were produced by just 15 countries. Taking into account the historical, economical, and cultural links of the diverse countries within this sample, at least three groups can be distinguished: (a) European countries (Spain and Portugal); (b) Latin American countries; and (c) developing small-island states in the Caribbean. The scientific productivity of the different states in this sample relates to the national levels of industrialization, higher education, and research policies developed over the last five decades.

In this article, the focus of the study is on the production of citable scientific articles in all fields of knowledge. To study the distribution of publications in each

---

This work and the information contains within is the responsibility of the author and is not necessarily the opinion of UNESCO. The designations employed and the presentation of material do not imply the expression of any opinion whatsoever on the part of UNESCO concerning the legal status of any country, territory, city, area, or its authorities, or the delimitation of its frontiers or boundaries.

G.A. Lemarchand (✉)
Consultant, Global Observatory of STI Policy Instruments (GO-->SPIN), Division of Science Policy and Capacity Building, United Nations Educational, Scientific and Cultural Organization (UNESCO), 7 Place de Fontenoy, Paris 75007, France
e-mail: galemarchand@gmail.com

H. Horta et al. (eds.), *Trends and Challenges in Science and Higher Education*, Knowledge Studies in Higher Education 3,
DOI 10.1007/978-3-319-20964-7_2

Table 2.1 World publication rankings using WoS data for a selected group of Ibero-American countries (1970–2012)

Country	1970[a]	1975	1980	1985	1990	1995	2000	2005	2010	2011	2012
Spain	29	26	19	15	13	10	10	9	10	10	10
Brazil	30	27	27	27	24	21	17	17	13	13	13
Portugal	46	46	47	45	42	39	37	34	32	27	26
Mexico	35	35	34	35	35	31	29	28	29	30	31
Argentina	27	28	29	29	32	35	33	39	39	39	40
Chile	33	36	36	36	40	44	42	43	45	46	46
Colombia	51	89	55	52	56	59	56	56	51	50	49
Venezuela BR	32	41	41	43	45	50	51	53	62	65	65
Cuba	84	68	62	59	57	55	52	58	65	63	64
Peru	53	59	63	68	59	88	81	78	72	73	75

Source: Estimations by the author based on raw data provided by WoS

[a]Based on information provided by De Solla Price (1986)

Ibero-American and Caribbean country,[1] Thomson-Reuters' Web of Science[2] (WoS) is used as the information source. Particularly, articles listed in the Science Citation Index Extended (SCI), Social Science Citation Index (SSCI), and the Arts and Humanities Citation Index (A&HCI) were examined. These databases constitute an appropriate and qualified indicator to investigate knowledge–production patterns within the region.

Table 2.1 shows the evolution of the world rankings in terms of number of publications listed in the WoS for the ten most prolific countries in the sample.

Within the analyzed period (42 years) shown in Table 2.1, Colombia moved up 40 places in the world ranking for number of publications, with Portugal moving up 20, Spain 19, Brazil 17, and Mexico and Cuba 4. In contrast, Venezuela fell 33 places, Peru 25, Argentina 17, and Chile 13. The relative position within the world ranking depends on the endogenous behavior of each individual country and the growth dynamics of other countries worldwide. For this global ranking, the same measurement method was used for countries with very different population,

[1]The countries included here for analysis within the Ibero-American and Caribbean region are: Antigua and Barbuda, Argentina, Bahamas, Barbados, Belize, Bermuda, Plurinational State of Bolivia, Brazil, Chile, Colombia, Costa Rica, Cuba, Dominica, Dominican Republic, Ecuador, El Salvador, Grenada, Guatemala, Guyana, Haiti, Honduras, Jamaica, Martinique, Mexico, Panama, Paraguay, Peru, Portugal, Spain, St. Kitts and Nevis, St. Lucia, St. Thomas, St. Vincent and the Grenadines, Suriname, Trinidad and Tobago, Turks and Caicos, Uruguay, and the Bolivarian Republic of Venezuela. These countries were studied to estimate the total number of regional publications per year (1966–2013). A number of nations were excluded from the study: Puerto Rico (with 18,442 publications over the period) because it is a territory of the United States, French Guiana (with a total number of 1,179 publications over the period), and the Caribbean colonial states (Anguilla, Aruba, British Virgin Islands, Cayman Islands, Curaçao, Guadeloupe, Martinique, Montserrat, Saint Martin, and Turks and Caicos). Most of the smaller Caribbean islands had no mainstream scientific publications during this period.

[2]http://thomsonreuters.com/thomson-reuters-web-of-science/

economic, and development levels. This is the main limitation of such scoreboards. The results can be misleading if this restriction is not taken into account; most scoreboards and world rankings do allude to this issue.

In the present chapter, it is made clear that each individual country has its own dynamics. To test this hypothesis, the different growth rates of the scientific articles published in mainstream journals will be analyzed.

Figure 2.1 shows the distribution of published articles between 1966 and 2013 with a log-linear scale for the 18 Ibero-American nations with the highest number of mainstream scientific publications. This sample includes more than 99 % of the total number of publications published within the same period. The distribution

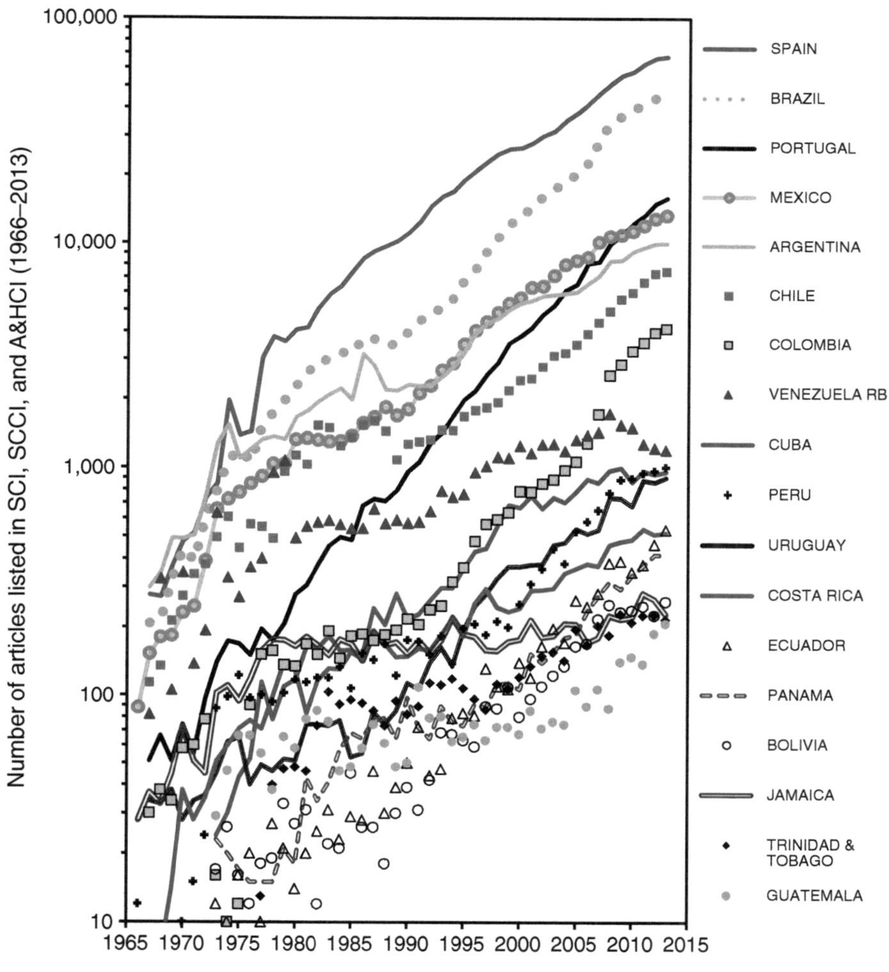

Fig. 2.1 Publication of peer-reviewed articles listed in SCI, SSCI, and A&HCI for the 18 most productive Ibero-American and Caribbean countries (1966–2013) (exponential growth); Portugal shows the highest growth rate

follows an exponential growth behavior, which can be described by $P_k(t) = \varphi e^{\gamma t}$, where $P_k(t)$ is the total number of publications of country k at time t, and φ and γ are empirically determined constants.

Table 2.2 shows the 30 most productive Ibero-American and Caribbean countries in terms of scientific publications listed in SCI, SSCI, and A&HCI between 1966 and 2013. The first column indicates the country ranking according to the total number of articles published (1966–2013). The third column shows the total number of articles published (Δ_1: 1966–2013) while the fourth shows the shares of each country in the same period. The fifth column shows the total number of articles published in the last decade (Δ_2: 2004–2013) and the sixth column the corresponding shares. The seventh column represents the size of publications over the last decade as a percentage of the total number of publications over the whole period (Δ_2 / Δ_1). The eighth column covers the number of publications during 2013 and the ninth the corresponding shares. The following six columns present the exponential growth rates (ρ, σ, λ) and their corresponding determination coefficients ($R^2_{\rho,\sigma,\lambda}$) to measure: (1) the growth in the number of citable publications over time (ρ and R_ρ^2); (2) the growth in the number of citable publications per million inhabitants over time (σ and R_σ^2); and (3) the growth in the ratio between the number of citable publications and GDP per capita in 2005 constant US dollars (λ and R_λ^2) over time. The population and GDP data used for these computations for each country and each year between 1966 and 2013 were sourced from the UN Statistics Division.

Interestingly in this calculation, the data analyzed cover a period close to five decades (48 years). For the most productive countries in the sample, the critical mass of researchers—according to their population size—was reached in the early 1980s (Lemarchand 2007, 2012). By including data between 1966 and 1980, the fitting coefficients are diminished because the system was not operating in the self-organizing regime owing to the lack of critical mass of researchers. However, by extending the analyzed period, other features can be detected. Changes in the slope of the different growth curves are usually associated with changes in economic, social, and political conditions within the different countries. Some of these changes are also associated with the application of specific policies for the promotion of scientific research and technological development (Lemarchand 2010).

Portugal has the highest ρ, σ, and λ growth constants of the sample. This is consistent with previous results (Lemarchand 2012). In these three cases, they show exponential growth with fitting determination constants $R^2_{\rho,\sigma,\lambda} > 0.99$. Within a period of 48 years, Portugal increased its publications by a factor of 311, its publications per million inhabitants by a factor of 247, and the ratio between the publications and GDP per capita in constant 2005 USD by a factor of 92. In 1966, Portugal ranked seventh for publications in the region; by 2013 it had moved to third place, overtaking Mexico, with a population 10 times the size of Portugal.

Spain has the second-highest growth rate for the number of publications and publications per capita, and the fourth-highest growth rate in terms of the ratio between publications and GDP per capita. Brazil has the third-highest growth rate in publications, the fourth for publications per capita, and the fifth for the ratio of publications/GDP per capita.

Table 2.2 Scientific productivity expressed by a set of selected indicators for three periods: 1966–2013, 2004–2013, and 2013

		Δ_1 : 1966–2013		Δ_2 : 2004–2013		Δ_2/Δ_1	2013		1966–2013		1966–2013		1966-2013	
Rank	Country	No. of articles	% Regional	No. of articles	% Regional	[%]	No. of articles	% Regional	ρ	R_ρ^2	σ	R_σ^2	λ	R_λ^2
1	Spain	914,613	41.64	523,614	40.58	57.25	67,439	39.58	0.1134	0.94	0.1082	0.93	0.0935	0.92
2	Brazil	508,482	23.15	333,489	25.85	65.59	44,949	26.38	0.1073	0.97	0.0898	0.96	0.0914	0.98
3	Mexico	187,601	8.54	106,052	8.22	56.53	13,299	7.80	0.0925	0.96	0.0730	0.94	0.0799	0.96
4	Argentina	170,252	7.75	80,644	6.25	47.37	9,909	5.82	0.0650	0.93	0.0521	0.90	0.0628	0.90
5	Portugal	150,550	6.85	106,156	8.23	70.51	15,848	9.30	0.1280	0.99	0.1240	0.99	0.1032	0.99
6	Chile	98,679	4.49	52,794	4.09	53.50	7,441	4.37	0.0710	0.91	0.0558	0.86	0.0405	0.66
7	Venezuela	37,829	1.72	13,774	1.07	36.41	1,193	0.70	0.0446	0.74	0.0209	0.40	0.0505	0.78
8	Colombia	35,071	1.60	25,330	1.96	72.22	4,118	2.42	0.1055	0.90	0.0850	0.84	0.0868	0.85
9	Cuba	17,770	0.81	8,915	0.69	50.17	955	0.56	0.0821	0.92	0.0785	0.92	0.0732	0.87
10	Peru	12 865	0.59	7,604	0.59	59.11	1,003	0.59	0.0849	0.85	0.0646	0.76	0.0798	0.83
11	Uruguay	11,756	0.54	6,844	0.53	58.22	900	0.53	0.0791	0.95	0.0738	0.94	0.0623	0.93
12	Costa Rica	9,741	0.44	4,446	0.34	45.64	517	0.30	0.0688	0.85	0.0453	0.71	0.0515	0.75
13	Jamaica	7,598	0.35	2,200	0.17	28.95	228	0.13	0.0312	0.62	0.0224	0.47	0.0885	0.96
14	Ecuador	5,276	0.24	3,454	0.27	65.47	534	0.31	0.0968	0.97	0.0742	0.94	0.0885	0.96
15	Panama	5,120	0.23	3,024	0.23	59.06	416	0.24	0.0800	0.94	0.0587	0.88	0.0630	0.88
16	Trinidad & Tobago	4,482	0.20	1,996	0.15	44.53	216	0.13	0.0664	0.63	0.0611	0.59	0.0506	0.44
17	Bolivia	3,605	0.16	2,136	0.17	59.25	259	0.15	0.0988	0.84	0.0775	0.75	0.0975	0.83
18	Guatemala	3,311	0.15	1,279	0.10	38.63	207	0.12	0.0630	0.51	0.0389	0.28	0.0098	0.57
19	Barbados	1,915	0.09	782	0.06	40.84	94	0.06	0.0681	0.72	0.0652	0.69	0.0006	0.61
20	Nicaragua	1,183	0.05	733	0.06	61.96	79	0.05	0.0898	0.88	0.0691	0.81	0.1028	0.89
21	Paraguay	1,124	0.05	670	0.05	59.61	105	0.06	0.0707	0.84	0.0543	0.67	0.0555	0.72
22	Dominican Republic	1,118	0.05	528	0.04	47.23	88	0.05	0.0665	0.77	0.0465	0.61	n/a	n/a

(continued)

Table 2.2 (continued)

		Δ_1 : 1966–2013		Δ_2 : 2004–2013		Δ_2/Δ_1	2013		1966–2013		1966–2013		1966-2013	
Rank	Country	No. of articles	% Regional	No. of articles	% Regional	[%]	No. of articles	% Regional	ρ	R_ρ^2	σ	R_σ^2	λ	R_λ^2
23	Honduras	1,066	0.05	558	0.04	52.35	82	0.05	0.0515	0.64	0.0244	0.26	0.0411	0.53
24	Bermuda	776	0.04	309	0.02	39.82	41	0.02	0.0455	0.69	0.0396	0.60	0.0281	0.46
25	El Salvador	735	0.03	425	0.03	57.82	54	0.03	0.0406	0.32	0.0268	0.15	0.0290	0.23
26	Haiti	724	0.03	436	0.03	60.22	78	0.05	0.0715	0.76	0.0506	0.60	n/a	n/a
27	Grenada	654	0.03	595	0.05	90.98	115	0.07	0.1058	0.57	0.0956	0.51	n/a	n/a
28	Guyana	611	0.03	226	0.02	36.99	19	0.01	0.0297	0.53	0.0295	0.53	0.0213	0.34
29	Bahamas	353	0.02	191	0.01	54.11	40	0.02	n/a	n/a	0.0312	0.28	n/a	n/a
30	Belize	308	0.01	184	0.01	59.74	25	0.01	0.0819	0.73	n/a	n/a	n/a	n/a

Following this first group (Brazil, Portugal, and Spain), it is interesting to observe that Grenada, Colombia, Bolivia, and Ecuador have the next highest growth rates in terms of publications and publications per capita.

In contrast, Guyana, Jamaica, and the Bolivarian Republic of Venezuela have the lowest ρ, σ, and λ constant rates. In these cases, the number of citable publications per million inhabitants had increased by less than a factor of two over the last 40 years.

In terms of the shares of publications over the whole region, three different periods are studied (1966–2013, 2004–2013, and 2013). Spain remained relatively constant, producing approximately 40 % of the publications, while the shares of Brazil, Portugal, Colombia, Ecuador, and Grenada increased. In contrast, those of Mexico, Argentina, Venezuela, Cuba, Costa Rica, Jamaica, Trinidad and Tobago, Bermuda and Guyana decreased. The shares of the rest of the countries remained constant across the three periods.

A very important feature is the fact that in several countries, the majority of the scientific production occurred in the last decade (see Δ_2/Δ_1). The most extraordinary case was that of Grenada with 91 %, followed by Colombia (72 %), Portugal (71 %), Brazil (66 %), Ecuador (65 %), Nicaragua (62 %), Haiti (60 %), Belize (60 %), and Paraguay (60 %). Bolivia, Peru, Panama, Uruguay, El Salvador, Spain, Mexico, Bahamas, Chile, Honduras, and Cuba present shares between 59 % $\geq \Delta_2/\Delta_1 \geq$ 50 %. The remainder of the countries have shares below 50 %.

These bibliometric measurements belong to the science and technology (S&T) output indicators set (Freeman 1969). The study of these measurements represents, in many ways, the performance of the national production of knowledge stemming from research and development (R&D) activities. The absence of growth for more than four decades (i.e., Venezuela and Jamaica) implies the failure of the applied S&T policies. In contrast, the research and innovation policies applied in Spain, Brazil, and Portugal present relative success, as shown by the temporal evolution of most of their main S&T indicators, such as R&D funding, number of full-time equivalent (FTE) researchers, higher education policies, generation of new PhDs, and by the increase in the number of scientific publications (Lemarchand 2010).

At this point it is important to consider that during the analyzed period, the number of journals and consequently the total number of published articles included within the WoS database substantially grew. Mabe (2003) showed that journal growth rates have been remarkably consistent over time with average rates of 0.034 since 1800 to the present day. That study presents evidence that during the whole twentieth century, this growth phenomenon appears to show a system that is self-organizing and in equilibrium with a 0.032 growth constant. Considering that the WoS database only includes a small fraction of all the new journals, the database growth rate would be even smaller than the one estimated by Mabe (2003). The values of the ρ-growth constants presented in Table 2.2 clearly show that $\rho > 0.032$ (with the exception of Jamaica and Guyana). Consequently, the productivity of scientific articles, published by the countries of our sample during the last 48 years, still experiences exponential growth ($\rho - 0.032 > 0$).

Mabe (2003) also analyzed the high coincidence between journal growth rates and world scientist growth rates over the last 50 years $\left(\gamma_{scie} \approx 0.03\right)$. According to that study, the phenomenon causing the increase in journal growth is the rise in the number of scientists. That hypothesis is consistent with the results of Table 2.2, and with the fact that recent growth rates for the number of scientists in Latin America have been well above the world average (Lemarchand 2010, p. 56). The rates for Spain and Portugal were even higher (see Fig. 2.2).

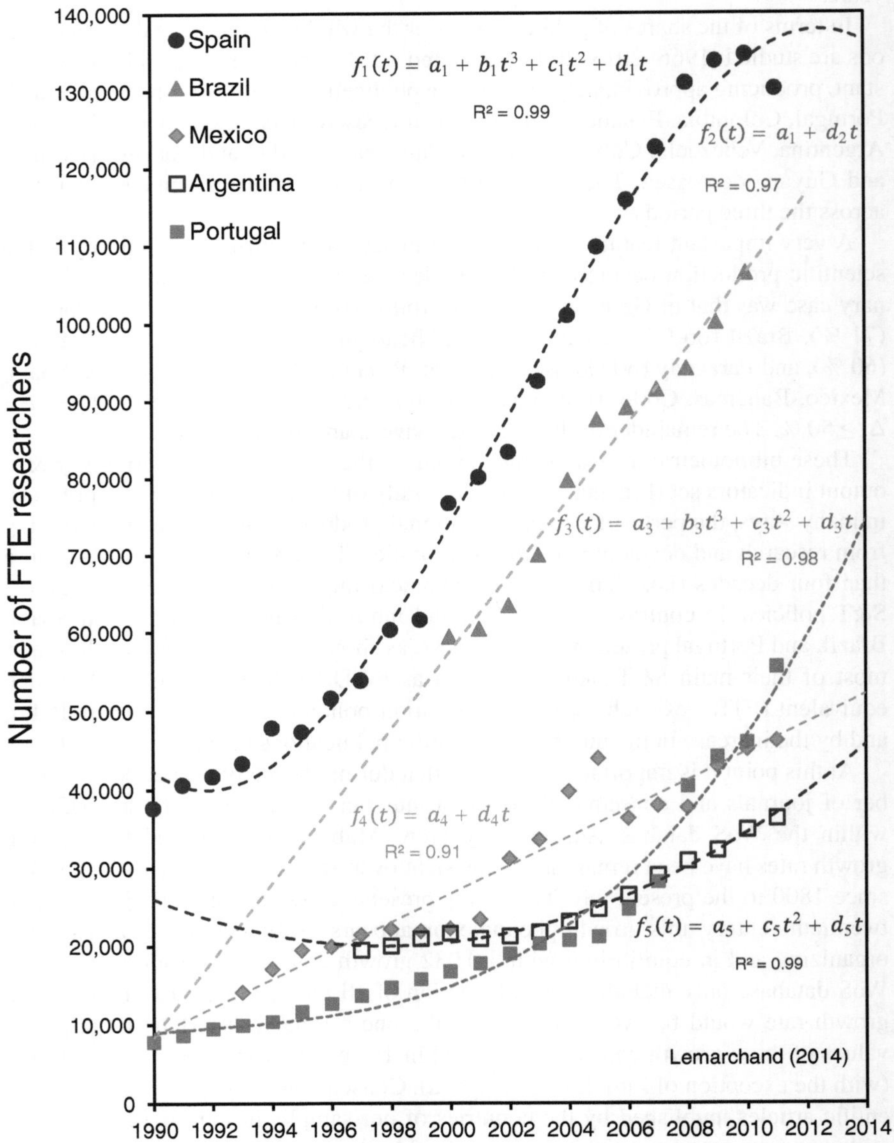

Fig. 2.2 Evolution of the number of FTE researchers (1990–2012) for Argentina (quadratic growth), Brazil (linear growth), Mexico (linear growth), Portugal (cubic growth), and Spain (cubic growth)

Since the early days of scientometrics, it has been suggested that a correlation exists between the total number of publications and national GDP (De Solla Price 1986). A recent study showed a high correlation between the number of publications over a 10-year period within the different regions of the world (Africa, Arab states, Asia-Pacific, Europe, North America, and Latin America and the Caribbean) and the shares of global GDP within the same regions (Lemarchand 2010, 2012). This correlation is tested here by analyzing the number of publications in relation to national GDP per capita expressed in constant 2005 USD over time (1966–2012). Table 2.2 shows these results. Only 57 % of the 30 countries analyzed present a high correlation ($R^2 > 0.8$). At this point, it is not possible to generalize the hypothesis that GDP is the only driver for scientific publications. However, these results are still consistent with previous findings (Lemarchand 2012).

2.2 Ibero-American Self-organizing Co-authorship Networks (1973–2013)

Co-authorship is one of the most tangible and well-documented forms of scientific collaboration. In a recent study, Lemarchand (2012) showed that scientific co-authorship among countries follows a power law and behaves as a self-organizing scale-free network, where each country appears as a node and each co-publication as a link. In that study, a mathematical model was developed to study the temporal evolution of co-authorship networks. The model shows how the number of co-publications $P_k^i(t)$ between country k and country i, against the coupling growth coefficients a_k^i, follow a power law mathematical relation.

To show that the number of co-publications among countries also grows quadratically against time, Lemarchand (2012) assumed a linear growth for the national scientific network and a preferential attachment strategy for co-publications. The quadratic growth constants for 352 different co-authorship networks were empirically determined for the period 1973–2006 for a group of 12 Ibero-American countries (responsible for 98 % of the total regional publications between 1973 and 2010) with a group of 46 other different nations.

The developed model predicted that the connectivity of Ibero-American countries with larger scientific networks (hubs) is growing faster than that of other less connected countries. Lemarchand (2012) determined that 70.4 % of the 352 analyzed cases linked their cooperation with hubs or larger co-authorship networks. These social webs were responsible for 39.6 % of the total number of articles published between 1973 and 2006 by the 12 countries in that sample. The rest of the co-authorship networks (smaller ones) accounted for only 8.8 % of the total regional production; Spain accounted for 6.1 %.

Lemarchand (2012) was also able to determine the dates, t_0, at which the co-authorship connectivities trigger the self-organizing scale-free network for each of the 352 cases. It was found that the latter follows a normal distribution around year 1981.4 ± 2.2, and this effect was associated with a brain-drain process generated within Ibero-American countries during the previous decade. The examined data showed a time lag of approximately 15 years, between the peak when most Ibero-American populations were living under dictatorships (massive brain drain) and the peak when most of the co-authorship networks were triggered. Emigrant scientists living abroad need a period of time to develop a wide range of S&T human capital assets, to enhance R&D knowledge, craft skills and know-how, to publish mainstream articles, to develop their ability to structure and plan research, and of course, to increase contacts with other scientists, the industry, and funding agents. After expanding these potentialities, emigrant scientists become "visible" to the scientific networks of their homeland. After this time lag, they may be in a position to transfer part of their accumulated knowledge and experience to their home country, via periodic visits and participating in knowledge and co-authorship networks. In this way, the preferential attachment links are enhanced and the self-organizing behavior triggered.

Based on the mathematical properties of the model, Lemarchand (2012) was able to develop a methodology to use the empirically determined growth constants for each co-authorship network to predict changes in the relative intensity of cooperation among countries. These predictions for the period 2007–2010 were tested within the 352 networks.

2.3 A Generalization of the Self-organizing Model of Scientific Networks

The original model developed by Lemarchand (2012) assumed that new researchers join the national scientific network (nodes) following a linear growth process described by the following equation $f_i(t) = d_i t + a_i$. Here, $f_i(t)$ represents the total number of nodes for country i, d_i the growth constant, and a_i the total number of nodes when $t = 0$. The model has the following assumptions: (a) nodes join the network in a linear way $(f_i(t))$; (b) incoming nodes link to already present nodes following preferential attachment; and (c) nodes already present in the network form new internal links following preferential attachment. Lemarchand (2012) defined $P_i(t)$ as the number of links (co-publications) at time t and country i. Thus, it is possible to show that the number of co-publications among pairs of countries over time follows the following equation:

$$P_i(t) = \alpha \left[t - t_0 \right]^2 + \gamma \tag{2.1}$$

Here, α is the growth constant, t_o is the time at which the system becomes self-organized, and γ is the minimum number of co-publications, which triggers the self-organizing process. The three constants (α, γ and t_o) can be empirically determined.

Because in the majority of countries the national scientific network grows linearly, the model worked extraordinary well for almost all nations, but there were some exceptions (i.e., China and India).

To generalize the model (Lemarchand and Barrera-Lemarchand 2016), no special assumptions were made regarding the type of mathematical functions for $f_i(t)$. The mathematical functions of $f_i(t)$ are: (1) constant ($f_i(t) = a_i$); (2) linear ($f_i(t) = d_i t + a_i$); (3) quadratic ($f_i(t) = c_i t^2 + d_i t + a_i$); or (4) cubic ($f_i(t) = b_i t^3 + c_i t^2 + d_i t + a_i$).

Consequently, the dynamics placed by the preferential attachment effect will induce a different set of mathematical functions for $P_i(t)$. If $f_i(t)$ is constant, then $P_i(t)$ will have a linear growth; if $f_i(t)$ is linear, then $P_i(t)$ will be quadratic; if $f_i(t)$ is quadratic, then $P_i(t)$ will be cubic; and if $f_i(t)$ is cubic then $P_i(t)$ will be quartic.

The behavior of the national scientific networks can be analyzed by studying the number of scientists in a particular nation. Figure 2.2 shows the long-term behavior of the number of FTE researchers between 1990 and 2012 for the five most productive Ibero-American countries. The definition of FTE researchers follows the one proposed by the Frascati Manual (OECD, 2002). The raw data were provided by the UNESCO Institute for Statistics.

The evolution of FTE researchers constitutes the national scientific network $f_i(t)$, where $i = 1$ (Spain), $i = 2$ (Brazil), $i = 3$ (Portugal), $i = 4$ (Mexico), and $i = 5$ (Argentina). To test the model, the behavior of 25 scientific networks was analyzed using data covering 41 years.

Figures 2.3, 2.4, 2.5, 2.6 and 2.7 show the evolution (1973–2013) of the number of scientific co-publications between the five countries of this sample (Spain, Brazil, Portugal, Mexico, and Argentina) with their five main scientific partners. The quadratic, cubic, or quartic fitting curves are in total agreement with the predictions made by the generalized model described here, improving the results obtained in Lemarchand (2012).

The values obtained for the determination coefficients (R^2) between the number of co-publications and the curves predicted by this model show that in 72 % of cases, $R^2 > 0.99$, in 16 %, $R^2 > 0.98$, in 8 %, $R^2 > 0.97$, and within the remaining 4 %, $R^2 > 0.96$. These are extraordinary results taking into account that this analysis covers the social behavior of co-publications among 25 different pairs of countries in a 41-year period.

2.4 Conclusions

This study analyzed the long-term evolution of scientific articles published in mainstream journals over a period of 48 years. The focus was on all Ibero-American and Caribbean countries. Within this period, the ten most productive Ibero-American and Caribbean countries accounted for 97 % of all publications, the 15 most productive for 99 % and the 30 most productive countries for 99.9 %.

The analysis also included the estimation of the various ranking positions of the top 10 Ibero-American within the world ranking since 1970 (see Table 2.1). Spain, Brazil, Colombia, and Portugal showed the most dramatic ranking improvements, whereas Venezuela, Argentina, and Chile had the greatest falls.

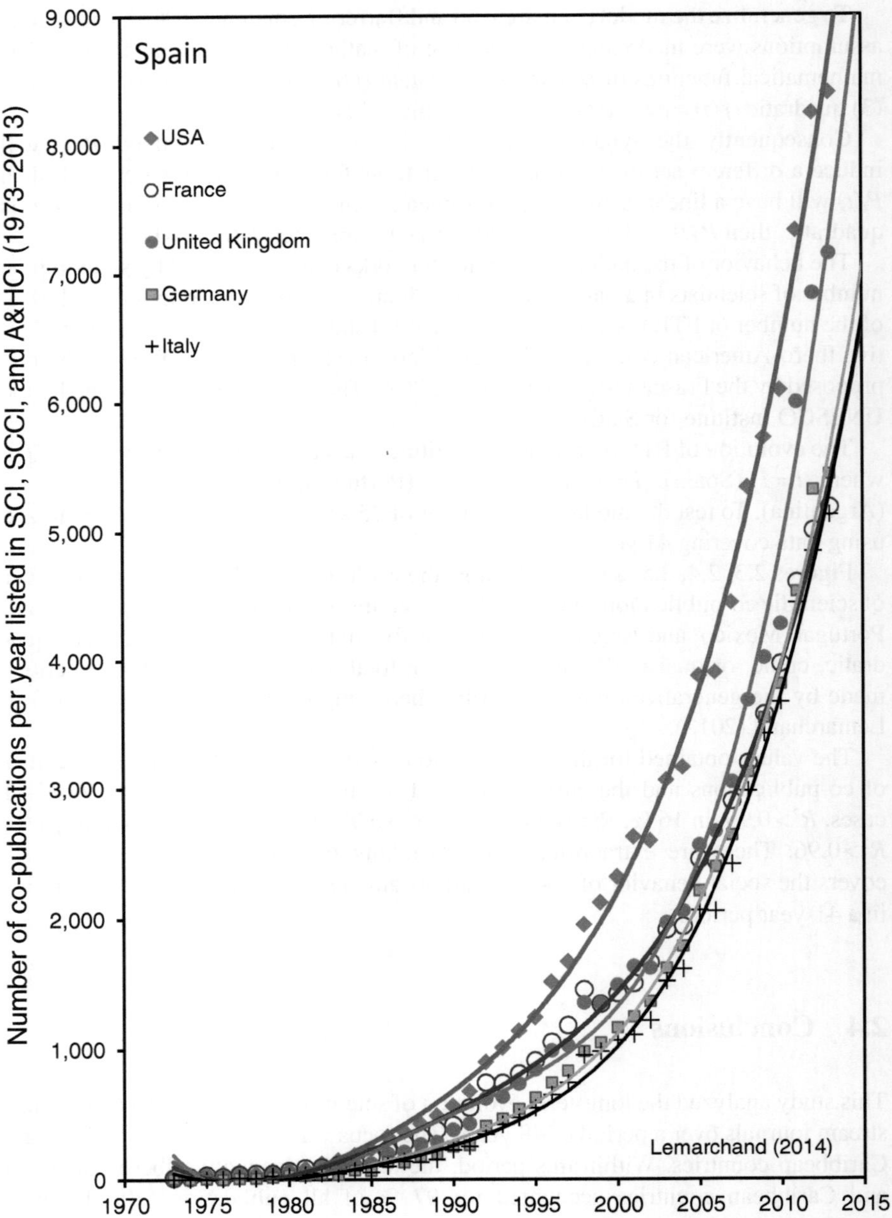

Fig. 2.3 Temporal evolution (1973–2013) of the co-authorship social network for Spain; the model predicts a quartic growth in the number of links (publications) over time because the FTE researchers in Spain shows cubic growth (the *solid lines* represent the quartic fit)

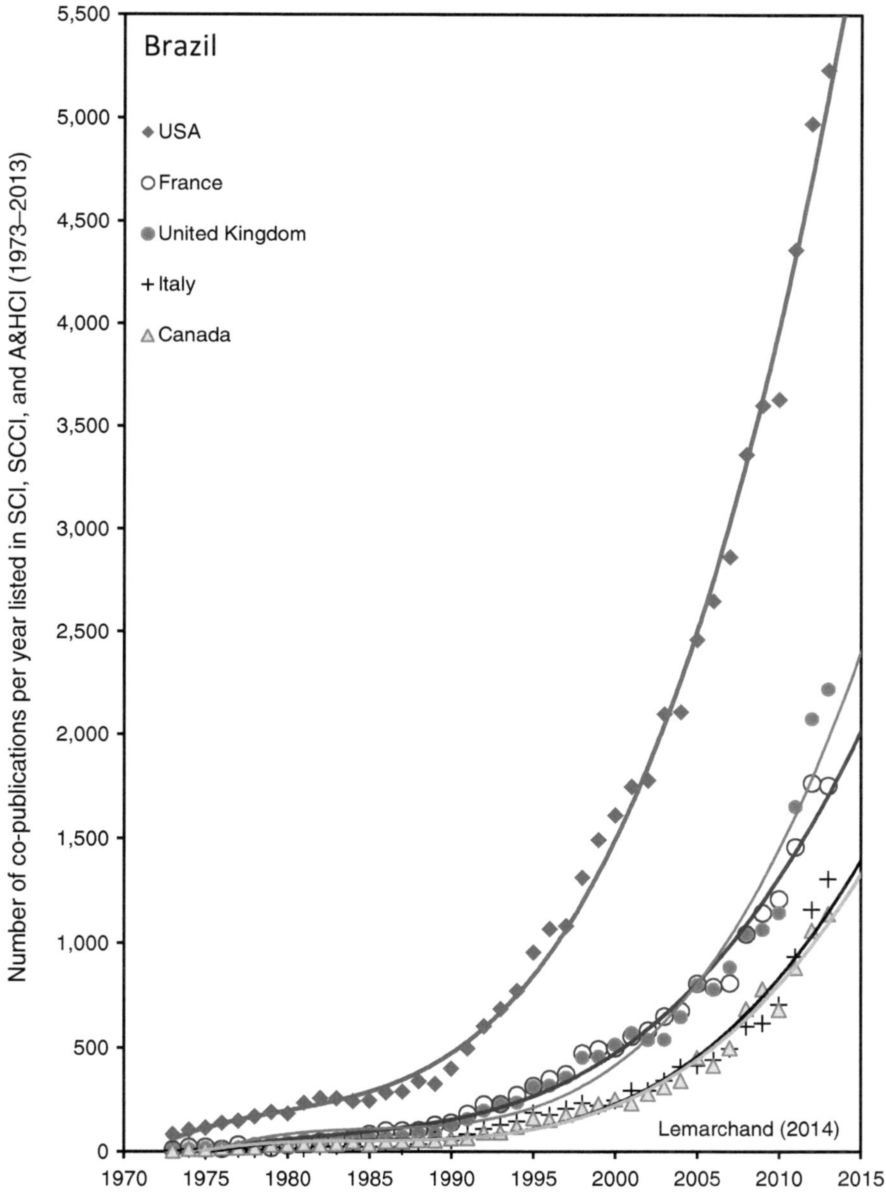

Fig. 2.4 Temporal evolution (1973–2013) of the co-authorship social network for Brazil; the model developed predicts a parabolic growth in the number of links (publications) over time (the *solid lines* represent the quadratic fit)

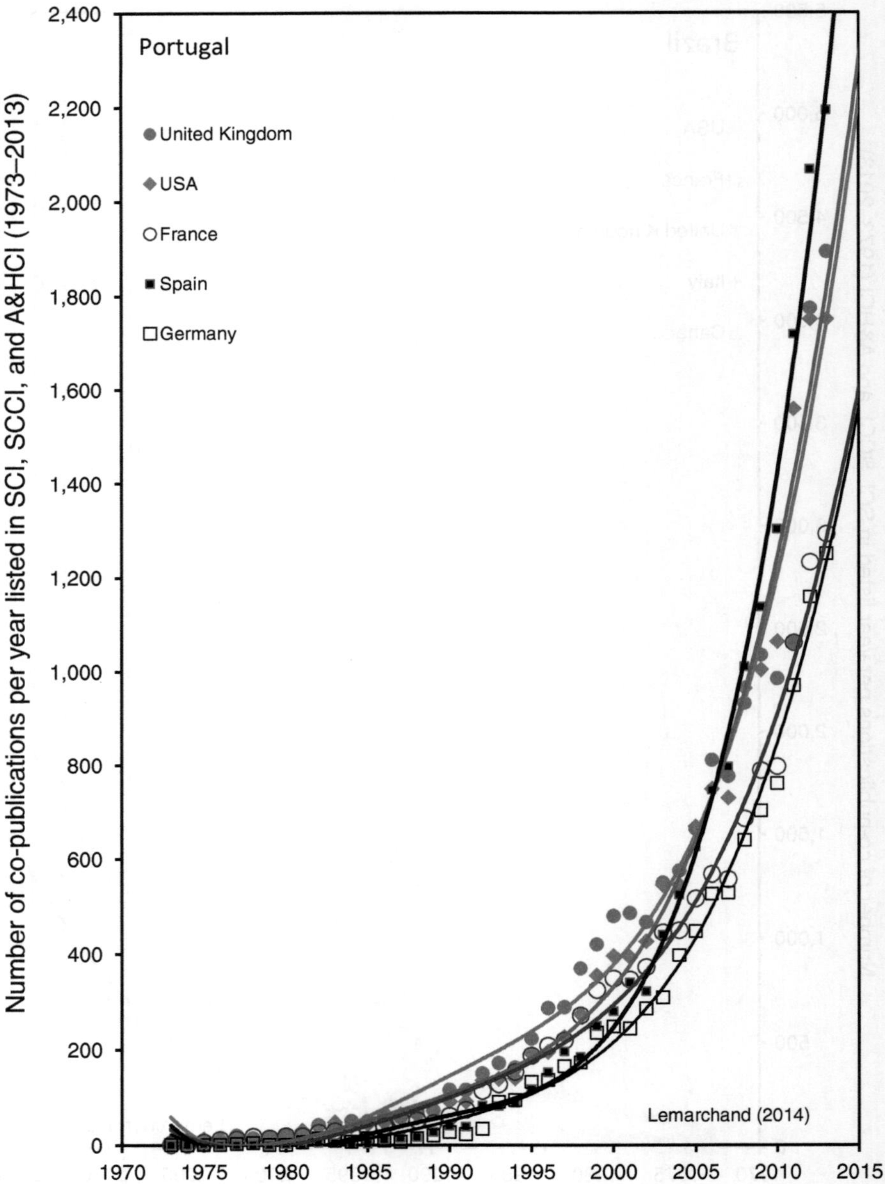

Fig. 2.5 Temporal evolution (1973–2010) of the co-authorship social network for Portugal; the model developed predicts a quartic growth in the number of links (publications) over time because the number of FTE researchers shows cubic growth (the *solid lines* represent the quartic fit)

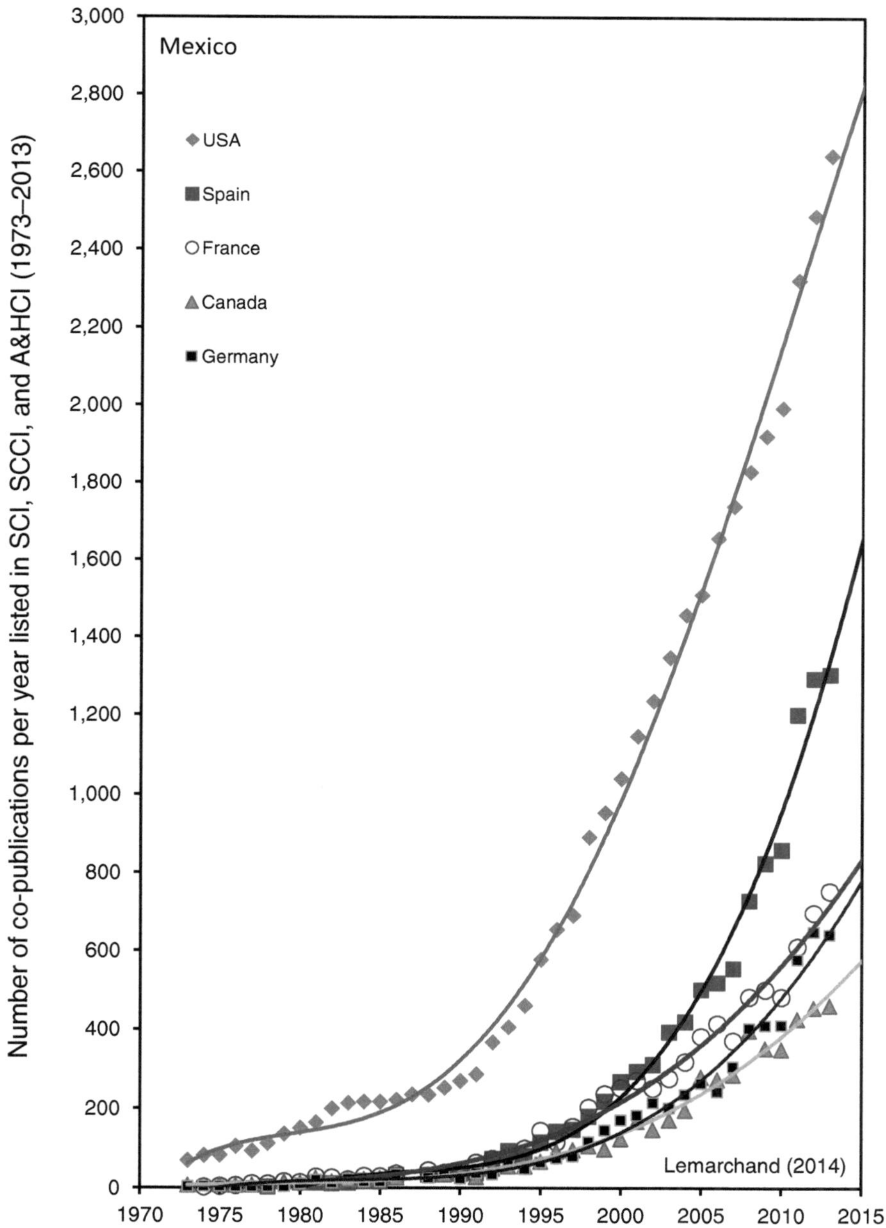

Fig. 2.6 Temporal evolution (1973–2013) of the co-authorship social network for Mexico; the model developed predicts a parabolic growth in the number of links (publications) over time (the *solid lines* represent the quadratic fit)

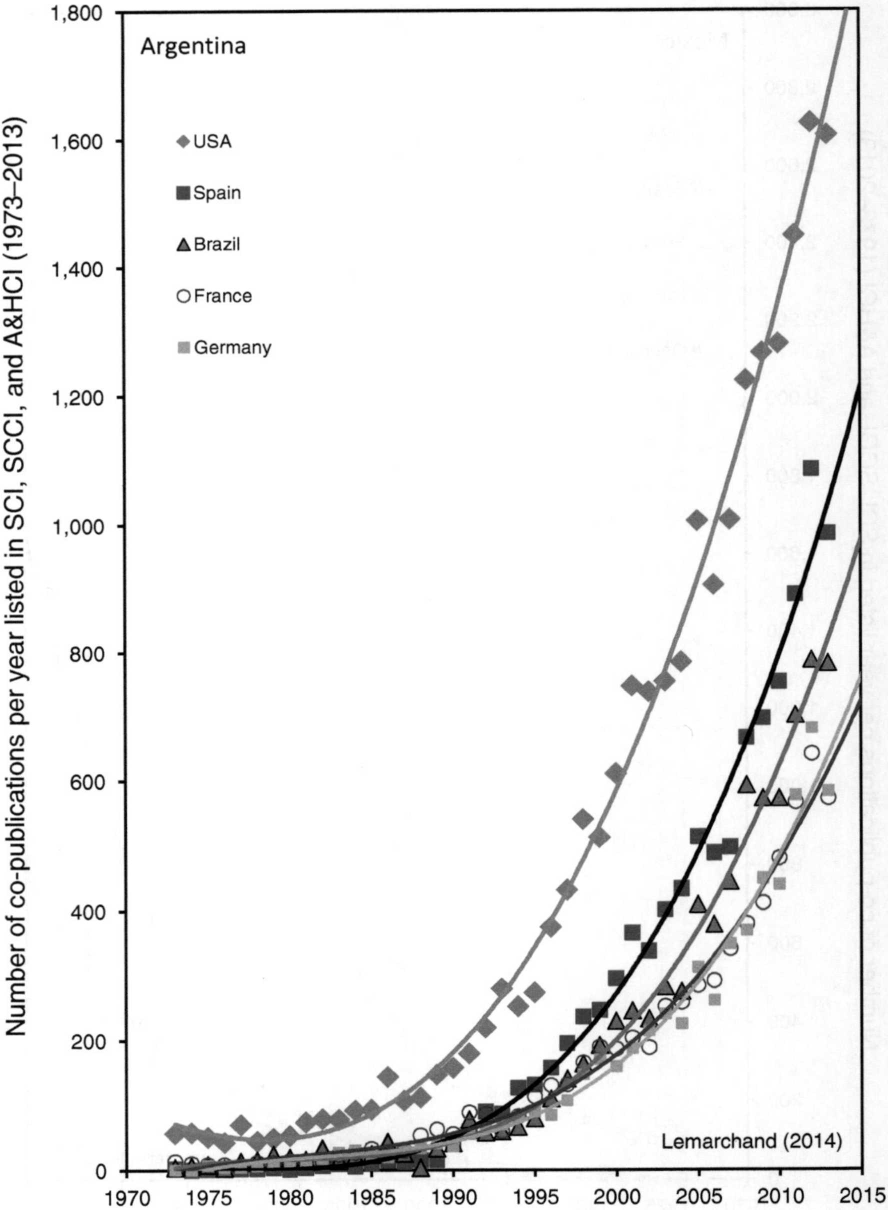

Fig. 2.7 Temporal evolution (1973–2013) of the co-authorship social network of Argentina; the model developed predicts a cubic growth in the number of links (publications) over time because the number of FTE researchers shows quadratic growth (the *solid lines* represent the cubic fit)

Participation in publication was also estimated, for 30 countries in three different periods (see Table 2.2). The last decade was extraordinarily productive, accounting for 59 % of all publications produced in the last 48 years.

The long-term growth (1966–2013) of scientific articles in mainstream journals for 18 Ibero-American countries was represented, accounting for 99.4 % of all publications (see Fig. 2.1). This analysis enabled the observation of slope changes, which may relate to modifications to national science, technology, and innovation ecosystems, or with the application of new public policies (Lemarchand 2010, 2012). In some cases productivity increased while in others the new policies failed and consequently productivity diminished.

The growth-rate constants for 30 countries were determined for three categories: (a) number of publications; (b) number of publications per million inhabitants; and (c) ratio of the number of publications and GDP per capita. In all three categories, Portugal showed the highest growth rates with determination coefficients $R^2 > 0.99$.

Furthermore, the characteristics of the self-organizing model of co-authorship networks (Lemarchand 2012) were analyzed. To ensure generalization, no limits were placed on the evolution of the national scientific networks over time. By representing the long-term evolution of the number of FTE researchers for Spain, Brazil, Portugal, Mexico, and Argentina (see Fig. 2.2) the type of mathematical function that describes the evolution of each national scientific network $(f_i(t))$ could be determined. With that information, it was possible to predict the mathematical behavior of the number of co-publications among pair of countries $(P_i(t))$. This model was tested against 25 different scientific networks using data between 1973 and 2013 and showed that 100 % of the determination coefficients had values $R^2 > 0.96$ while 72 % had values $R^2 > 0.99$.

The results obtained in this study are remarkable, especially taking into account that the variable described in this model is associated with co-authorship behaviors among groups of individual scientists working in different countries, with different environments and levels of political stability, over a period of 41 years.

The reliability of the original and generalized model of self-organizing scientific networks enables their application for foresight studies and for the design of specific research and innovation policy instruments.

Acknowledgements This research was supported by the Swedish International Development Cooperation Agency (SIDA) through the UNESCO extrabudgetary project 503GLO2002.

References

De Solla Price, D. J. (1986). *Little science, big science ... and beyond*. New York: Columbia University Press.

Freeman, C. (1969). *Measurement of output of research and experimental development: A review paper*. Paris: UNESCO.

Lemarchand, G. A. (2007). Science indicators of cooperation in Iberoamerica: The co-authorship of main-stream scientific articles. A long-term temporal evolution's study (1966–2006).

Presented at the *VII Congreso Iberoamericano de Indicadores de Ciencia y Tecnología*—Sao Paulo (Brazil), 23–25 May 2007.
Lemarchand, G. A. (2010). Science, technology and innovation policies in Latin America and the Caribbean during the past six decades. In G. A. Lemarchand (Ed.), *LAC, Vol. 1, National science, technology and innovation systems in LAC* (pp. 15–139). Montevideo: UNESCO. See: http://unesdoc.unesco.org/images/0018/001898/189823e.pdf
Lemarchand, G. A. (2012). The long-term dynamics of co-authorship scientific networks: Iberoamerican countries (1973–2010). *Research Policy, 41*, 291–305.
Lemarchand, G. A., & Barrera-Lemarchand, F. U. (2016). A generalization of the self-organizing model of scientific co-authorship networks. *Physica A* (submitted)
Mabe, M. (2003). The growth and number of journals. *Serials, 16*(2), 191–197.
OECD. (2002). *Frascati manual. Proposed standard practice for surveys of research and experimental development*. Paris: Organisation for Economic Co-operation and Development.

Chapter 3
Towards a Knowledge-Based Economy in Brazil

Anne-Marie Maculan

3.1 Introduction

Brazil, as an emerging economy, is reformulating its economic development strategy. So far, its ability to export commodities or low-technology industrial products has ensured high GDP growth rates. However, Brazil is not immune from competition with more dynamic and innovative economies, or from the negative impacts of the global crisis that have sparked in recent years and brought new challenges.

The Brazilian economy needs to move to the next level and become capable of generating scientific advancements and significant technological innovations. Its ability to respond to growing competition from China, whose performance in research and innovation is far superior to that of Brazil,[1] is a necessary condition that affects its sustainable economic growth.

Brazilian firms now face the challenge of going beyond the capacity to assimilate advanced technologies; they now need to generate their own product and process innovations. This implies a major departure from the 1980s and 1990s. For this to occur, the country needs highly qualified human resources, firms investing in research and development (R&D), and public policies to foster research and innovation.

Research institutions and universities are facing the need to accept new undertakings, not only to train more qualified human resources but also to take a prominent role in local economic development. This involves transforming its relations with firms, civil society, and government agencies.

[1] China's investment in R&D in relation to its GDP is now approximately 40 % higher than Brazil's.

A.-M. Maculan (✉)
Federal University of Rio de Janeiro, Rio de Janeiro, RJ 21941-901, Brazil
e-mail: amdmpep@gmail.com

H. Horta et al. (eds.), *Trends and Challenges in Science and Higher Education*, Knowledge Studies in Higher Education 3,
DOI 10.1007/978-3-319-20964-7_3

Between 2000 and 2010, Brazil sought to strengthen its national innovation system and create the foundations for economic sustainable growth. The main goal, as explained in many government programs, was to promote the transition from an industrial Taylorist-based economy, until then characterized by a weak capacity to innovate and the passive incorporation of technology, to an economy based on knowledge, research, and innovation. Brazilian governments have used the concept of the knowledge economy as the central pillar of their economic development policies, a concept that emphasizes investments in the area of science & technology (S&T).

After a decade of the renewal of Brazil's S&T policy, is it possible to make a first assessment of the progress made and its impacts? Is it legitimate to assume that in the last decade, the implemented interventions in the innovation system were an appropriate response to the crisis, limiting the decline in GDP to 0.6 % in 2009? Would that mean that Brazil was able to establish solid foundations for a knowledge economy? Are there new challenges to be faced? In the context of this global crisis, can the Brazilian economy keep growing without incorporating new knowledge in various sectors of the economy? First, what is the country's ability to maintain a high rate of growth and face international competition with an economy based on exporting commodities, heavily dependent on demand from China? Second, is it possible to be a knowledge-based economy with industrial firms that do not develop radical innovations? Third, how can a nation deal with the need to promote new technological solutions that also meet the demands of environmental protection?

This chapter presents some of the Brazilian government's responses. These responses lead to a reformulation of S&T policies, an increase in the capacity to generate knowledge. The changes aim for a more dynamic innovation system and to ensure that firms are more capable of facing the global economic crisis. This new political orientation is characterized by a steady increase in research investment, the expansion of higher education, and greater financial support for innovative firms; this is the mission given to universities in technology transfer and the formation of a true knowledge market.

Our hypothesis is that Brazil has identified knowledge as a strategic ingredient of economic development, firm competitiveness, and capacity to innovate. As a result, it attempts to organize the production and transfer of knowledge to the main economic actors. The new regulatory framework aims to strengthen interactions and partnerships between producers and users of knowledge. It intends to structure a knowledge market that serves as a key element in the transition to an economy able to withstand the current crisis.

First, we list the most significant actions that seek to complete the creation of an integrated and efficient national innovation system. Then, an evaluation of the impacts of these measures is presented, identifying the existence of bottlenecks that limit the progress achieved. Finally, we point out the challenges to be faced to ensure the sustainability of economic growth.

3.2 Funding the Research

Several studies have identified the weakness of the Brazilian innovation system as a significant and important barrier to the capacity of firms to innovate. Thus, these barriers present a negative factor in the global economic context.[2] Important reforms were implemented with regard to the organization and financing of scientific research and technology, and Brazil's system of innovation became the main conceptual reference.

The new resolutions reflected the view that it was necessary to structure a real national system of innovation because research activities should enable the generation of new knowledge to serve as inputs for the building of a firm's innovative capacity. In 2001, the Ministry of Science & Technology (MCTI) incorporated the "I" of innovation into its title and adopted a new policy that created conditions for further knowledge generation and its dissemination in economic activities.

The government resumed investments in R&D and deployed a new funding system. Most research remained funded by the federal government, with over 51 % of the investments being in S&T. However, a small increase in the investment in research by firms could be observed. Research infrastructure was substantially expanded and human resources training grew also increased. With a regular and permanent allocation of funding secured, it became possible to plan a new research system (Fig. 3.1).

The creation of Sector Funds in 1999 represented the first step of a new system of research funding.[3] The introduction of industrial sectors as a reference represented an effort to link the allocation of research resources to the applicability of results and specific economic sectors. Sector Funds are fuelled by tax contributions from firms that exploit natural resources belonging to the state or act in some specific sectors of the economy. For the first six funds, revenue and financing were fully attached to the sectors that provided the resources. The funds were intended for exclusive application in the new strategic plans set out in the Industrial, Technological, and Foreign Trade Policy adopted in 2003.[4]

Sector Funds have two remarkable features. First, the resources come from the activities of the infrastructure industry. Second, much of the resources should be allocated to projects undertaken in cooperation with universities and firms. The definition of research projects is made in formal announcements that express the government's priorities and place different groups of researchers in competition. The search for the practical applicability of results in industrial activities becomes the rule and the projects to be funded must be submitted by a team that brings together researchers from universities or R&D institutions and firms. These conditions pro-

[2] Da Motta e Albuquerque (2003); Viotti (2002); Cassiolatto and Lastres (1999).

[3] The first Sector Funds (Oil and Natural Gas was set up in 1998, and Information Technology, Mineral Resources, Electricity, Water, and Aerospace created in 2001) were financed by funds from the sectors themselves and were used to invest in those sectors.

[4] http://www.mct.gov.br/index.php/content/view/27646.html

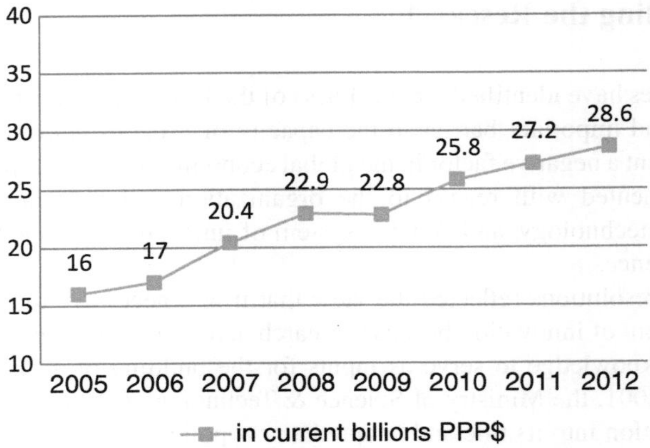

Fig. 3.1 Brazil's gross expenditure on R&D in 2005–2012 (Source: www.mct.gov.br/index.php/content/view/300803/)

mote industry–research relations and partnerships, and strengthen ties between the actors in the innovation system.

The new funding system not only allows the revitalization of research activities but also introduces a significant change in the public funding of R&D in firms. Until now, public funding was intended for academic research and could not directly benefit firms. Moreover, the Sector Funds system aims to promote cooperation in the research of firms, universities, and R&D institutions. In particular, the so-called Green–Yellow Fund[5] serves to fund the University–Industry Cooperation Innovation Support Incentive Program, which welcomes extremely diverse demands from different industries. However, although the diversity of the target public enhances firm participation, it also hinders the definition of priority areas and criteria for the selection of research proposals because of the heterogeneity of the object, its scope, and relevance. In this way, it can be a source of ineffectiveness.

The use of Sector Funds to finance cooperative research and the Innovation Law of 2004 are both examples of the government's decision to bring together firms and universities. The government is guided by the idea that an increase of university–firm relationships will create a well-structured environment that is more conducive to innovation. The new criteria to allocate funds are based on the idea that an increase in the capacity of firms to innovate fundamentally depends on interactions with universities or public research institutions. From the joint development of research projects, it is expected that in addition to increasing the potential applicability of the results, this can also promote the beginning of R&D activities within firms. The objective is to promote a more intense process of learning and technological training in firms. Thus, it may further spread the culture of research and innovation to different industry segments.

[5] Created in 2001, the name symbolically refers to the green and yellow of the Brazilian flag.

3.3 Universities and the Formation of Human Resources

The first task, which is essential for a more dynamic innovation system, is to secure human resource training, to achieve both a critical mass of researchers and to enable the dissemination and use of knowledge in the industry. The average length of schooling of people aged over 18 years still does not exceed 7.5 years. In 2012, 12 million people aged 24 years and older were illiterate. According to The Brazilian Institute of Geography and Statistics (IBGE) (2010), between 2002 and 2009, the number of people who completed tertiary education rose from 6.9 million to 12.2 million, an increase of 74.8 % (although this represents only 13 % of the whole workforce).[6]

From the post-war period until the 1980s, companies' manpower needs were mainly for low-skilled labor. The resumption of economic growth in the late 1990s, and the changes in the structure of Brazilian economy have generated a huge demand for skilled human resources. The education system's poor performance—insufficient places and low quality at all levels of education—is becoming one of the main bottlenecks for sustainable growth and the creation of a knowledge economy. Recently, the country has faced a shortage of engineers and medical doctors.

Of the 24.8 million workers seeking jobs in 2009, only 19.3 million had any qualifications or professional experience. The importance of the number of unskilled workers (22.2 %) requires policies to combat a form of social and economic exclusion that distinguishes Brazil from South Korea or China. Even with the steady growth of GDP between 2004 and 2013, there are still a significant number of unemployed and unskilled workers who will be unable find their place in a knowledge economy. This vulnerability of the Brazilian economy is the result of its poor education system, which for many years received no attention, either from government agencies or businesses, but now has a strong impact on conditions for competitiveness.

The biggest challenge is to improve the quality of Brazil's education. It is necessary to correct educational flow to make primary education available to all and to ensure the universal completion of studies. Furthermore, the number of places available in high schools is not able to accommodate eligible 15–17-year-olds. Significant investment is needed to improve and expand the physical capacity that would ensure access and permanence. This bottleneck affects higher education where the challenges are even greater because of low enrollment rates and extreme social and regional disparities.

Universities are responsible for the formation of qualified human resources who are essential intermediaries of knowledge dissemination in an economy. While the number of places available in higher education increased by 47 % between 2004 and

[6]According to IBGE (2013) there were 101 million employed and 6.5 million unemployed in Brazil between 2011 and 2012.

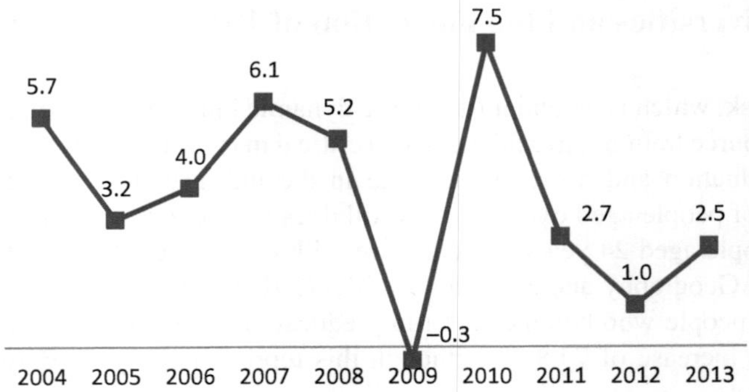

Fig. 3.2 GDP annual growth (Source: www.IBGE.gov.br)

2013,[7] the rate of enrollment in higher education institutions for those aged between 18 and 24 does not exceed 14 %.[8]

The expansion of universities has enabled the training of more qualified human resources and researchers. Between 2001 and 2012, the number of PhD and master holders more than doubled (Fig. 3.2). PhD programs correspond to the government's policy to seek greater integration between research and innovation systems (Center for Strategic Studies and Management in Science, Technology and Innovation [CGEE] 2010). Most PhD holders (80 %) still work for universities and R&D institutions. However, an analysis of employment trends shows a growing dispersion of PhD holders in geographical terms, and a flow between sectors of activity (from research to industry) (CGEE 2008) (Fig 3.3).

The goals of the National Education Plan for 2011–2020 predict that 33 % of the population aged 18–24 will be enrolled in higher education; this represents a 50 % increase over 10 years. The number of master graduates should increase from 35,000 a year to 60,000 and the number of PhD graduates from 11,000 to 25,000.

However, there is a lack of professionals in all areas of science and especially engineering. As a result, firms face a shortage of skilled professionals. The lack of engineers is probably the most notable example of the demand for more qualified human resources. Engineers play a key role in technological development. They are the key professionals required to introduce improvements in processes and products, organization and production management, and to conduct research and development in firms.

This shortage of engineers strongly affects firms' competitiveness and the capacity to innovate, and only recently have firms begun to express concern about this situation. A few years ago, Brazil had to seek foreign engineers to build a steel mill in Rio de Janeiro (Saavedra Durao 2010). According to statistics from the Ministry of Education (INEP 2013), of all graduate students, only 6.7 % are engineers. In

[7] See http://portal.inep.gov.br/superior-censosuperior-sinopse.

[8] According to data from IBGE (PNAD 2010), Brazil had a population of 24 million aged between 18 and 24 in 2010.

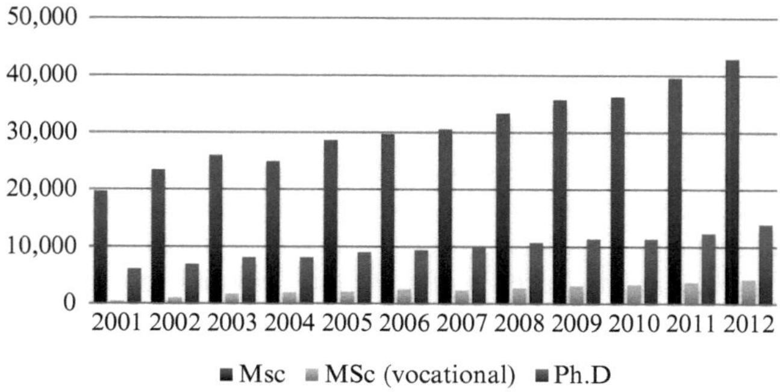

Fig. 3.3 Conferment of MSc and PhD degrees (2001–2012) (Source: www.mct.gov.br/index.php/content/view/6629.html

2013, Brazil graduated only 76,000 engineers, less than Mexico (88,000), China (400,000), and India (220,000).

In the 1980s during the debt crisis, there was a sharp reduction in the number of graduate engineers because of the low demand from firms. Today, firms have a higher demand for engineers because of their fundamental role in assimilating new technologies and innovating. A recent report from the Institute for Applied Economic Research (IPEA 2010) estimated the demand for skilled human resources and predicted that Brazil would have a major shortfall of engineering knowledge and technological skills.

3.4 Universities and the Research–Industry Relationship

Universities and research institutions are the main sources of (and are almost exclusively responsible for) advances in knowledge creation, either through human resource training or research. They are considered to play a pivotal role in the formation of the knowledge market. A second line of action by the government since 2003 has been to bring together industry and research and to encourage universities to seek the applicability of their research and transfer technology to firms.

The mission of universities, embodied in various Ministry of Science, Technology, and Innovation programs, is to increase the frequency of interactions between research institutions and industrial firms, that is, cooperation in research or commercializing technology. The need to promote closer collaboration between universities and industry and to define mechanisms of economic valuation for research, together with the issue of innovation, is a concern of the government.

Although Brazilian universities and R&D institutes have formed a consolidated and competent infrastructure for research, this infrastructure remains very distant from industry, and therefore the results achieved in research are rarely transformed into innovation for the market (Maculan and Mello 2009). Firms' capacity to innovate remains very low, and this has negative consequences for the competitiveness of the Brazilian economy. Thus, the government must promote closer collaboration between universities and industry as well as define mechanisms of economic valuation for research projects.

Two laws—the Innovation Law (2004) and the Law of Goods (2005)—have created more favorable conditions for cooperation between research institutions and firms. These laws encourage the exchange of technology and collaborative research, and also reduce much of the legal uncertainties that hindered collaboration between the two. Furthermore, the acts legitimize the informal entrepreneurial mission undertaken by some universities with greater research capacity and infrastructure. Universities with a longer tradition in research[9] took the initiative to create environments that were favorable to the creation of spinoff firms by teachers, researchers, or students, and to the provision of consultancy services to firms. These initiatives are now incorporated into the responsibilities of all universities. Thus, the entrepreneurial activities of universities regarding entrepreneurship, with the creation of business incubators and technology parks, now take on a new dimension.

In addition to teaching and research, S&T institutions should nowadays take on the task of transferring knowledge acquired from research to actors in the innovation system, especially firms. This means cooperating in research or commercializing technology. Science, Technology, and Innovation (ST&I) policies favor procedures such as patent licensing, contractual transfer of technology, and creating spinoff firms, procedures that concretize the institutional capacity for technological push. These policies strongly encourage the use of intellectual property to protect knowledge to turn it into an object of exchange.

Some universities invest in the possibility of capturing additional financial resources where they can attract the interest of firms with patented technologies. The new regulatory framework encourages universities to participate effectively in the creation of technology-based firms using new tools offered by the Innovation Law.

3.4.1 Incubators and Science Parks

Regarding universities' third mission, the expected impact on the capacity to innovate and firm competitiveness start to become visible. Many successful incubators have been created in recent years. More recently, major universities have begun to establish technology parks. Some parks have proven to be very dynamic and have attracted large multinational firms, such as the park of the Federal University of Rio de Janeiro, which benefited from being close to an important oil hub led by Petrobras. Multinational firms such as L'Oreal, Schlumberger, and General Electric are placing research centers on the university campus.

The Brazilian states of Minas Gerais and São Paulo have a policy to create a number of technology parks around state universities, to support local industry.

[9] These are mostly public universities located in the south and southeast of the country.

However, the complexity of negotiations required to define the profile of these parks, the importance of funding, and the period in which they are developed do not enable us to consider parks as a simple tool for the valuation of research. There are no studies to date that assess the impact of these experiments on the widespread creation of technology-based firms. The results of some surveys (Santos 2010; Parreiras 2010) show that these new undertakings require universities to make organizational changes, develop new skills, set goals and strategies, and to make use of performance indicators.

The Funding Authority for Studies and Projects (FINEP)—the MCTI's agency to foster innovation—has recently launched its PRIME Program. This program seeks to use the experiences gained by incubators at universities for the promotion of entrepreneurship to expand the impact of innovation funding programs. The goal is to invest R$1.3 billion (approximately US$700 million) over the next 4 years in emerging technology-based firms. In 2011, it was expected that more than 1800 applications would be received for new projects.[10] Eighteen incubators have been selected to prepare the bid documents and select the proposals. It is likely that most of these 1800 firms are university spinoffs, that is, firms created to explore the knowledge created in research activities. The successful experiments by incubators in creating small technology-based firms—although modest in global quantitative terms—raise the interest of local governments (municipalities) to explore the potential of creating spinoffs together with universities.

These programs are based on the understanding that connecting mechanisms are missing between the research capability of universities and R&D institutions on the one hand, and firms' ability to innovate on the other. This situation could be improved by creating administrative structures specializing in technology transfer and by educating researchers on the benefits of the intellectual property system.[11] The expectation is that universities and R&D institutes can gather a portfolio of patents that will enable the subsequent commercialization of that particular technology. Therefore, it is important that universities and research institutions have a department specializing in intellectual property and technology transfer.

3.4.2 *Experience of the Technological Innovation Nuclei—NITs*

These new governmental orientations have encouraged universities to apply economic value to their research results, protect their creations, patent the technologies they develop, and support the creation of technology-based firms. To do this,

[10] According to FINEP, 1,380 emerging firms were selected to receive financial and management support.

[11] Until recently, Brazilian companies appear to have been unaware of the possible benefits of intellectual property rights.

they must institutionalize their entrepreneurial activities and create a technology transfer office (so-called NITs)[12] or an agency for innovation.

NITs have the power to commercialize knowledge, apply for patents, and manage contracts with firms (there are currently more than 160 active NITs), and replace the first technology transfer offices (TTOs).[13] The larger universities created various TTOs in the last decade. Their primary mission was to encourage researchers to make use of intellectual property. The creation of these TTOs was the final stage in the institutionalization of universities' technology and knowledge transfer mechanisms. The performance and capabilities of these new structures vary greatly depending on the research capacity of the academic institution. In addition, they should offer the possibility of closer cooperation with firms with the expected sharing of laboratory facilities or the exchange of researchers.

NITs should act as strategic intermediaries to define and implement university innovation support programs. Initially, it is expected that firms will become interested in the licensing of patents generated from research findings and that universities will be able to obtain additional budgetary resources from the payment of royalties. Although the legislation emphasizes the role of NITs in intellectual property management, the scope of NITs is broader, incorporating not only the commercialization of technology and patents, but also contract management, provision of services, projects, and consultancies. In some cases, they also handle the management of university incubators and parks. They offer a legal support to the universities that acted, in a semi-formal manner, as entrepreneurs. This can be an important element to stimulate technological entrepreneurship; unfortunately, there is no evaluation of their performance.

Some studies conducted at universities with a tradition in research[14] show that there is a quantitative increase in their use of intellectual property. In 2012, 25 Brazilian universities applied for patents. However, this is insignificant if there is no licensing and the transfer of new knowledge. The main responsibility of R&D institutions is to intensify research activity to ensure results can be transferred to productive sectors. This is essential to maximize innovation potential. It is still not possible to gather accurate data on this trend.

Moreover, not all universities have adequate research (both in quality and number) to generate high-potential technology patents that can be licensed to firms. Many universities are more apt at technology transfer via consulting services. Investing in patenting should be a priority only when there are good prospects of getting some return by licensing. Patenting all research results is probably an unnecessary expenditure. There is no doubt that patents have an impact for firms that operate in strategic markets, as it facilitates interaction with the industry, particularly in some specific sectors (e.g., pharmaceuticals, electronics). Statistics from the National Intellectual Property Institute, World Intellectual Property Organization,

[12] According to the Innovation Law of 2004.

[13] The MCTI should check the compliance of this clause in the Innovation Law via information provided by universities and research institutes.

[14] An example is the State University of Campinas or the Federal University of Minas Gerais.

and United States Patents and Trademarks Office show that patenting in Brazil remains relatively uncommon and a comparison with other newly industrialized countries like Korea or China is not favorable. Furthermore, it is worrying that, contrary to what occurs in developed economies, Brazilian firms patent even less than universities.

NITs do not evaluate the effectiveness of the patent system on the capacity to innovate. While all claim to have signed technology transfer contracts, none can provide numbers as to how many contracts were signed, nor can they evaluate the cost-benefit relation of the patents. There are a very limited number of professionals with skills in managing and licensing patents.[15] With the exception of some larger universities that have strong and traditional activities in research, there is no evidence that NITs will be able to perform their role as expected by the Innovation Law (Maculan and Neder 2010).

3.5 Funding Innovation in Industry

In addition to increasing investment in research, the government has expanded the funding of innovation. The goal was twofold: to encourage firms to develop innovation projects, and to conduct internal research activities.

Two types of public funding for innovation prevailed until 2006: loans at lower interest rates than bank interest rates, and tax exemptions on imported equipment used for R&D activities. No assessment has yet been made of the effectiveness and impact of these two types of funding provided by FINEP[16] and the National Bank for Economic and Social Development. Some studies show that only a few firms took advantage of this funding, either because they were unaware of it or they did not meet the legal requirements. It can be concluded that these modalities were not appropriate to foster innovation.

This perception is reinforced by the Technological Innovation Survey (PINTEC). The IBGE conducted four PINTEC surveys between 2000 and 2010 (IBGE 2002, 2005, 2007, 2013). The surveys analyzed the innovative practices of a sample of 12,000 Brazilian industrial firms with 10 or more employees. Data on the practices of firms that effectively innovated raise the fundamental question of the efficacy of support policies. The results show that firms innovate very little and in a very hesitant manner. Firms rarely develop new products but give preference to the modernization of equipment and machines. Moreover, few firms make use of public

[15] These activities require not only legal knowledge (negotiating agreements, contracts) and administrative expertize (general organization of the NIT), but also skills in technology areas such as chemistry, biology, and engineering, which are all essential to address issues of technology and patents.

[16] The support of the Technological Development in National Firms (ADTEN) program, which ran between 1996 and 2003, benefited only 0.07 % of Brazilian industrial firms. The loaned funds accounted for between 1.6 % and 3.0 % of the benefited firms' spending on R&D (de Negri et al. 2008).

funding. Very few firms actually invest in developing new products or processes, or have internal R&D activities. As a source of funding, firms use their own resources (for 75 % of costs). No more than 20 % of innovation development costs come from government agencies (IBGE 2010).

These surveys were very influential in defining new forms of public funding for innovation.[17] In an attempt to change this negative scenario, the government rethought the financial measures to foster innovation in firms, measures that included the Innovation Law and the Law of Goods. Both laws provide new financial tools to promote innovation.

The Law of Goods provides incentives for the development of new products and has simplified the use of tax incentives. More firms now take advantage of tax incentives; since 2006, the number of firms that have benefited from tax incentives for innovation has grown exponentially (Araujo 2010). Although the size of the firm is not a criterion to benefit from the law, the way the tax exemption is calculated tends to favor larger firms.

The modality labeled "economic subvention" represents a genuine break from the previous system because the financial aid does not need to be repaid. This "subvention" modality is designed to finance innovative activities, which can be broadly understood as the purchase of raw materials or equipment, the hiring of consultants or experts, and patent requests. Between 2007 and 2012, more than 700 firms received this grant, and over 70 % of them are small businesses or start-ups.

The grant award is put into operation by FINEP in partnership with each research state agency to expand the reach of the program. The firms that receive support work must belong to industrial sectors such as electronics, software, pharmaceuticals, capital goods, bio fuels, aeronautics, nanotechnology, biotechnology, and alternative energy. Those sectors are considered priorities by the industrial policy presented by the government in 2003. The results have not yet been fully analyzed and therefore the actual impacts are unknown. Some results appear to indicate that the adoption of the two laws has provided an environment more conducive to innovation and cooperation between firms and universities.

Several studies show the importance of the accumulation of internal skills in firms for the efficient external transfer of technology. Therefore, one objective of the program is to get firms to hire researchers to leverage the internal research capacity and optimize the transfer of technology acquired externally. A provision of the law allows the partial funding of the salaries of researchers hired by the firms. Thus, firms are able to collaborate with R&D institutions and develop joint research. Based on this premise, it is also expected that university researchers could work in firms for a limited time.

Other programs destined for start-up firms complete the picture of public funding for innovation. Special attention is given to technology-based firms that are university spinoffs. State innovation agencies and university business incubators are mobilized to run these programs. This initiative aims to increase the reach of funding programs that until now had been almost completely unknown to smaller firms.

---

[17] In previous decades, traditional S&T policies have paid little attention to the innovation process, because of (a) the lack of research on the nature and intensity of innovation activities of firms, and (b) the priority given to scientific research conducted in universities.

The PRIME program[18] was implemented in early 2009. Its goal is to offer financial support to start-up firms and to consolidate the initial phase of projects without the need for entrepreneurs to have parallel activities (in general consultancy or provision of services) as a makeshift solution to ensure the maintenance of the firm. PRIME allows new entrepreneurs to fully devote themselves to their major project, the development of products, and their integration into the market. PRIME's first offer received over 3000 requests, and approximately 1300 firms obtained financial support. The proliferation of start-ups with better chances of survival is expected to leverage the growth of Brazilian economy.

Tax incentives, funding with low interest or no need for reimbursement used to support projects in cooperation with research institutions, the economic grant program and financial assistance for hiring researchers are all tools that aim to reduce the costs of the innovation process. Although they serve as an incentive for firms to be able to develop innovations, these programs do not increase user demand for new products.

3.5.1 *Firms Investing in In-House Research*

The IBGE surveys show that firms still undertake few innovative activities. According to data from PINTEC, the few firms that have internal R&D activities maintain on average 9.4 people dedicated solely to research. Of this group, 60 % have a university degree (2.5 % are doctorate holders and 6.6 % master holders).

The Law of Goods has provisions facilitating the hiring of master and doctorate holders so that firms have employees able to identify new knowledge and incorporate new technology into the production layout. This is also a prerequisite to develop research projects in cooperation with universities and to use patent protected technology. However, industrial firms show little interest in developing radical innovations based on the results of academic research.

Data from the most recent IBGE survey (IBGE 2013) show a small increase in the number of people devoting themselves solely to research activities within a firm (an average of 11.3 employees). The absorption of people with skills and qualifications in R&D remains very low. This is an important indicator of the industry's limited R&D efforts, and reflects the small impact that government incentives have to encourage firms to recruit PhD holders. It remains a challenge for the government to make firms enhance their technological strategies. The decrease in high-tech exports shows that to withstand competition from China and India, export rates must increase as soon as possible.

The availability of new funding possibilities for R&D activities means firms require a greater capacity for organization and planning, and to formulate clearer strategies in developing their innovation projects. Strengthening firms' capacity to innovate, in many cases, involves significant changes in their organizational culture

[18] PRIME: Program for New Innovative Firms.

and, more specifically, in the structuring of R&D activities. University incubators have an important role in introducing modern methods of business management and training new entrepreneurs. They yield a class of entrepreneurs who are more efficient but limited in number. In the vast majority of small firms, managers must acquire new skills to plan innovation processes or to enter the international market.

All legal instruments of innovation support aim to reduce the costs of innovation. In this way, firms are encouraged to invest more in internal R until now, this has been a rare occurrence. This is crucial to develop more radical innovations. However, these laws do not seem to be enough to persuade firms to take bigger risks in developing new products or new technologies. Although there are significant advances in the process of technological training and learning, firms are still limiting themselves to modest incremental innovations. To achieve a significant gain in competitiveness, firms must introduce significant changes in their adaptive technological pathways.

There is a need, as suggested by a study conducted by the National Association of R&D Firms (ANPEI) and CGEE/MCTI (2009), to define new modalities of innovation support that reduce risks, and integrate into a bolder industrial policy with specific sectorial targets and an assured inflow of government orders. A successful example in the past was the creation of Embraer. The firm's competitiveness in the international market enabled aircraft exports to rank first in high-tech exports.

3.6 Regional Interconnectedness of Research and Innovation-Fostering Policies

Historically, the formulation and implementation of Brazilian R&D policies have occurred at the federal level.[19] The state of São Paulo was the only one to create a decision-making structure of funding and support for R&D. However, the adoption of new state constitutions in 1988 produced a strong regional presence in universities and research institutions, thanks to state research agencies and S&T secretariats.

Today, research capacity is still concentrated in the south and southeast of the country, where there are more R&D institutions with greater experience to raise funds for large research projects. In other regions, R&D institutions have fewer qualified human resources and scientists, and research activities are more modest. To reduce regional disparities, federal research funding programs now have a rule to

[19] Federal institutions such as the National Council for Scientific and Technological Development (CNPq), Funding Authority for Studies and Projects (FINEP), and the National Bank for Economic and Social Development (BNDES) are primarily responsible for the decision making and governance of the ST&I system. Only the state of São Paulo, because of its historical background and strong local industrial base, was able to create an autonomous system, although it is part of and funded by the federal system.

put aside a percentage of resources available to fund projects in other regions because of the limited research capacity and smaller industrial base.

To enforce such measures, agencies fostering research and innovation are creating a network of local partners to overcome the drawbacks of geographic distance, heterogeneity of regional industrial bases, and inequalities in educational and social development. The MCTI has introduced a certain degree of regional interconnectedness and state research institutions are asked to cooperate in the selection of projects. Major research programs—PRONEX, Millennium, and INCT—have been developed in networks formed by researchers belonging to scientific institutions located in various regions. The goal is to reduce regional disparity in the ability to create knowledge and in human resources training. The participation of state scientific institutions allows for greater diversity in defining research programs, allocating resources, and in the capacity to attend to the specific needs of local firms.

As firms start to be considered the main actors in the innovation system and the main target of ST&I policy, it is essential to approach them and provide a regional dimension for the funding programs. In many cases, the firms that are the true targets of the programs have no information about possible funding; in general, micro and small firms are unaware of such programs. The MCTI is aware of the difficulties of coordinating a large number of programs for institutions located all over the country. It has selected regional institutions as partners to work in cooperation with state S&T institutions. It streamlines intermediary structures, incorporating other actors in the national system of innovation (state R&D agencies, universities, technology centers, and firms) to carry out the ST&I policy and ensures greater interconnectedness in the innovation system. The decisions are made by federal agencies along with representatives of industry associations, members of other federal departments, committees of scientists, and state R&D agencies. This new configuration diminishes the bureaucratic burden of the governance of the whole system. It streamlines the selection process and increases access to programs that foster innovation. The positive impact is visible in the case of PRIME and the Program for Research Support in Firms (PAPPE) programs designed to fund small business innovation projects.

The regionalization of the ST&I policy encourages a greater geographic spread of R&D investments. Data show that 35 % of funds are now going to north, northeast, and midwest regions in Brazil, ensuring the participation of the least developed regions. This trend also directly strengthens regional innovation systems. The most notable example is the implementation of PAPPE in partnership with the Foundation for Research of the Amazonas State. The supported projects are related to local activities such as boat building, the phytopharmaceutical and phytocosmetics industries, fishing, and ecotourism. The objectives of PAPPE are to encourage partnerships between researchers and firm entrepreneurs, increase career opportunities for researchers, contribute to research and innovation in small technology-based firms, and improve the geographical distribution of investments in R&D. Thus, PAPPE demands that researchers and firms work in partnership. In this way, a culture of research and innovation can be introduced in firms. PAPPE currently operates with the help of research support agencies in 12 states.

3.7 Bottlenecks

Brazil has established a policy of incentives that will likely encourage the emergence of a knowledge-based economy to serve as an engine of growth. It follows what may be regarded as the current paradigm of global research and innovation fostering policies, seeking to encompass those dimensions that constitute a dynamic national innovation system. The new instruments to promote innovation and the efforts of government agencies to transform the Brazilian economy into a knowledge economy are very broad. The expectation is that in this way, the economy will be more competitive in the face of the global recession.

However, several measures adopted recently have encountered those structural characteristics of the innovation system that hinder the achievement of targets, act as bottlenecks, and affect the actors of the Triple Helix and Brazil's capacity to base its economic and social development in a true knowledge economy.

First, the trajectory of Brazilian S&T policy has always favored scientific development at the expense of innovation fostering in firms. Government agencies, responsible for implementing technology policy guidelines, have a long history of supporting academic research. What prevails in these institutions is the paradigm of prioritizing research in the innovation process. This is evident in the types of analysis of innovation funding projects. According to a study conducted by ANPEI and CGEE/MCTI (2009), the system to foster technological development is focused on scientific research support, although the general approach adopted by the goals of ST&I policies is to give more weight to innovation itself.

Second, the development of a strategic forward-looking vision and the role of planning in technological development-fostering institutions are poorly structured. In this context, new instruments used to encourage innovation lack a clear strategic proposal and continue to be too broad. The absence of clearly defined goals and sector prioritization make it impossible to assess compliance to industrial policy goals.

Third, the lack of prospective vision is also linked to the absolute absence of performance analysis and result indicators. The evaluation of the programs' effectiveness is completely thwarted by the lack of reports on the activities of the various fostering institutions, the investments made by each state, the stages achieved by the researchers working together with firms, and regarding the results concerning new products or services. There are no data available on the number, origin, or academic profile of the researchers benefiting from the new provisions, or on their areas of expertise in the firms.

One flaw is the lack of permanent instruments to evaluate and monitor the supported projects. It is essential to introduce a systematic evaluation of results, taking into consideration the strong growth in the number of firms that receive loans, grants, or venture capital based on resources generated by sector funds, the Innovation Law, and the provisions of Tax Incentives for Innovation (Morais 2007). This is important to define new targets for information on the success rates, costs, or timelines of projects that result in new products, on factors that prevent the completion of abandoned projects, and on successful experiences that could serve as references for projects that fail to have satisfactory results. To better characterize the profile of firms that receive subsidies and to determine their returns, it is crucial to know the impact of the programs on the recipient firms, in terms of innovations obtained, jobs created, the nature and number of links with universities, and the increase in their technology exports.

3.8 Challenges for a New ST&I Paradigm

The solutions for continuous growth must have global dimensions and be within the realm of sustainability. These two characteristics are directly related to the environmental crisis and the exhaustion of a certain mode of industrialization, which is not concerned about the finiteness of natural resources. Both the environmental crisis and the financial crisis are of global concern, and their effects have no borders. The profound changes that must be introduced into the productive paradigm require significant advances in knowledge and thus research investments. This will only be possible if the state has a strategic policy of innovation. In an interdependent and globalized economic system, innovation has an important role because it determines the capacity of each nation's economy to find its own niche.

Brazil is experiencing a boost in terms of innovation, from an understanding that there is a need to diversify the economy, expand exports, and increase productivity and competitiveness. The country is in a crucial stage of its economic development. Brazil has come through the global financial crisis with less harm than most countries and has had satisfactory rates of growth over the last decade. Still, it faces serious challenges, such as competition from cheaper imported products—which threaten some industrial sectors—and the relative decline of the export of high-tech products (Fig. 3.4).

A decade after the renewal of Brazil's S&T policy, it is possible to make a first evaluation of the progress thus far. With the goal to form a truly national system of innovation, and the expectation to strengthen the competitiveness of its firms, Brazil has adopted measures to encourage research and innovation that are common in the ST&I policies of most OECD countries. There is no doubt that the interactions between the actors in the Triple Helix have intensified markedly and that this should continue.

However, despite significant advances in terms of research, doubts remain as to the government's ability to change Brazil into a knowledge economy, which would ensure sustainable growth. Although Brazil has made significant progress in recent years, there is still a long way to go.

Some questions remain: why do Brazilian companies show so little innovation? Why does the Brazilian economy have so much trouble incorporating new knowledge and developing more radical innovations? Is it necessary to seek a new set of policies and instruments capable of guiding Brazil to positively globalize its economy? How to turn knowledge into the true backbone of the economic system?

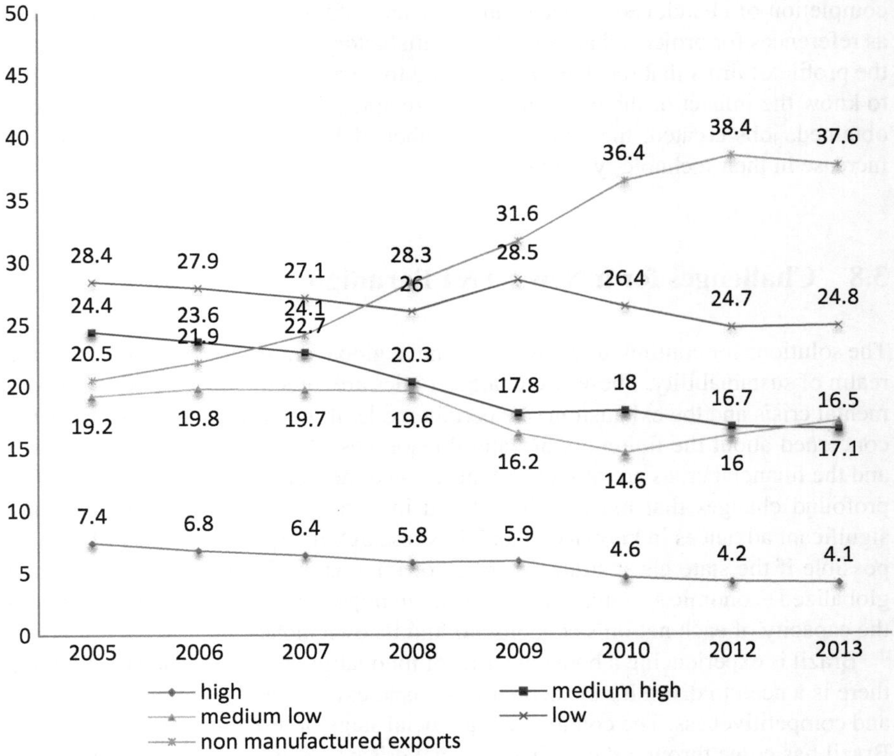

Fig. 3.4 Brazilian exports according to their technology intensity (%) (Source: http://www.desenvolvimento.gov.br/sitio/interna/interna.php?area=5&menu=1113)

Brazil, as an emerging country, is increasingly engaging in the global economy and in a diversified manner. The redefinition of the main trends in the global economy will be a test to verify the effectiveness of these advances.

Acknowledgment We would like to thank CAPES/MEC for financial support for this work.

References

Araujo, B. C. (2010). *Incentivos fiscais a pesquisa e desenvolvimento e os custos de inovação no Brasil* [Tax incentives for R-D and innovation costs in Brazil] (pp. 3–11). Brasilia: IPEA 46. RADAR 9, agosto de 2010.

Cassiolatto, J. E., & Lastres, H. M. M. (1999). *Local, national and regional systems of innovation in the Mercosul*. In Paper prepared for the DRUID's summer conference on national innovation systems, industrial dynamics and innovation policy, Rebild, 9–12Jun 1999.

CGEE. (2008). *Características do Emprego dos Doutores Brasileiros: Características do emprego formal no ano de 2004 das pessoas que obtiveram título de doutorado no Brasil no período*

1996–2003 [Employment characteristics of Brazilian PhD holders. Features of formal employment in 2004 of persons who hold doctorate in Brazil in the period 1996–2003.] http://www.cgee.org.br/publicacoes/doutores.php. Accessed 6 Apr 2011.

da Motta e Albuquerque, E. (2003). *Immature systems of innovation: Introductory notes about a comparison between South Africa, India, Mexico and Brazil based on science and technology statistics*. Belo Horizonte: UFMG/Cedeplar.

de Morais, J. M. (2007). *Políticas de apoio financeiro a inovação tecnológica: Avaliação dos programas MCT/FINEP para as empresas de pequeno porte* [Financial support of technological innovation: Evaluation of MCT/FINEP programs for small business]. TD 1296. Brasilia: IPEA. http://www.ipea.gov.br. Accessed 13 May 2008.

de Negri, J. A., de Negri, F., Lemos, M. B. (2008). *O impacto do programa ADTEN sobre o desempenho e o esforço tecnológico das empresas industriais brasileiras* [The impact of ADTEN program on performance and technological effort of Brazilian industrial companies]. IPEA. http://www.ipea.gov.br/portal/index.php. Accessed 28 Mar 2013.

dos Santos, C. M. C. (2010). *A Transferência de Conhecimento segundo o conceito de Universidade Empreendedora: Algumas Experiências Na Universidade Federal Fluminense (UFF)* [The knowledge transfer under the concept of entrepreneurial university: Some experiences of the Federal Fluminense University]. *Dissertação de mestrado*. PEP/COPPE/Universidade Federal do Rio de Janeiro.

IBGE. (2010). *PNAD. Pesquisa Nacional por Amostra de Domicílios*. [National Survey of Households]. Volume 30. http://www.ibge.gov.br/home/presidencia/noticias/noticia_visualiza.php?id_noticia=1708. Accessed 24 Jul 2011.

IBGE. (2002, 2005, 2007, 2013). PINTEC. Pesquisa de Inovação Tecnológica. [Technological innovation survey]. http://www.ibge.gov.br/home/estatistica/economia/industria/pintec/2011/default.shtm. Accessed 9 Dec 2013.

IBGE. (2013) *PNAD Pesquisa Nacional por Amostra de Domicílios. Síntese de indicadores 2012* [National Survey of Households. Indicators Synthesis 2012*]*. Rio de Janeiro 2013. Accessed 30 Sep 2013.

INEP. (2013). *Sinopse 2013 [Synopsis 2013]*www.inep.gov.br/superior/censosuperior/sinopse/default.asp. Accessed 4 Nov 2013.

IPEA. (2010). *Comunicados do IPEA N° 41 Emprego e oferta qualificada de mão de obra no Brasil: impactos do crescimento econômico pós-crise* [Employment and qualified supply of labor in Brazil: Impacts of post-crisis economic growth]. http://ipea.gov.br. Accessed 4 Mar 2011.

Maculan, A.-M., & Mello, J. M. (2009). University start-ups for breaking lock-ins of the Brazilian economy. *Science and Public Policy, 36*, 109–114.

Maculan, A. M, & Neder, A. R. (2010). *Projeto PIBIC—Núcleos de Inovação Tecnológica em Universidades Empreendedoras*. Relatório de pesquisa [PIBIC project—Technological innovation offices in entrepreneurial universities]. COPPE/Universidade Federal do Rio de Janeiro.

Parreiras, V. M. A. (2010). *Processos de Transferência de Conhecimento e Interações entre Universidades e Empresas: o caso da PUC-Rio. Dissertação de mestrado*. [Knowledge transfer and interactions between universities and firms: The case of PUC-Rio.] PEP/COPPE/Universidade Federal do Rio de Janeiro.

Saavedra Durão, V. (2010). *Empresa alemã trouxe 600 Chineses para obra de usina* [German company hires 600 Chinese workers]. *Valor Econômico*. http://www.valoronline.com.br/impresso/brasil/97/124112

Viotti, E. B. (2002). National learning systems: A new approach on technological change in late industrializing economies and evidences from the cases of Brazil and South Korea. *Technological Forecasting & Social Change, 69*, 653–680.

Chapter 4
Current Challenges in Higher Education and Their Implications for Research and Development in *Colombia*

Marta Losada

4.1 Introduction

The relevance of educational attainment to achieve economic development and human wellbeing has been comprehensively studied. Recently there have been updated studies and analyses regarding key issues in higher education and the advanced training of human resources in Latin America (see Heitor and Horta 2014 and the references therein). A main conclusion from these studies emphasizes the need to broaden the base of human resources involved in knowledge activities and to promote a much stronger research system. There are several other reports that contain useful reference material, benchmark data, and policy recommendations (OECD 2014a; European Commission 2014a; OECD/International Bank for Reconstruction and Development/The World Bank 2012).

This chapter presents the key indicators for Colombia in both research and development (R&D) and higher education benchmarked at the international level. These reference values illustrate the significant gaps that exist and provide the basis for the suggestion of new policies. To address the main topic of this chapter, we start with a quick overview of the status of education in Colombia. The data presented in this chapter are mostly taken and adapted from the original statistical sources of UNESCO Institute for Statistics (UIS), Organisation for Economic Co-operation and Development (OECD) Science and Technology Statistics, the National Ministry of Education of Colombia (MEN), Colombian Institute for the Promotion of Higher Education (Instituto Colombiano para el Fomento de la Educación Superior; ICFES), the Colombian Observatory for Science and Technology (OCyT), and Network Indicators of Science and Technology (Red de Indicadores de Ciencia y

M. Losada (✉)
Universidad Antonio Nariño, Bogota, Colombia
e-mail: rector@uan.edu.co

H. Horta et al. (eds.), *Trends and Challenges in Science and Higher Education*, Knowledge Studies in Higher Education 3,
DOI 10.1007/978-3-319-20964-7_4

Tecnología; RICyT) (UIS2 2014, UIS 2014; OECD 2014a, b; MEN 2014a; OCyT 2014; RICyt 2014).

As can be seen in Table 4.1, Colombia has obtained significant results in getting close to universal primary and secondary education for the corresponding age ranges. The gross enrollment ratio at the tertiary level has increased significantly, and the gender gap has been essentially overcome at the undergraduate level.

It should be noted that there have been increased efforts to achieve greater student enrollment in higher education, given the poor results that were the norm until recently. Data from the 1993 census (National Administrative Department for Statistics; DANE) show the difference in educational attainment levels for men and women aged 20–24 and 55–64 years (Table 4.2). In addition, in the early 1990s, the gross enrollment ratio in higher education in Colombia was among the lowest in Latin America and the Caribbean (LAC). Enrollment in higher education has increased from close to half a million students in 1990 to approximately 2,100,000 in 2013 (MEN 2014b).

More recent data enable us to see the evolution of educational attainment levels in Colombia and to compare these with world averages. Data from 2011 show that for the population aged 25 years and older, 7.9 % have no schooling, 35.7 % have primary-level, 36.2 % have secondary-level (both ISCED[1] levels), and 19.7 % have tertiary-level education. In addition, these data show that 19.2 % of men and 20.2 % of women have tertiary-level education. Table 4.3 shows the educational attainment of Colombia in comparison with other countries. From this we can infer that signifi-

Table 4.1 Gross and net enrollment ratio for primary, secondary, and tertiary education for 2012

	Gross enrollment primary (%)	Net enrollment primary (%)	Gross enrollment secondary (%)	Net enrollment secondary (%)	Gross enrollment tertiary (%)
Women	106.9	83.5	96.8	76.6	47.7
Men	108.7	86.2	89	70.8	42.4
Total	106.9	83.9	92.8	73.6	45

Source: UIS, Country profiles

Table 4.2 Distribution of education attainment levels for given age groups in the 1993 census

	None (%)	Primary (%)	Secondary (%)	Higher (%)	Total (%)
Women aged 20–24	3.8	36.7	48.3	13.2	100
Men aged 20–24	6.9	38.3	45.6	11.2	100
Women aged 55–64	21.4	58.6	17.4	2.6	100
Men aged 55–64	19.2	58.7	16.2	5.9	100

Source: DANE 1993 Census

[1] ISCED: International Standard Classification of Education.

Table 4.3 Educational attainment levels for population aged 25+ for Colombia (2011) and other countries (2010)

	Below upper secondary (%)	Upper secondary (%)	Tertiary (%)	Total (%)
Colombia	58	22	20	100
OECD average	26	44	30	100
EU21 average	24	48	28	100
G20	41	33	26	100

Source: UIS, Education Database

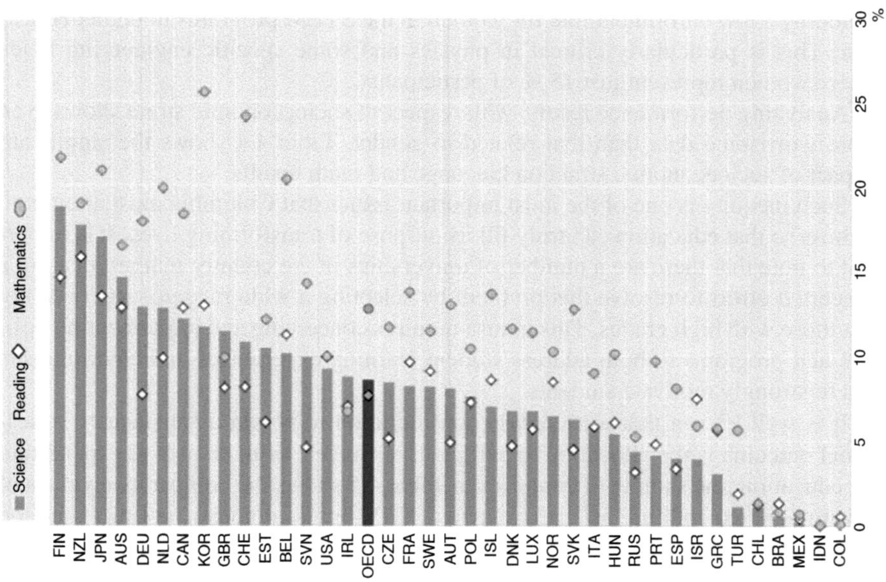

Fig. 4.1 Science, reading, and mathematic top performers at age 15, 2009 PISA results (Source: Adapted from OECD, PISA 2009 database)

cant efforts must still be made to enable a larger proportion of the population to acquire at least upper-secondary and tertiary-level schooling. It is important to keep in mind that EUROPE 2020 (European Commission 2014b) defined a specific goal to increase educational attainment at the tertiary level to reach 40 % for the 30–34 age group by 2020 (the 2012 value is 37 % for the 30–34 age group for EU27 countries).

Although these certainly are improvements over a period of almost two decades, there are a number of critical issues that should be mentioned. Performance results in language, math, and science are poor in comparison with other countries, in particular those countries that are part of the OECD. Figure 4.1 shows the top performer percentages for the 2009 Program for International Student Assessment

(PISA) for different countries including Colombia. The graph shows a negligible number of top performers from Colombia.

Similar to many other countries, Colombia has noticeable gender gaps in education performance. There are some significant gender gaps present in language and math, and these increase specifically between fifth and ninth grades (ICFES 2014). For example, the average scores from national math exams show little difference between fifth grade boys and girls; however, there is a noticeable difference by the ninth grade. There are significant gender differences in reading, math, and science in Colombia in the 2012 PISA assessment results: the average score for girls is nearly 10 % lower than the average score for boys. This is an important issue to address in general. Furthermore, this difference has a number of consequences including a low enrollment rate for women in the STEM programs in higher education. This is particularly critical in physics and some specific engineering fields where women represent just 15 % of participants.

Analyzing performance results with respect to socioeconomic strata shows even more worrisome data than that related to gender. Table 4.4 shows the significant impact of socioeconomic strata on language and math results.

Such inequity is one of the most important issues that Colombia must attempt to address so that education can truly fill its purpose of transforming lives. It is important to note that there are a number of universities in the country that are making a concerted effort to resolve this problem by selecting a wide range of students, not just those with high grades. This in turn requires conceiving and implementing policies and programs to both address student learning experiences and environments and to strongly motivate students.

It is well known that a necessary consideration to obtain a satisfactory educational outcome is the allocation of sufficient resources. With respect to expenditure for education, the sample of countries in Table 4.5 shows the comparative values of government expenditure as a percentage of GDP (gross domestic product) and the amount per student for primary, secondary, and tertiary education levels. The OECD (2014) reported that the expenditure per student at the tertiary level in Colombia was significantly higher than that reported by UIS for 2012. An important aspect to note is that there are significant differences with respect to private expenditure in education among these countries, given the variations in the educational systems. While Spain, Portugal, and Denmark are dominated by public education, in Colombia and Korea especially, there is a strong component of private higher education institutions.

Table 4.4 Results in language and math on national tests for fifth and ninth grade; average scores for four different socioeconomic levels NS1, NS2, NS3, NS4

Average score	NS1	NS2	NS3	NS4	Average
Language fifth grade	255	277	301	353	291
Language ninth grade	252	273	294	344	293
Math fifth grade	246	270	294	343	282
Math ninth grade	247	267	289	344	289

Source: ICFES, www.icfes.gov.co extracted 2014

Table 4.5 Government expenditure on education as % of GDP and per student (in USD PPP) for different levels of education for selected countries

	Colombia 2012	Spain 2010	Portugal 2010	Korea 2011	Denmark 2009
% of GDP	6.4	5.0	5.6	5.2	8.7
Per student in primary education	1609	6816	5656	6641	11,169
Per student in secondary education	1589	8430	9394	7010	12,738
Per student in tertiary education	2449	9057	7685	3491	21,971

Source: UIS, Country Profiles

4.2 Higher Education

As mentioned above, Colombia presents an interesting result with regard to gender parity for gross enrollment at the tertiary level. However, it is important to point out that gender parity varies for undergraduate, master, and doctoral levels. There are also significant differences across the fields of knowledge. For example, women only represent less than 30 % of all graduates in engineering. Table 4.6 shows the number of graduates from university undergraduate, master's degree, and doctoral degree programs by gender for 2002–2011. With respect to the total faculty in 2013, 74,031 university professors were men and 41,746 women.

Table 4.7 illustrates the distribution of students enrolled in tertiary-level programs according to ISCED 5A, 5B, and 6 levels as defined by UIS. There are two striking differences between the distributions by tertiary ISCED level in Colombia and those of other countries. First, Colombia has lower enrollment rates in ISCED 5A than most developed countries, but even more importantly, for those that have similar values of enrollment in ISCED 5A level, there is huge difference in the enrollment of ISCED 6. This extremely low proportion of advanced research degrees, doctoral or the equivalent, is one of the main challenges to be addressed.

It is even more worrisome if we take into account the evolution of the distribution among tertiary ISCED levels for 1999–2012. Tables 4.8 and 4.9[2] show this evolution for Colombia and Korea, respectively.

The increase in ISCED 5B enrollment in Colombia has been due to a strongly promoted public policy since 2002. This is unfortunate given that essentially, as discussed above, the lower socioeconomic population largely has this option for tertiary education because of their low performance outcomes at the end of secondary school. In Colombia, as in most countries, there is a strong correlation between

[2] We use data as reported in the UNESCO database, although the value for 1999 of percentage enrolled in ISCED 6 level is not consistent with other national data sources.

Table 4.6 Gender distribution of graduates for 2002–2011 from undergraduate, master, and doctoral programs

Program level	Men	Women
University undergraduate	477,279 43.4 %	624,082 56.7 %
Master	23,930 56.3 %	20,173 45.7 %
Doctoral	926 66.8 %	504 35.2 %

Source: MEN

Table 4.7 Distribution of enrollment by tertiary ISCED level for 2012

Country 2012	% ISCED 5A	% ISCED 5B	% ISCED 6
World	77.9	20.7	1.4
Spain (2011)	80.9	15.6	3.5
Portugal	95.0	0.01	6.9
Chile	56.9	46.7	0.4
Colombia	68.0	31.8	0.2
Korea	76.0	26.1	1.9
Denmark	86.7	12.0	3.3

Source: UIS, Education dataset, extracted June–September 2014

Table 4.8 Evolution of enrollment in ISCED levels from 1999 to 2012 in Colombia

Level	1999	2004	2009	2012
ISCED 5A	77.9	81.8	67.6	68.0
ISCED 5B	16.9	18.1	32	31.8
ISCED 6	5.2	0.1	3.2	0.2

Source: UIS, Education dataset, extracted June–September 2014

Table 4.9 Evolution of enrollment in ISCED levels from 1999 to 2012 in Korea

Level	1999	2004	2009	2012
ISCED 5A	58.7	59.7	76.8	76.0
ISCED 5B	40.3	39.1	23.6	26.1
ISCED 6	1.0	1.2	1.6	1.9

Source: UIS, Education dataset, extracted June–September 2014

level of education and increased wages and lower unemployment rates. Thus, the current framework constrains the effective social mobility of the population.

One of the consequences of relatively low proportions of the population with advanced research degrees is the low number of university professors with doctoral or equivalent degrees. For 2013, of the 116,819 university professors in Colombia, only 6803 had PhD degrees, which corresponds to just 5.8 % (MEN 2014b). These data represent teaching staff at all higher education institutions including those that offer 2-year programs.

4.3 Research and Development

This section includes an in-depth analysis of current R&D expenditure in Colombia in comparison with other countries. This analysis will clearly show that a modification in public policy with respect to R&D expenditure is crucial. The gross expenditure on R&D (GERD) includes the expenditure on R&D from business enterprises (BERD), higher education (HERD), government (GOVERD) and private non-profit organizations (PRIVRD). South America contributes only 2.5 % of global R&D expenditure, corresponding to 36 billion USD PPP for 2011 (National Science Foundation 2014).

The amount of R&D expenditure for 2012 as a percentage of the GDP for Colombia is in the order of 0.18 %. To illustrate and benchmark Colombia's R&D ecosystem, we compare a few key R&D indicators with those of a number of other countries. The countries selected for more detailed comparisons fulfill criteria such as similar population size (Spain, Korea), regional benchmark (Chile, Mexico, Brazil), transforming their science and technology sectors over last several decades (Korea, Portugal, Spain), and public–private composition of the higher education sector (Chile, Korea), among others. First, we show in Tables 4.10 and 4.11 the highest and lowest expenditures of OECD countries for GERD, BERD, HERD, and GOVERD as percentage of the country's GDP. It is clear that there is a large spread in these values, and variations in the proportions from each sector are characteristics of each country's economy and strengths or weaknesses.

We present in Table 4.12 comparative reference data of each country's GDP, GDP per capita population, and GERD per capita population. The OECD average value for GERD per capita population is 882.91 current PPP USD in 2012.

Table 4.10 Highest and Lowest GERD and BERD as percentage of GDP of OECD countries for 2012

Highest GERD (% GDP)		Highest BERD (% GDP)	
Korea	6.35	Korea	3.39
Finland	3.55	Japan	2.56
Sweden	3.40	Finland	2.44
Japan	3.35	Sweden	2.30
OECD	2.39	OECD	1.62
Lowest GERD (**% GDP**)		Lowest BERD (**% GDP**)	
Turkey	0.92	Turkey	0.41
Poland	0.89	Slovak Rep.	0.34
Slovak Rep.	0.82	Poland	0.33
Greece	0.69	Greece	0.23
Chile	0.35	Chile	0.11

Source: OECD Science and Technology Indicators, extracted June 2014

Table 4.11 Highest and Lowest HERD and GOVERD as percentage of GDP of OECD countries for 2012

Highest HERD (% GDP)		Highest GOVERD (% GDP)	
Denmark	0.95	Korea	0.49
Finland	0.77	Germany	0.42
Austria	0.72	Czech Rep.	0.36
Netherlands	0.70	Slovenia/USA	0.36
OECD	0.43	OECD	0.27
Lowest HERD (**% GDP**)		Lowest GOVERD (**% GDP**)	
Slovenia	0.29	Portugal	0.097
Slovak Rep.	O.28	Ireland	0.08
Greece	0.27	Israel	0.07
Hungary	0.24	Switzerland	0.023
Chile	0.12	Chile	0.014

Source: OECD Science and Technology Indicators, extracted June 2014

Table 4.12 GDP in billions of USD 2012, GDP per capita (current PPP USD) 2012, GERD per capita population (current PPP USD) 2012

Country	GDP	GDP per capita	GERD per capita
Spain	1,481	32,043	423.6
Chile	391	22,363	75.4
Colombia	498	10,436	18
Portugal	267	25,389	386.9
Korea	1,540	30,801	1,307.8

Source: RICyT and OCyT

We present the GERD as a percentage of GDP for the OECD average and a group of countries in Table 4.13 and the corresponding portion from BERD, HERD, and GOVERD.

The total R&D expenditure for 2011 by funding source (business enterprise, government, higher education, private non-profit, and abroad) is shown in Table 4.14, clearly showing how Colombia contributes a significantly higher proportion of its own higher education resources to R&D expenditure.

It is useful to understand how each sector contributing to the total GERD of a country is funded. Table 4.15 shows the total R&D expenditure in business enterprise, higher education, and origin of funding (business enterprise, BE; higher education, HE; government, GOV; private, PRIV; international, INT) for Portugal, Korea, Spain, and Chile. The table also shows the percentage per funding source for each segment and the average values. On average, 87 % of business enterprise expenditure on R&D is financed by the same sector, while government funds average 9 %. For the higher education sector, the average government funding is 71 % of HERD, with HE sector funds at 17.5 %, and business enterprises at 4.3 %. It should be noted that the inclusion of Chile significantly lowers the average value of government contribution to HERD funding.

Table 4.13 GERD, BERD, HERD and GOVERD as a percentage of GDP for selected countries

Country	GERD (% GDP)	BERD (% GDP)	HERD (% GDP)	GOVERD (% GDP)
OECD	2.39	1.62	0.43	0.27
Spain	1.31	0.69	0.36	0.26
Portugal	1.5	0.69	0.55	0.11
Chile	0.35	0.11	0.12	0.014
Colombia	0.18	0.043	0.068	0.006

Source: OECD Science and Technology Indicators, extracted June 2014, RICyT, OCyT

Table 4.14 R&D expenditure in 2011 by source of funds

Country	Government (%)	Business enterprise (%)	Higher education (%)	Private (%)	Abroad
Spain	46.47	46.31	3.98	0.55	6.67 %
Colombia	41.85	30.85	16.98	6.54	3.76 %
Chile (2010)	37.26	35.37	10.28	1.34	15.73 %
Portugal	41.80	46.06	6.22	2.02	5.87 %
USA	31.01	62.43	3.02	3.52	–
LAC	56.03	41.62	3.28	0.25	0.80 %
Ibero-America	49.16	41.07	6.55	0.69	6.50 %

Source: RICyT

In comparison, the distribution of expenditure in Colombia for science, technology and innovation activities (STIA), which include the corresponding expenditure for R&D, is shown in Table 4.16 for both business enterprises and higher education.[3] The table also shows the origin of funding for total R&D expenditure in Colombia for 2011. We can conclude that the government clearly underfunds the expenditure in HERD (covering only 30.7 %) compared with the average values shown for Portugal, Chile, Spain, and Korea (average of 71.1 %).

As will be shown in more detail, the dominant cost of R&D expenditure in higher education is labor costs. Thus, there is a significant correlation between the number of researchers and the GERD of a country. Figure 4.2 illustrates this point clearly; Colombia is located on the lower left side of the graph. This view is much more useful because a direct comparison of GERD per researcher can be misleading: if both values are small, the ratio is not necessarily small (see Table 4.17). In contrast, there are large differences in the number of researchers per 1000 labor force, shown both for headcount and full-time equivalent (FTE)[4] in Table 4.18.

Another important difference arises in the distribution of researchers by performance sector in each country. Figure 4.3 shows the distribution of researchers by

[3] STIA includes other types of costs, some of which are not directly related to R&D. The total expenditure on STIA as a percentage of GDP for Colombia was 0.46 % for 2012 OCyT. (n.d.). *Indicadores de Ciencia y Tecnología de Colombia.* Retrieved June 2014, from Observatorio Colombiano de Ciencia y Tecnología: ocyt.org.co.

[4] The exchange rate has fluctuated between 1800 and 2000 COP per USD in recent years.

Table 4.15 Amount of R&D expenditure (in 2005 PPP millions and %) in the BE and HE sectors and origin of funding for selected countries

SOURCE	PORTUGAL		KOREA		SPAIN		CHILE (2010)		AVERAGE %	
	BE	HE	BE	HE	BE	HE	BE	HE	BE	HE
TOTAL	1,626.1	1,315.0	42,401.6	5,589.9	8,767.6	4,743.9	400.4	316.3		
BE	1,491.3 91.7%	23.9 1.8%	39,755.4 93.7%	614.9 1.1%	6,835.1 77.9%	379.9 8.0%	339.8 84.8%	20.2 6.3%	87%	4.3%
GOV	65.5 4.0%	998.5 75.6%	2,567.8 6.0%	4,480.5 80.1%	1,260.8 14.4%	3,410.2 71.8%	59.9 14.9%	180.2 56.9%	9%	71.1%
HE	0.16 0.01%	210.1 15.9%	16.2 0.04%	368.1 6.9%	2.7 0.03%	657.2 13.8%	0.2 0.04%	105.5 33.3%	0.03%	17.47%
PRIV	0 --	14.5 1.1%	8.7 0.02%	104.3 1.8%	16.32 0.18%	45.1 0.9%	0.07 0.001%	1.8 0.5%		
INT	69.1 0.06%	68.1 5.1%	53.5 0.1%	22.1 0.4%	652.6 7.4%	251.52 5.3%	0.29 0.07%	8.6 2.7%		

Source: Adapted from OECD, Science and Technology Indicators 2011

Table 4.16 Distribution of R&D and STIA expenditure and source of funding for Colombia; the STIA amounts are given in millions of Colombian pesos and the corresponding percentage distribution

	R+D (%)	STIA BE	STIA HE
Total	100	943.5	636.24
BE	30.85	716.3 76 %	111.30 17.4 %
GOV	41.85	227.2 24 %	195.96 30.7 %
HE	16.98	–	282.65 46.4 %
PRIV	6.54	–	–
INT	3.76	–	46.31 7.2 %

Source: OCyT and RICyT

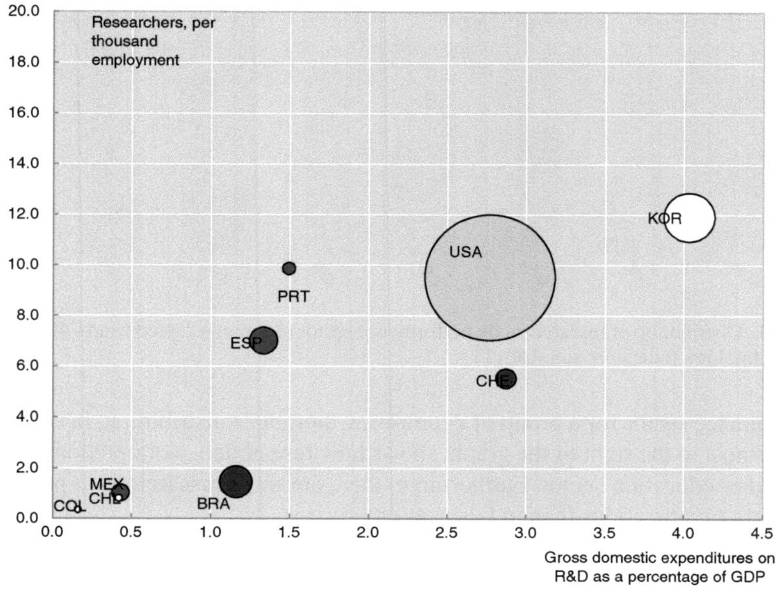

Fig. 4.2 Researchers per 1000 employment versus GERD as a percentage of GDP; the size of the circle for each country quantifies the total R&D expenditure (Source: Adapted from OECD Science and Technology Indicators, extracted June 2014)

Table 4.17 GERD per researcher (1000 s of PPP USD) in 2011

Country/region	GERD/ headcount	GERD/FTE
Spain	89.72	151.74
Colombia	53.33	100.24
Chile (2010)	122.02	212.03
Portugal	41.19	82.63
LAC	92.98	161.32
Ibero-America	85.56	149.72
Brazil (2010)	110.45	87.04

Source: RICyT

Table 4.18 Number of researchers (headcount and FTE) per 1,000 labor force in 2011

Country/region	Headcount	FTE
Spain	9.53	5.63
Colombia	0.69	0.37
Chile (2010)	1.19	0.68
Portugal	18.26	9.10

Source: RICyT

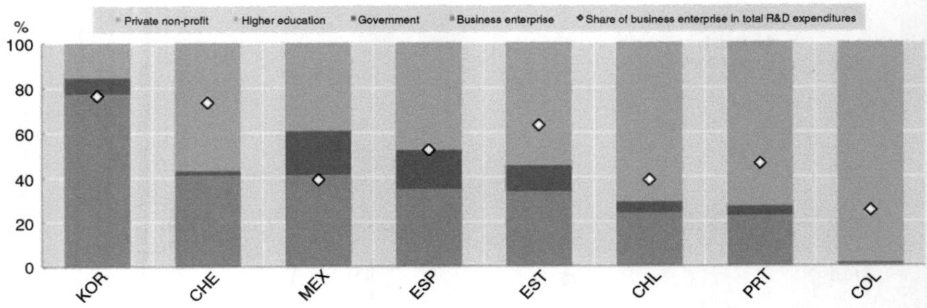

Fig. 4.3 Distribution of researchers by performance sector (Source: Adapted from OECD Science and Technology Indicators and RICyT)

performance sector for a group of countries. Colombia's distribution, in the column furthermost to the right of the graph, shows how researchers work predominantly in the higher education sector. Furthermore, there are more researchers in private not-for-profit organizations than in business enterprises.

4.3.1 Research and Development in Higher Education

R&D has been a significant activity in higher education over many decades with profound implications for the advancement of human knowledge, changing paradigms, establishing new theories, developing new technologies, and in general, impacting not only the learning process but society as a whole. Recent decades have also seen increased levels of technology transfer that are ultimately integrated into the knowledge economy. Looking at the R&D expenditure in higher education as a percentage of GDP, Colombia has the smallest HERD of the group of our sample countries. If we break down total HERD costs, standardized to the methodology of the Frascati Manual, we see in Table 4.19 the distribution of the costs among current costs, land and buildings, and equipment. In addition, current costs are split as current labor costs and other current costs. Clearly, current labor costs are the most significant cost for Spain, Portugal, and Chile with respect to the total amount. For Colombia, the breakdown is not available and it could be that current labor costs are undervalued in the amount that is officially included as HERD. Given that there has been an increase in the number of active researchers, it is not clear whether the full labor costs are being properly accounted for as HERD.

Given the large proportion of HERD costs arising from associated labor costs, we analyze the number of personnel involved in R&D activities. In 2012, Spain had 127,129 FTE R&D staff in all sectors, of whom 39 % had doctoral degrees. In higher education there were 59,775 FTE R&D staff and 68,865 PhD students (2011). The average number of PhD graduates over the 2007–2011 period was close to 8000 per year, of which nearly 3400 were graduates in science and engineering. Korea, in 2011, had more than double the number of Spain, with 288,900 FTE R&D

Table 4.19 Distribution of HERD costs per country in 2005 PPP millions USD according to the accepted usage for disclosure of R&D expenditure using the Frascati Manual

	Spain	Portugal	Chile	Korea	Colombia
Total costs	4,743.95	1,315.0	316.3	5,589.9	294.2
Current	4,188.3	1,252.7	262.0	5,189.2	
Current labour	3,227.6	915.3	189.6	1,930.1	
Current other	960.7	337.4	72.4	3,259.0	
Land and buildings	202.4	15.1	15.3	4.8	
Equipment	353.1	47.2	38.9	395.9	

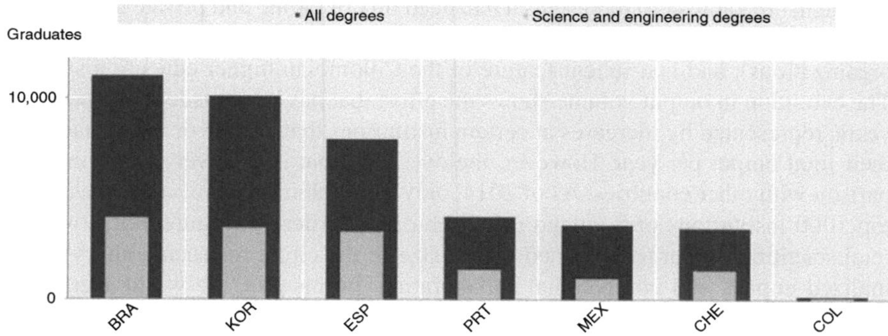

Fig. 4.4 Average numbers of doctoral graduates per year for selected countries (Source: Adapted from OECD, Science and Technology Indicators)

staff. Of these, 40,844 were FTE staff working in higher education institutions, which is significantly less than in Spain. Korea had 62,312 PhD students in 2012 with about 10,000 graduates per year. In 2012, Portugal had 50,694 FTE R&D staff, of which 30,185 were FTE R&D staff in HE. Of the total FTE, 27.2 % had doctoral degrees with 19,227 PhD students that same year.

In contrast, in 2010 Chile had 5440 FTE R&D staff and 3274 (FTE) R&D staff in HE. Furthermore, 43 % of R&D staff in all sectors had doctoral degrees. There were 4328 PhD students in 2012 and the average number of PhD graduates was 355 per year, of which 221 were in science and engineering. Finally, Colombia had a FTE R&D staff of 8675, with 35 % holding doctoral degrees, a FTE R&D staff in HE of 7535, and 3063 PhD students in 2012. The number of PhD graduates in 2011 was 258 over all areas of knowledge, and 156 in science and engineering.

Figure 4.4 shows the average number of PhD graduates per year for seven countries and the number of graduates in science and engineering. It is evident that Colombia must make a concerted effort to increase its number of doctoral students. Furthermore, up until 2014, no formal initiatives existed to fund postdoctoral positions in Colombia. The 2014 call for the funding of postdoctoral positions was very limited, especially with respect to the institutions that could take part in the process.

4.3.2 Scientific Output, Impact and International Collaboration

The number of scientific articles published in international indexed journals is a standard measure of scientific output. Others measures include impact, which can be measured by citations and excellence. International collaboration is also a traditional benchmark to analyze research products. Table 4.20 shows the scientific output of the top 25 Colombian universities ordered by the total volume accumulated up to August 2014 (note: Colombia has 81 universities and almost 280 higher education institutes).

The list in Table 4.20 represents a balanced mix of public and private universities (15 and 10, respectively; by law all higher education institutions are not-for-profit organizations), and is a salient feature of the Colombian higher education system. The evolution in output volume shows the effort that has been made in the past few years, represented by increases at certain institutions that double or even quadruple their total output per year. However, the overall output is relatively small in comparison with other countries. As of 2014, only one Colombian university was in the top 1000 institutions of Scimago using the output criterion. Figure 4.5 shows the total scientific output (determined by the size of the circle for each country), normalized impact, and international collaboration. In this plot, the world average of normalized impact is fixed to equal 1. Thus, research from Colombia must produce a greater scientific impact. More detailed analyses (Scimago 2014) show that there is a strong correlation between international collaboration and scientific impact on an institutional level. The proportion of international collaboration is high for Colombian institutions, caused in part by the many active Colombian researchers who obtained their doctoral degrees abroad and continue to participate in joint international research projects. There are seven Colombian universities in the top 1000 higher education institutions for international collaboration in the Scimago database. Regarding the Scimago excellence criterion for 2014, only two Colombian universities appear in the top 1000: Universidad Antonio Nariño and Universidad de los Andes.

4.3.3 A Case Study in Colombia: Universidad Antonio Nariño

We briefly present the case study of the Universidad Antonio Nariño (UAN), which is a relatively young institution founded in 1976. The mission of the university explicitly describes its efforts to provide access to higher education for a broad profile of students, thus enabling those from a lower socioeconomic background to take advantage of new possibilities for university-level education. In addition, over the last decade or so UAN has engaged in a strong policy for capacity building to perform R&D that is measurable at the international level.

The results of these efforts are impressive, with nearly 90 % of its students coming from the three lowest socioeconomic groups, and with a high labor insertion

Table 4.20 Scientific output of Colombian universities 2011–2014

Position					Output			
2014	2013	2012	2011	University	Aug 2011	Aug 2012	Aug 2013	Aug 2014
1	1	1	1	Universidad Nacional de Colombia	6515	7951	9633	12,263
2	2	2	2	Universidad de Antioquia	4151	4890	5755	6739
3	3	4	4	Universidad de Los Andes	2758	3335	3924	4806
4	4	3	3	Universidad del Valle	3086	3434	4027	4423
5	5	5	5	Universidad Industrial de Santander	1734	1983	2242	2584
6	6	6	6	Universidad Javeriana	1448	1763	2112	2514
7	8	8	7	Universidad Pontificia Bolivariana	548	653	824	1052
8	7	7	8	Universidad del Rosario	536	692	846	1044
9	9	9	10	Universidad del Norte	381	482	666	844
10	10	12	12	Universidad de Cartagena	334	425	573	736
11	12	13	13	Universidad EAFIT	314	410	533	694
12	11	11	11	Universidad de Caldas	365	435	564	679
13	13	10	9	Universidad del Cauca	390	448	524	659
14	14	14	14	Universidad Tecnológica de Pereira	285	349	457	586
15	15	17	22	Universidad Antonio Nariño	117	246	378	506
16	16	20	–	Universidad Francisco Jose de Caldas	–	204	353	456
17	17	16	17	Universidad de La Sabana	191	250	327	428
18	20	19	18	UPTC	159	224	286	382
19	19	18	16	Universidad El Bosque	201	235	299	372
20	18	15	15	Universidad del Quindio	213	255	302	362
21	21	23	22	Universidad Militar Nueva Granada	117	165	260	341
22	23	25	21	Universidad del Magdalena	121	157	210	271
23	22	21	19	Universidad del Tolima	150	184	224	261
24	27	26	24	Universidad CES	105	141	186	254
25	24	24	23	Universidad de Nariño	112	162	198	250

Source: Scopus Database

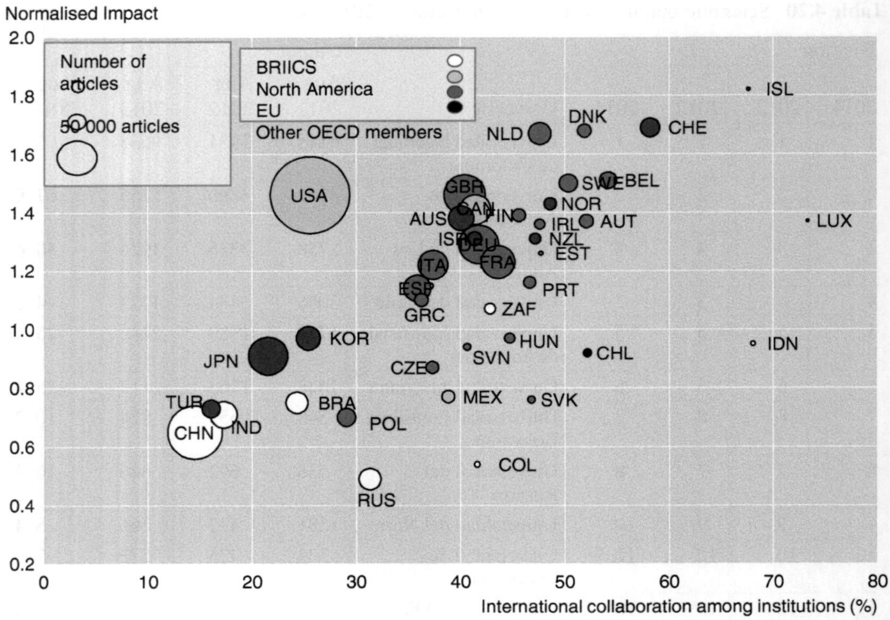

Fig. 4.5 Comparison of scientific impact, output, and international collaboration (Source: Adapted from OECD, Science and Technology Indicators)

after graduation measured through employability rates. One of the main mechanisms implemented by the university was the placement of its campuses in multiple cities in Colombia (28 in 2014), with several located in city districts that previously had no higher education institutions.

The various policies implemented in UAN to strengthen R&D capacity include:

- A substantial increase in the number of professors with doctoral degrees. This policy was implemented via a twofold strategy: funding a selection of current professors to obtain advanced degrees, and with special calls for new professor positions only available to doctorate holders. As discussed earlier, while this continues to be an issue at the national level, UAN has clearly surpassed the national average for faculty with PhD degrees.
- A constant and significant investment in R&D with a large proportion coming from direct university funds.
- An international perspective and benchmark of activities and goals.
- Awareness that the results of these policies are obtained at in the medium and long term.

Some of the results and consequences of these policies can be easily verified. University rankings that measure research results and impact do not typically have impressive results for Latin American institutions with a few exceptions. Some of these rankings focused on Ibero-America or Latin America give interesting results for UAN.

Table 4.21 Ranking of Colombian universities from the 2013 Ibero-American Extended Shanghai Ranking

University	Ranking 2013 COL	Ranking 2013 Ibero-America
Universidad Nacional de Colombia	1	52
Universidad de los Andes	2	58
Universidad de Antioquia	3	85
Universidad del Valle	4	131
Pontifica Universidad Javeriana	5	139
Universidad Antonio Nariño	6	146
Universidad de Cartagena	7	159
Universidad del Rosario	8	182
Universidad Industrial de Santander	9	195

Source: Ibero-American Extended Shanghai Ranking 2013

The Shanghai Ranking extended to Ibero-America (Docampo 2013) ranks the top 400 universities in Ibero-America. The focus of this ranking is excellence in research. Table 4.21 shows the results for Colombian institutions in the top 200 universities for 2013. It is important to note that there are very significant gaps in the research performance of Colombian institutions when compared with the top institutions in the world.

Similarly, other university rankings also show UAN in a privileged position. For example, it ranks 15th among Colombian universities in both the QS ranking for Latin America and 15th, as shown before, in the Scimago ranking (when measuring total output). When specific indicators are analyzed, for example, the student/professor ratio, it is ranked fourth in Colombia in the QS 2012 Latin America ranking and first for excellence, international collaboration, and impact in the Scimago indicators for 2014.

These results show the immense effort undertaken by UAN, following clear goals and strategies, albeit with limited resources to support them.

4.4 Discussion and Conclusions

The data presented in this chapter show that Colombia needs to make a qualitative change in its policies for higher education as well as for R&D in all performance sectors.

It is therefore pertinent to propose some specific policy alternatives to significantly speed up and strengthen the higher education system and its relation with R&D. The main policy recommendations are as follows:

1. Set concrete quantitative goals concerning higher education attainment levels, employability rates, and the national R&D expenditure to close the gap with more developed countries. The goals should have a well-defined timeframe with corresponding funding resources to attain such goals.

2. Build stronger universities, taking into account that greater funding per student is required and that a larger student demand is likely to occur. Strong universities also require better support to achieve higher quality education. Current governmental actions focus on control rather than supporting diverse and flexible initiatives. A clear tension exists between external evaluation and autonomy.
3. Devise more effective strategies to broaden participation and the completion of tertiary level education.
4. Build universities that are visible at the international level. One way to do this while also strengthening research quality and output is to be active participants in international research projects. There are diverse strategies that can be implemented, in particular fostering collaboration with top institutions worldwide with real peer interactions on both sides.
5. Implement a balanced approach in which both the stronger parts of the research system continue to increase their performance levels and simultaneously further elements of the system should be included and results consolidated.
6. Reduce the rate of low performers in reading, science, and math. This goes beyond the scope of higher education but is essential for the consolidation of the full national system of education.

It is necessary to implement those strategies that will accompany the vision of having an economy based on more highly qualified human resources with a high proportion of tertiary-level graduates. This will in turn foster competitiveness and economic growth.

The GDP of Colombia has significantly increased over the last few years, positioning the country's economy among the top 30 in the world. However, several other international comparisons do not confer such a privileged position. In international education assessments exercises, as mentioned earlier, Colombia does not perform well. The Global Competitiveness Index positions Colombia in 69th place in its 2012–2013 ranking. One pillar measured by this index is higher education and training; Colombia is ranked 67. In part, this reflects the nature and structure of the Colombian economy. There is a clear need for better qualified professors at universities with advanced research degrees, given that the national average of 5.8 % leaves much room for improvement. In addition, researchers should be employed not only in the higher education sector but also by business enterprises. This could help to transform the economy and positively impact social wellbeing.

Capacity building requires long-term sustained efforts. The resources that must be allocated to increase advanced research qualifications must also be accompanied by an increase in grant resources to carry out R&D projects. Innovation is fueled in different ways but it is with no doubt that a human talent base is one of the most essential. Thus, a greater effort must be made.

Trends in higher education have become more dynamic, and the consolidation of a highly competitive sector at the international level is required in Colombia. Graduates will be global citizens and require strong skills in critical thinking as well as being both versatile and resolute. Significant efforts must also be made by higher education institutions to ensure their relevance on an international scale. At the same time, a good balance is required to reduce the inequities that are currently present in the system.

A recent promulgation of Colombia's public policy for education envisions the ambitious goal of becoming "the most educated" population in Latin America. However, it is unclear what this means in terms of specific goals with respect to educational attainment, educational performance results, the contribution towards advancing knowledge, and the identification and planning of all that is necessary to achieve these specific goals.

References

Docampo, D. (2013). Shanghai ranking expanded. https://ec3metrics.com/wp-content/uploads/2013/09/Shanghai-Ranking_EXPANDED_Iberoamericanas_2013-.pdf. Accessed Sept 2014.

European Commission. (2014a). Education and training strategic framework. http://ec.europa.eu/education/policy/strategic-framework/index_en.htm. Accessed Sept 2014.

European Commission. (2014b). Europe 2020 in a nutshell. http://ec.europa.eu/europe2020/europe-2020-in-a-nutshell/targets/index_en.htm. Accessed 7 Oct 2014.

Heitor, M. V., & Horta, H. (2014). Further democratizing Latin America: Broadening access to higher education and promoting science policies focused on the advanced training of human resource. *Journal of Technology Management and Innovation, 9*(4), 64–82.

ICFES. (2014). Resultados pruebas saber. http://www.icfes.gov.co/investigacion/informes-de-resultados-de-evaluaciones-nacionales/saber-3o-5o-y-9o. Accessed 10 Aug 2014.

MEN. (2014a). SNIES módulo de estadísticas. http://bi.mineducacion.gov.co:8380/eportal/web/snies1/docentes. Accessed 25 Jul 2014.

MEN. (2014b). Estadisticas de educacion superior. Ministerio de Educacion Nacional. http://www.mineducacion.gov.co/sistemasdeinformacion/1735/articles-212350. Accessed 25 Jul 2014.

National Science Foundation. (2014). Science and engineering indicators 2014. http://www.nsf.gov/statistics/seind14/. Accessed 12 Jun 2014.

OCyT. (2014). Indicadores de ciencia y tecnología de Colombia. Observatorio Colombiano de Ciencia y Tecnología. http://ocyt.org.co/es-es/InformeAnualIndicadores/ArtMID/542/ArticleID/223/Indicadores-de-Ciencia-y-Tecnolog237a-Colombia-2013 Accessed 9 Jun 2014.

OECD. (2014a). *Education at a glance 2014.* http://www.oecd.org/education/eag.htm

OECD. (2014b). Main science and technology indicators. OECD Statistics. http://stats.oecd.org/. Accessed 9 Jun 2014.

OECD. (2014c). Science, technology and industry scoreboard 2013. http://www.oecd-ilibrary.org/science-and-technology/oecd-science-technology-and-industry-scoreboard-2013_sti_scoreboard-2013-en. Accessed 10 Jun 2014.

OECD/International Bank for Reconstruction and Development/The World Bank. (2012). Reviews of national policies for education: Tertiary education in Colombia 2012. http://dx.doi.org/10.1787/9789264180697-en. Accessed Jan 2014.

RICyT. (2014). Indicadores. Red Iberoamericana de Ciencia y Tecnología. http://www.ricyt.org/indicadores

Scimago. (2014). SIR Iber Colombia 2014. http://www.scimagoir.com. Accessed 2014.

UIS. (2014). Education database. UNESCO Institute of Statistics. http://www.uis.unesco.org/Education/Pages/default.aspx. Accessed 7 Jul 2014.

UIS2. (2014). Country profiles. UNESCO Institute for Statistics (UIS). http://www.uis.unesco.org/DataCentre/Pages/country-profile.aspx. Accessed Aug 2014.

Chapter 5
Internationalization of Chinese Higher Education in Latin American Campuses

Jae Park

5.1 Introduction

The first university in the American continent, San Marcos, was established by a 1551 royal charter in the Viceroyalty of Peru, and the founding of the University of Mexico followed a few months later, a century ahead of Harvard University (Marsiske 2006; Robles 2006). Borrowing a term coined by Hopkins (2002), the establishment of early Latin American universities occurred during a period of *archaic globalization*, a time characterized by the wars and colonization waged by monarchs before 1600.

The currencies and modus operandi of this older type of globalization and its universities have not remained unchanged. On the contrary, Latin American higher education underwent a thorough modernization as shown by the reform that followed the *Manifiesto de Córdova*, the student-led modernization of the administration and management of higher education in Argentina, leaving behind the then prevalent Napoleonic model (Mariátegui 2008 [1928]; Tünnermann 2008).

This is not to say that today's Latin American universities are not without their difficulties. Salmi (2014) points out in his article The Governance Challenge in Ibero-American Universities that Ibero-American universities "suffer from serious limitations because of their populist traditions, characterized by the democratic election of university presidents or rectors by the entire academic community, high levels of inbreeding, and a lack of international outlook, as can be seen in Spain and most of Latin America." Nevertheless, these problems and the corresponding efforts of the academic community to constructively problematize them could also be seen as a reflection and sign of the struggle for growth and change.

J. Park (✉)
Department of International Education and Lifelong Learning, Hong Kong Institute of Education, Hong Kong, China
e-mail: jpark@ied.edu.hk

H. Horta et al. (eds.), *Trends and Challenges in Science and Higher Education*, Knowledge Studies in Higher Education 3,
DOI 10.1007/978-3-319-20964-7_5

What is relevant to this chapter is that it is within this backdrop that Chinese universities are carrying out a Chinese version, so to speak, of the internationalization of higher education. Chinese universities have different backgrounds and are relatively recent, the oldest being the Imperial University of Beijing (1898), which was established by imperial edict in an effort to modernize Qing China with Western science and technologies during the British Colonial Globalization (1800–1940s), a sub-era of modern globalization for Hopkins (2002).

This chapter looks at the emerging internationalization of higher education and the transnationalism of China, as experienced by the protagonists of Confucius Institutes (CIs) in Peru. This is relevant because (1) most studies in sociology or political science about Chinese international relations treat the Latin American region in its entirety and on a macro level, thereby neglecting the contexts of the individual countries and the experiences of the people, and (2) reports on CIs in Peru are almost non-existent despite the broad recognition of the fast expanding Chinese–Latin American relations. This chapter also offers new theoretical perspectives to understand the intersection of international relations and the internationalization of higher education where both competition and collaboration of higher education occur on the global stage.

CIs constitute a global network of Chinese cultural and language centers. These institutions are similar to other international language–cultural centers such as the Alliance Française (1883) and the British Council (1934), which are subsidized and monitored by governments or state agencies. The first CI was only established in 2004. As of December 2014, the number of CIs worldwide is 475; this number excludes the 851 Mandarin-speaking classrooms called Confucius Classrooms at elementary and secondary education levels. The total number of full-time and part-time faculty and staff now exceeds 10,000, with 3.45 million registered students (Meng and Cao 2014). The Chinese state agency *Hanban*, which oversees CIs, dispatches more than 1000 Mandarin-language instructors every year to the 126 countries and regions with established CIs. Between 2009 and 2010, the number of instructors doubled from 2000 to 4000 (Liu 2009, 2010).

Extant studies consider CIs to be homogenous worldwide and have investigated them using cross-sectional analysis (Gil 2008; Paradise 2009; Yang 2010). However, given that Chinese transnationalism in Latin America has been present for almost two centuries, the consideration of only contemporary data in studying the transnational relationship between Peru and China would result in an incomplete and distorted analysis. Thus, this chapter also looks at the longitudinal aspects of international relations and examines the Sino–Peruvian ties mediated by education as soft power. Furthermore, it focuses on the relational aspects of power instead of an exclusively substantive dimension, and examines the Sino–Peruvian ties using the historiographic method, which has been dubbed by Foucault as "archaeology," namely a "differential analysis of the modalities of discourse" (Foucault 1972, p. 139). In addition to historiography, a later section of this chapter uses an empirical data set from 2010 interviews with staff and students at a CI in Peru, direct observations and written information from a CI conference held in Beijing, December 2010, and, tacitly, the author's personal experience during the years of Maoist civil war in Peru.

5.2 Early Sino–Peruvian Relations

Sino–Latin American relations started and matured during the "Hundred Years of weakness and poverty" (1840s–1949) of the impoverished Qing Empire, followed by a republic divided by civil wars and invaded by Japan (Wang 1993). The Peruvian abolitionist law[1] caused a serious shortage of manpower in the land of the Incas, which was experiencing fast-paced industrialization after its independence in 1821. The scarcity of manpower was more glaring in sugar plantations, copper mines, and guano beds because of their harsh working conditions. It was then that laborers from the impoverished Qing Empire appeared as an alternative manpower source—to that of British Indian coolieism—to Peruvian planters, politicians, and traders (cf. Rodríguez Pastor 2001).

Tens of thousands of Chinese men were systematically lured and deceived during their recruitment by local crimps and foreign dealers, with promises of, for example, a voyage to great fortunes at "Gold Mountain" (the idealized Chinese name for San Francisco) during the Gold Rush (Cuba Commission 1970). As a result of this and other strong motives, such as selling out local revolutionaries instead of executing them or disgraced gamblers selling themselves to pay debts, some 90 to 100 thousand Chinese laborers were shipped to Peru. Between 1849 and 1874, the 3-month transpacific passage had an average mortality rate of 11 %, with its peak at 16 % between 1860 and 1869 (Meagher 2008).

Typically, Chinese coolies to Peru signed a contract for 8 years, twice as long as British indentured servants and seventeenth to eighteenth century French *engagés* in North American and Caribbean territories. They were often called "Chino Macao" or "Chino Manila," depending on their shipping ports. Their working and living condition in Peru was appalling, "Their situation was so horrible that they were generally called 'slaves'; only, if scarcely, in legal documents were they defined by the term 'free laborers', which was the official designation for them. In the haciendas and farms, they were treated much worse than black slaves" (Pérez de la Riva 1976, p. 20).

The reason for their comparably terrible treatment was that African slaves were owned, and therefore better cared for. Although Chinese coolies were called free bonded labor or indentured servants, reports and literature on the matter do not do justice to this assessment (Irick and Center 1982; Meagher 2008; Stanley and American Council of Learned Societies 1998; Stewart 1951; Yen 1985). Only young, strong males survived their first 8 years of indenture, particularly on guano islands (Stewart 1951). Furthermore, upon the culmination of the contracts after 8 years, instead of being freed, they were usually thrown back into a cycle of never ending re-indenture under financial and legal coercions such as the inability to pay their own right of abode or passage back to the homeland (Meagher 2008).

[1] African slavery in Peru was abolished in 1854, about 20 years after Britain and 10 years earlier than the 13th Amendment to the US Constitution.

Peruvian newspapers of that time denounced these abuses as being against what would be termed, a century too late, as human rights: "The greatest part of those being newly contracted by the plantations are coming to replace, not those who are completing their contracts, but rather those who died fulfilling them" (Stewart 1951, p. 105). They witnessed striking similarities between African slavery and Chinese coolieism. *El Nacional*, a Peruvian newspaper, defined Chinese coolieism as "another African slave trade (except that now) the trade was not in ebony, but in copper" (Stewart 1951, p. 117).

Research on Anglo-American labor law in the nineteenth century refers to this period as an era in which "unfree serf labor gave way to regulated wage labor and finally to free wage labor" (Steinfeld 2001, p. 1). However, English, and to a lesser degree American, wage workers of that time were subject to non-pecuniary pressures, that is, criminally charged and sent to gaols, masterfully described by Charles Dickens. It can be safely stated that certain aspects of nineteenth century Western wage-labor constituted coerced or "unfree" labor. Chinese coolieism in Peru was synchronous with these legal and social developments, which defined coolies' legal, social, and juridical identities. Borrowing from McKeown (1999), the history of Chinese coolies in Peru can be described as a chapter of "diaspora-as-exile" and not "diaspora-as-diversity".

Although agonizingly slow, such a physical violence and flat denial of human dignity eventually produced a response from the Qing Empire (Irick and Center 1982; Yen 1985). The treaty of Tien Tsin (1874) between the Qing Empire and Peruvian republic marked the start of their formal diplomatic relationship (Rodríguez Pastor 2001) as well as the official end of coolieism in Peru and Latin America.

As stated in an official site of a CI in Peru, CIs arrived in this part of Latin America after "a historical contact of more than 160 years against hair" (PUC 2012, author's translation). Katz (1965) categorizes the sources of international conflicts over resources, ideologies, and power. They are not mutually exclusive, for example, "power is not only a means for securing economic advantage or ideological maximization, it is also a source of conflict in its own right" (1965, p. 374). The conflicts in the early transnationalism between China and Peru hinged principally on national and global economics.

5.3 Maoism in Peru

It is widely recognized that more so than the French Revolution and its aftermath, the independence of Peru from Spain in 1821 represented a change of power from monarchist to nationalist *señores feudales* (feudal lords). Latin-Americanist and Peruvian thinker Mariátegui argues in his 1928 opus magnum that Chinese coolies were to the new Peruvian republic what black slaves were to viceroyalty—working limbs and not human (Mariátegui 2008 [1928]). By and large, Peruvian lands and their industrial exploitation remained in the hands of the same landowning ruling class until the 1969 land reforms and expropriation by a Marxist military regime.

The latter also brought about a considerable growth of leftist political movements in the Andean highlands (Taylor 1983). One such movement, led by university activists and academics in the late 1960s, became known as the *Sendero Luminoso* (Shining Path) (Paredes 2009; Zucha 2007). Sendero's leaders called themselves a Maoist movement, and they were ideologically and militarily trained in China between 1964 and 1967 (Barnhurst 1991; Zucha 2007). This was a logical turn of events because when the Communist International or *Comintern* was dissolved in 1943, followed by the death of Stalin 10 years later, Mao became the de facto leader of the international communist cause. However, in the 1980s, the already established guerrilla organization Sendero was not on good terms with the modernizing Chinese state. Sendero regarded Deng Xiaoping (China's leader at that time) as a reactionary promoting a market economy in China, which clashed with Mao Tse-tung's ideals and undermined Sendero's ideological foundations. Soon after, the Peruvian version of the People's War began, nurtured by Maoist ideology but weaned from Deng Xiaoping's Chinese state; this movement has been engaged in implementing a "socialist market economic system" since 1978 (Evans 1995).

The social cost of this typically terrorist war was enormous: "Since it began operating in 1980, Sendero had mobilized some eighty thousand subversives in a war that killed over twenty-five thousand people" (Soto 2002, xiv). Sendero's theory and praxis for power, its propaganda and actions, were met by the armed forces speaking the same language—one of violence. The casualties (on both sides) and daily suffering of ordinary people became the norm in Peru. The war rapidly subsided after 1991 with the arrest and imprisonment of Abismael Guzman, the leader of Sendero, also known as "President Gonzalo." From the perspective of Katz's (1965) conflict theory, the war between Sendero and the Peruvian state could be attributed to disagreements over values and ideology, although the struggle over economic and political hegemony also played a role: "A relationship of violence acts upon a body or upon things; it forces, it bends, it breaks on the wheel, it destroys, or it closes the door on all possibilities" (Foucault 1983, p. 220).

No sooner had the Sendero leaders fallen, another news story about China made to the front pages of Peruvian newspapers. A state-backed company from Beijing purchased an important iron mine that was once owned by Americans prior to its nationalization by a dictatorial regime; the buyers were hailed as heroes (Romero and Zárate 2010). From then onwards, China has continued to purchase and sell in Peru, becoming its greatest trade partner and investor. Chinese investments in Peru amounted to US$10 billion in 2010, matching those in the United Kingdom (Minperu 2010). Chinese investments are focused on copper and iron mine megaprojects such as Toromocho, Galeno, Rio Blanco, Mina Justa, Marcona, and Pampa del Pongo (Minperu 2010). What makes these companies different from other international mining companies stationed in Peru is that their entire production is exclusively shipped to Mainland China.

It is these cultural, ideological, and economic backgrounds on which Chinese CIs started their Peruvian operations in 2008.

5.4 Chinese Higher Education in Peruvian Campuses

Various facets of CIs have been previously investigated. Zhang and Liu (Zhang and Liu 2006, 2008) studied the early years of CIs regarding the transnational collaboration in promoting language instruction. The recent popularity of Mandarin-language instruction is consistently attributed to commercial potential rather than an interest in Chinese culture (Gil 2008; Lo Bianco 2007).

CIs utilize preexisting universities that are time-honored global institutions, which assure from the outset institutional prestige, administrative functionality, and an appropriate infrastructure. By and large, this phenomenon is the most significant state-sponsored projection of Chinese universities abroad, with impacts on the internationalization of universities not only in Mainland China but also those in the hosting countries.

Interviewed by a researcher during a yearly CI conference, the vice-chancellor of an important university in the United Kingdom used the term "Confucius Institute Movement," and justified the use of this phrase stating, "I am a linguist. Movement has several meanings. The first one is change. CIs are mechanisms for change in mutual understanding in language, business and politics. A second meaning is direction. CIs as movements mean a next step with a direction and towards a global recognition. Third, movement also means shared understanding in works being done, exchange, sense of purpose and change."

The implications and scope of CIs as a movement depend on the intentions and actions taken by both sides of the agreement. Such intentions, sense of direction, and actions converge into the quest for international recognition for the culture, language, and social status of both the Chinese state and host institutions.

Although the struggle for recognition is not always explicit, to consider CIs as mere charm displays would be mistakenly simplistic. It should be remembered that such intentions and actions at a macro level are translated into those at a micro level. Thus, CIs worldwide have a basic set of personalized goals, namely to help individuals learn both the Chinese language and culture, and increase their competitiveness in upward social mobility.

The Peruvian case is no exception. There are four CIs in Peru, two in the capital city of Lima and two in provincial capitals. All four are located on campuses of prominent private universities (three of these are publicly known to be founded on conservative religious ideals). The most recently erected CI at the Ricardo Palma University aims to produce future interpreters (Castro 2009), but all four CIs share comparable *Putonghua* (standard Mandarin) language instruction and cultural activities.

The Peruvian CI in this study is located in Piura, the capital city of the most populated northern province that carries the same name. As with other CIs around the world, a national Beijing university meets approximately half of the CI costs. This particular CI offers Mandarin-language programs at basic, intermediate, and advanced levels; it also functions as a national venue for international Mandarin proficiency tests. The courses are not credit bearing and accept off-campus students, accounting for half of all enrollments. This CI is located within the compounds of the university language center where English and French are also taught. The tuition fee is less than half that of other CIs in Lima. According to its director, the CI is expected to contribute to regional development by easing the language barriers for people in business with China and Asia under various trade frameworks such as Free Trade Agreements and Asian Pacific Economic Cooperation. Furthermore, the CI can facilitate students' access to scholarships (UDEP 2010).

For a 30-year-old female student with technical training in the garment industry, her main motivation to learn Mandarin language is as follows: "…to improve my professional future. The tie between Peru and China has strengthened in recent years and increasing business between the two countries will demand [the use of] Chinese language. I would like to be an interpreter or a language teacher. I know that the road to fluency is long but I am hopeful that I'll achieve it."

Social mobility may be one of the main motivations for studying the Mandarin language, but it is not the chief one at this particular CI. An administrative staff member from Beijing (who is also a Mandarin instructor) said that 10–15 % of her students have some Chinese ancestry. She believes that cultural interest is perhaps the most important motivation for her students, as their genealogical background truly motivates them and creates a friendly and hospitable environment for teaching and learning.

A 55-year-old mother of two is the daughter of a first-generation immigrant father from China and a Peruvian mother: "I was always interested in learning languages, since I was a girl. My father is [a] native of Guangdong and his conversations with his friends and stories about China have always fascinated me. I wish I could visit one day. Moreover, Chinese written characters intrigue me and I am very eager to learn them." Thus, the on-going globalization provides an altruistic element to the learning of Chinese culture and language. A website of a CI in Lima states that "We work to bridge understanding among people with different cultures contributing to friendship and construction of a harmonious world" (PUC 2012, author's translation).

CIs are almost always established in partnership between a Chinese university and one in the host country. This model contrasts with Western state-sponsored language and cultural centers that seldom operate on campus. CIs take advantage of the pre-existing university system, a time-honored global institution that instantly assures basic standards, administrative structure, and perhaps even prestige. CIs also aim to become instrumental in university-based research, Sinology in particular (PUC 2012). However, this is the least developed area, shown not only by the exclusive character of their yearly conference in Beijing (closed to the public and researchers) but also by the fact that most Sinology research centers in the host universities were established prior to CIs and run by researchers unrelated to CIs. Indeed, these clearly overlapping functions and interests between the two parties, CIs and pre-existing research and language education centers, have probably caused a good deal of tension, if not unnecessary complications.

A fourth CI with an emphasis on training interpreters was established in Peru in 2009 (Castro 2009). Its aim is to produce graduate interpreters to contribute to the mutual understanding between China and Peru, in particular for trade and business. However, the type of tasks waiting for them can be rather unsettling: "The Chinese see us as little more than slaves," said Hermilia Zamudio, 58, a resident of Ruta del Sol, whose husband was fired from the mine after working there for almost 30 years. "They deem it beneath them to talk to us, and when they need to address problems here, they do so with their thugs." Clashes with private security guards and with the police, who receive a monthly stipend paid by the Shougang Corporation, are common in Ruta del Sol, on land where Shougang says it has concessionary rights to exploit deposits of dolomite, a mineral it hopes to extract for smelting iron and steel (Romero and Zárate 2010).

Chinese coolies were described as being "little more than slaves" upon their arrival in Peru more than a century earlier. This term was also attributed to the degraded human condition of many Peruvians under the physical violence of the Maoist civil war. What is fascinating is the metamorphosis of power forms. What is perhaps even more fascinating is the complexity of power relations, their control, perpetuation and, above all, the reversal of the flow of power.

5.5 Higher Education as a State Apparatus

The current academic discourse on the internationalization of higher education is based on neoliberal paradigms. The global quest is for world-class universities, league tables, citation indexes, and for "high-status" research (Deem et al. 2008). They have created not only a new academic culture but have also reinforced existing national and linguistic hegemonies. Although the Chinese state has also joined the discourse with concrete actions to "internationalize" higher education (e.g., the 211 Project and 985 Project), it should be remembered that the concepts of world-class university rankings, marketization, and the brain gain/drain of higher education can be criticized (Altbach 2004; Philip Altbach and Salmi 2011) and are not entirely related to the phenomenon of CIs. The alternative suggested here looks at the unique phenomenon of the internationalization of Chinese higher education in Peruvian campuses using a historiography of power relations.

Substantive power discourses deal with the origin, nature, and ownership of power, particularly with its excesses and decadence. However, discourses that typically consider power as a threat that needs to be moderated with supra laws (Aristotle), something that is ruthlessly repressive yet eventually beneficial by necessity (Machiavelli), or as a taxonomy of state power (Montesquieu), are inadequate to investigate CIs. First, these discourses focus on the system of overt might, which is not exactly true of CIs; second, they usually look at power structures within a geographically delimited nation-state whereas CIs stand on a transnational stage.

The analysis of power relations in this chapter starts with a distinction between *auctoritas* and *potestas* by discussing an ancient text, the *Res Gestae Divi Augusti*, scholarly expounded by Adcock (1952). In this text, Emperor Augustus describes how, after turning the Roman Empire into a republic with a senatorial system, he himself possessed *auctoritas* but limited *potestas*. The meaning of *potestas* here is a form of power by imposing obedience, that is, by coercing others to act in a determined manner. In contrast, *auctoritas* is a form of power that is supported or justified by earned prestige and recognition, which enjoys a corresponding level of legitimacy. A feature of *auctoritas* is that it cannot be obtained with pecuniary pressure or physical violence.

The literature on CIs often uses the concept of "soft power" coined by Joseph Nye (2004). Soft power is the non-military and non-economic state coercion (*auctoritas*) used by China to cast an image of China into the international arena. While hard power is military and economic might to intimidate others in global politics, soft power is "getting others to want the outcomes that you want (and) rests on the ability to shape the preferences of others" (Nye 2004, p. 5). Nye's version of *auctoritas* is interestingly based on neoliberal assumptions. His main concern is international relations, which he articulates with a set of strategies, international policies, and actions without hard coercion and inducement (e.g., military force, economic sanctions, even bribes); that is, soft power.

Lo and Pan (2014) suggest a blended framework to investigate CIs: Nye's tripartite taxonomy of coercion, inducement, and attraction together with another tripartite taxonomy of resources, strategies, and outcomes (Tellis et al. 2000). While these frameworks add multiple dimensions and methodological sophistications, they are only moderately relevant to the present research. In fact, CIs should be regarded unambiguously as power structures and infrastructures, or loosely as outlets of state power configuration.

As a movement, CIs are inextricably related to the wide range of powers that China possesses. Most of these powers are blended between soft and hard forms. For example, China's international health aid has been noted as being related to access to natural resources (Wharton Health Economics 2011). However, softness evaporates from her rhetoric when it comes to international territorial disputes or secessionist movements from within. CIs have been empowered by direct state intervention and they count on robust financial and administrative support from the state. The Chinese state in turn is a state centered in executive power at the expense of legislative and judicial powers, and its participatory powers such as private social institutions, regionalisms, and syndicalisms are, to be fair, still rudimentary. Indeed, CIs little resemble a participatory power structure.

Executive power-centered states try to transition from *potestas* to *auctoritas* both on national and international stages. Indeed, it is in the global arena that CIs become most relevant to China. Beyond considerations of structure and origin of power, CIs appear as handy instruments to bridge the chasm between the state and its outreach points overseas, playing an ancillary role for the former. In the same way that Alliance Française has been serving the French state for more than a century, in recent years CIs have played an outstanding auxiliary role for the transnational projections of a state seeking *auctoritas*; that is, CIs are state apparatuses.

Althusser argues that state apparatuses often function through education and culture, and typically through schooling because "no other … state apparatus has the obligatory (and not least, free) audience of the totality of the children in the capitalist social formation, eight hours a day for five or six days out of seven" (2006, p. 97). Discourses on educational and cultural activities (schooling inclusive) are seldom neutral socially and politically. In fact, educational institutions are usually depicted as one of the most politically sensitive social institutions because of their "fundamental role in maintaining the existing society [by] cultural transmission, role socialization, and value acquisition around a preoccupation with the principles of consensus, cohesion, and stability" (Giroux 1983, p. 48) and to the point of equating pedagogic interventions to symbolic violence in the pursuit of social reproduction (Bourdieu and Passeron 1977). A perfect example from China could be the situation and role of schools and universities during the Cultural Revolution (Chung-kung chung-yang tang-shih yen-chiu-shih 1991).

With the ivory-towered *potestas* and *auctoritas* in precipitous decline, nineteenth century Qing China offered an interesting contrast to CIs. The transnationalism that followed Qing's weakened power caused great diasporas and human capital exports (Wang 1993, 2001). During this period, Chinese subjects in Peru and several other Latin American nations were at the mercy of private businessmen, planters, and coolie clipper owners in cahoots with state apparatuses: "All the State Apparatuses function both by repression and by ideology, with the difference that the (Repressive) State Apparatus functions massively and predominantly by repression, whereas the Ideological State Apparatuses function massively and predominantly by ideology" (Althusser 2006, p. 94).

The chief cause of the conflict during the nineteenth century was an earlier form of globalization, industrialization, regardless of underlying capitalist or socialist ideologies. Chinese indentured laborers, their culture, and language were introduced into the industrialization-blind power structures of Peru, encountering a repressive practice of power apparatuses with a utilitarian ideology and African slavery practices that reverberated in the memories of the ruling class.

Today, economic growth at all costs continues to be the leading raison d'être of state apparatuses. Economy has become the golden yardstick to gauge state *auctoritas*. This precise space is the landing pad of CIs in Peru. After having left behind 160 years of transnationalism loaded with coercion and violence, CIs in Peru stand tall as new state apparatuses in pursuit of new resources (language and culture), new values (multiculturalism and leisure), and new power (soft), and all are globally plausible.

5.6 International Relations and Higher Education

To some extent, every discussion on substantive power includes relational aspects because power transactions occur among people and institutions in "alterity" positions (Lévinas 1991; Park 2009). For this reason, dialectical arguments dominate power discourses in the field of education such as Freire's pedagogy of oppressor–oppressed (1972) or Bourdieu's education as a perpetuator of social strata through the reproduction of habitus, taste, or consumption (1977, 1984).

A relational approach to CIs (as a phenomena of power) enhances our understanding of their substantive characteristics and, more importantly, presents a novel perspective. Foucault calls this relational approach "a new mode of inquiry" consisting of a discursive method focused on relations among "subjectivized" people rather than on the phenomenon of power itself, and by intentionally avoiding the excesses of substantive and rationalistic kind of arguments (1983, pp. 208–210). His relational discourse on power focuses on the process of the centralization and rationalization of state institutions (apparatuses for Althusser): "power relations have been progressively governmentalized, that is to say, elaborated, rationalized, and centralized in the form of, or under the auspices of, state institutions" (Foucault 1983, p. 224). Foucault also argues that power relations only occur when the one over whom power is exerted has previously enjoyed freedom; as soon as power is exercised, her/his freedom vanishes.

A critical analysis of the historical background of CIs in Peru seems congruous with this power–freedom correlation. Chinese coolies were free when they signed their contracts but led slave-like lives because of the Peruvian state apparatuses that legally but unjustly enforced indentured bondage. However, the assumption that education is a byproduct of colonization, or that university is a part of Western cultural imperialism (Spring 2009) should be contextualized into the Peruvian case taking into account: (1) the spirit and context of the early stage of colonization can never remain static in time; and (2) universities have a life and dynamism of their own and tend to forge their own culture rather than holding to the dated subculture of colonial subjectivity. These claims can be validated with several cases of Latin American universities that renounced their pontifical and royal titles to become secular, including the two abovementioned universities and the ongoing case of the Pontifical Catholic University of Peru.

During the Maoist war in the 1980s, the freedom of the people was severely constrained, and then at gunpoint. Similarly, today, a robust functioning of CIs over various forms of social reproduction *à la Bourdieu* could well be seen as conditioning or delimiting freedom. This is because CIs, as like any other linguistic and cultural instruction system, provide and create a situation that Foucault describes as the privileges of knowledge, the "effects of power which are linked with knowledge, competence, and qualification" (1983, p. 212).

Indeed, nowhere else in the education sector are language and communication more critical than in language–cultural centers. For CIs, the Mandarin language is both the medium and the message (McLuhan 1994). If the Baconian dictum "knowledge is power" were to be applied, language is the power (Fairclough 1989). Thus, in the case of CIs, the medium, message, and power contribute to a system of differentiations, namely the "linguistic and cultural differences, differences in know-how and competence" (Foucault 1983, p. 223). The system of differentiations is the basis for laying down the objectives of a power relation, something that is at stake and intentionally sought after.

Furthermore, the objectives of CIs are explicitly aligned with the state strategy of cultural development via soft power (Hu 2007), as are most major state interests (e.g., growing trade, greater exploitation of natural resources, and international fraternity). For example, if it were true that there is an "over-dominance" of English language in global education (De Bary 2007), it is not unthinkable to regard the Chinese language education offered by CIs as a counter-hegemonic movement.

After identifying a system of differentiations and its objectives, the next steps in the analysis of power relations should be the determination of the means of bringing power relations into being and the ensuing forms of institutionalization (Foucault 1983, p. 223). Chinese transnationalism in Latin America has unarguably transitioned through all of the following stages: Chinese coolieism as an institutionalization of human capital trade in Latin America during the nineteenth century; the war of Sendero Luminoso as a process of juxtaposing a new order upon Peru's existing democratic order ("one man-one vote"); last, CIs as an institutionalized system of language and culture education.

In addition to hosting a nation's legal structures and established educational institutions, namely universities and primary and secondary schools, CIs possess distinctive hierarchical structures but with relative autonomy in their functioning; that is, techniques of power. Foucault articulates the techniques of power with a panopticon prison design analogy. He looks at structure rather than infrastructure, and at the operational mechanism of power rather than the nature of power. CIs resemble a panopticon because they do not only describe a symbolic presence (linguistic and cultural capitals) but also a way of reifying the power relation between the Chinese state and host country/university. CIs could be regarded as apparatuses of panopticism with their "connection between bodies, space, power, and knowledge" (Dreyfus et al. 1983, p. 192) that are central to the degrees of rationalization and perpetuation of any power relations.

If a wider view is taken and the entire discussion re-examined thus far, it is clear that Chinese transnationalism in Latin America has had several degrees of rationalization, with various layers of adjustments to the situation of every époque, unveiling a power exercise that is "elaborated, transformed, [and] organized" (Foucault 1983, p. 224).

5.7 Conclusions

During my years in Peru, I have often heard that Latin America is a land of contradictions and counterpoints. Its higher education institutions seem no exception. On the one hand, Latin America enjoys noted socialist thinkers such as Paulo Freire and Santos Souza, while others such as José Carlos Mariátegui, Víctor Raúl Haya de la Torre, and Luis Alberto Sánchez have made intellectual contributions to the Latin American university system. However, as other chapters in this volume would probably substantiate, Latin American higher education appears to be sieged by neoliberal values, which are doubtlessly far less historical and authentic.

Thus, with my bias that there could never be a cultural discourse without history and vice versa, I highlighted in this chapter the importance of historiography as a method to look into the internationalization of Latin American higher education in conjunction with the internationalization of Chinese higher education on Latin American campuses. I argued that such historiography should include an account of power relations, that is, a relational power discourse needs to be incorporated in the substantive power discourse, if we are to look at CIs in the context of global power transactions.

The manpower shortage of an industrializing Latin America and the dire poverty of Qing China fatefully blended and converged into coolieism in nineteenth century Chinese transnationalism: greed met need. Maoist transnationalism in the twentieth century ended in the institutionalization of violence as a currency for power relations. At the dawn of the twenty-first century, CIs and their sponsoring states seem to be deploying a new power technique, capitalizing on the internationally more plausible soft power, with a set of charm-offensive strategies such as cultural export and language education. Although CIs capitalize on a time-honored and innately global institution of universities, it would be rather unfair to consider their worldwide operations as repressive, physically violent, or psychologically coercive. It is fairer to argue that the historiography of the power relation between Latin America and China has ruggedly evolved with distinctive forms and techniques to culminate in a power flux inversion. CIs as Chinese-language education and cultural centers under the direct patronage of the state are the latest velvet-gilded state apparatuses in the pursuit of state *auctoritas* for a change in global imageries and recognition.

Finally, with the extra freedom usually allowed to a book chapter format, I would say that for Latin American universities, the three missions of higher education, namely teaching, research, and community engagement, should be prioritized in a more flexible manner than the rampant neoliberal paradigm elsewhere. Even if research were to be given priority for any overriding reasons, it ought to be research with a pedagogical soul and social concern rather than pragmatically seeking a high position in the global university rankings. I also believe (and without pretentions of science) that embracing neoliberal ideologies as the only strategy for higher education development would not only be unwise but also a grievous lack of justice to our beloved Latin America, which has too often been trampled in political and social turmoil. Addressing these challenges, while maintaining inward sustainability, should perhaps be the main priority of Latin American universities today.

References

Adcock, F. E. (1952). A note on Res Gestae Divi Augusti, 34, 3. *Journal of Roman Studies, 42*(Part 1 and 2), 10–12.

Altbach, P. G. (2004). Globalisation and the university: Myths and realities in an unequal world. *Tertiary Education and Management, 10*(1), 3–25.

Altbach, P. G., & Salmi, J. (2011). *The road to academic excellence: The making of world-class research universities*. Washington, DC: World Bank.

Althusser, L. (2006). Ideology and ideological state apparatuses (notes towards an investigation). In A. Sharma & A. Gupta (Eds.), *The anthropology of the state: A reader* (pp. 86–111). Malden: Oxford: Blackwell Pub.

Barnhurst, K. G. (1991). Contemporary terrorism in Peru: Sendero Luminous and the media. *Journal of Communication, 41*(4), 75–89.
Bourdieu, P. (1984). *Distinction: A social critique of the judgement of taste*. Cambridge: Harvard University Press.
Bourdieu, P., & Passeron, J. C. (1977). *Reproduction in education, society and culture* (2nd ed.). London: Sage.
Castro, P. (2009, December 18). Aprender chino Habrá una cuarta sede en el Perú del Instituto Confucio. El Comercio. http://elcomercio.pe/noticia/383173/noticia/444718/alan-garcia-sobre-indulto-crousillat-siento-que-he-sido-burlado. Accessed 21 Jan 2011.
Chung-kung chung-yang tang-shih yen-chiu-shih. (1991). *History of the Chinese communist party: A chronology of events, 1919–1990*. Beijing: Foreign Language Press.
Cuba Commission. (1970). *Chinese emigration: Report of the commission sent by China to ascertain the condition of Chinese coolies in Cuba*. Taipei: Cheng Wen Publ. Co.
De Bary, W. T. (2007). *Confucian tradition and global education*. Hong Kong/New York: Columbia University Press/Chinese University Press.
Deem, R., Mok, K. H., & Lucas, L. (2008). Transforming higher education in whose image? Exploring the concept of the 'world-class' university in Europe and Asia. *Higher Education Policy, 21*(1), 83–97.
Dreyfus, H. L., Rabinow, P., & Foucault, M. (1983). *Michel Foucault, beyond structuralism and hermeneutics* (2nd ed.). Chicago: University of Chicago Press.
Evans, R. (1995). *Deng Xiaoping and the making of modern China*. London/New York: Penguin.
Fairclough, N. (1989). *Language and power*. London: Longman.
Foucault, M. (1972). *The archaeology of knowledge*. London: Tavistock Publications.
Foucault, M. (1983). The subject and power. In H. Dreyfus & P. Rabinow (Eds.), *Michel Foucault: Beyond structuralism and hermeneutics* (2nd ed., pp. 208–226). Chicago: University of Chicago Press.
Freire, P. (1972). *Pedagogy of the oppressed*. London: Penguin.
Gil, J. (2008). The promotion of Chinese language learning and China's soft power. *Asian Social Science, 4*(10), 116–122.
Giroux, H. A. (1983). *Theory and resistance in education: A pedagogy for the opposition*. London: Heinemann Educational.
Hopkins, A. G. (2002). *Globalization in world history*. London: Pimlico.
Hu, J. T. (2007). *Hold high the great banner of socialism with Chinese characteristics and strive for new victories in building a moderately prosperous society in all. Report to the Seventeenth National Congress of the Communist Party of China on Oct. 15, 2007*. Beijing: Xinhua News Agency.
Irick, R. L., & Center, C. M. (1982). *Ch'ing policy toward the coolie trade, 1847–1878*. Taipei: Chinese Materials Center.
Katz, D. (1965). Nationalism and strategies of international conflict resolution. In H. C. Kelman & Society for the Psychological Study of Social Issues (Eds.), *International behavior: A social-psychological analysis* (pp. 356–390). New York: Holt, Rinehart and Winston.
Lévinas, E. (1991). *Totality and infinity: An essay on exteriority*. Dordrecht: M. Nijhoff Publishers.
Liu, Y. (2009, December 11). *Mutual collaboration, continuous innovation, and sustainable development for the future of Confucius Institutes around the world*. Paper presented at the Fourth Confucius Institute Conference, People's Republic of China.
Liu, Y. (2010, December 10). *Working together towards the sustainable development of Confucius Institutes*. Paper presented at the Fifth Confucius Institute Conference, People's Republic of China.
Lo Bianco, J. (2007). Emergent China and Chinese: Language planning categories. *Language Policy, 6*(1), 3–26.
Lo, J. T.-y., & Pan, S. (2014). Confucius Institutes and China's soft power: Practices and paradoxes. *Compare: A Journal of Comparative and International Education* (ahead-of-print), 1–21. doi: 10.1080/03057925.2014.916185.

Mariátegui, J. C. (2008 [1928]). *Siete ensayos de interpretación de la realidad peruana*: Linkgua ediciones S.L.
Marsiske, R. (2006). La Universidad de México: Historia y Desarrollo. *Revista Historia de la Educación Latinoamericana, 8*(1), 9–34.
McKeown, A. (1999). Conceptualizing Chinese diasporas, 1842 to 1949. *Journal of Asian Studies, 58*(2), 306–337.
McLuhan, M. (1994). *Understanding media: The extensions of man*. London: Routledge.
Meagher, A. J. (2008). *Coolie trade: The traffic in Chinese laborers to Latin America 1847–1874*. Bloomington: Xlibris Corporation.
Meng, Z., & Cao, P. (2014). 475 Confucius Institutes in 126 countries and regions worldwide established. *Xinhuanet*. http://english.hanban.org/article/2014-12/26/content_568333.htm. Accessed 31 Mar 2011.
Minperu. (2010). CHINA Necesita: Cobre, hierro, energía, alimentos. *Minperu*. http://mineriadelperu.com/2010/08/20/china-necesita-cobre-hierro-energia-alimentos/. Accessed 30 Dec 2014.
Nye, J. S. (2004). *Soft power: The means to success in world politics* (1st ed.). New York: Public Affairs.
Paradise, J. F. (2009). China and international harmony: The role of Confucius Institutes in bolstering Beijing's soft power. *Asian Survey, 49*(4), 647–669.
Paredes, M. (2009). Indigenous peasant mobilization and the Left in Peru. http://www.crise.ox.ac.uk/copy/mobilisation%20conference/paredes.pdf. Accessed 31 Mar 2011.
Park, J. (2009). *Leadership of recognition: A philosophical study*. Saarbrücken: VDM Verlag Dr. Müller Aktiengesellschaft & Co. KG.
Pérez de la Riva, J. (1976). *Para la historia de las gentes sin historia*. Barcelona: Editorial Ariel.
PUC. (2012). Quiénes somos. http://confucio.pucp.edu.pe/nosotros. Accessed 29 Mar 2012.
Robles, E. O. (2006). Origen de las Universidades más Antiguas del Perú. *Revista Historia de la Educación Latinoamericana, 8*(1), 35–48.
Rodríguez Pastor, H. (2001). *Hijos del Celeste Imperio en el Perú (1850–1900): Migración, agricultura, mentalidad y explotación = Zai Bilu de tian guo zi sun* (2nd ed.). Lima: Sur Casa de Estudios del Socialismo.
Romero, S., & Zárate, A. (2010). Tensions over Chinese mining venture in Peru. *The New York Times*. http://www.nytimes.com/2010/08/15/world/americas/15chinaperu.html. Accessed 6 Mar 2011.
Salmi, J. (2014). The governance challenge in Ibero-American universities. http://tertiaryeducation.org/2014/01/the-governance-challenge-in-ibero-american-universities/. Accessed 12 Dec 2011.
Soto, H. D. (2002). *Otro sendero. [The other pat: The economic answer to terrorism]*. New York: Basic Books.
Spring, J. H. (2009). *Globalization of education: An introduction*. New York: Routledge.
Stanley, A. D., & American Council of Learned Societies. (1998). *From bondage to contract wage labor, marriage, and the market in the age of slave emancipation*. Cambridge/New York: Cambridge University Press.
Steinfeld, R. J. (2001). *Coercion, contract, and free labor in the nineteenth century*. Cambridge: Cambridge University Press.
Stewart, W. (1951). *Chinese bondage in Peru: A history of the Chinese coolie in Peru, 1849–1874*. Durham: Duke University Press.
Taylor, L. (1983). *Maoism in the Andes: Sendero Luminoso and the contemporary guerrilla movement in Peru*. Cambridge: Centre for Latin American Studies, University of Cambridge.
Tellis, A. J., Bially, J., Layne, C., & McPherson, M. (2000). *Measuring national power in the postindustrial age*. Santa Monica: RAND Arroyo Center, Rand Corporation.
Tünnermann, C. B. (2008). La Reforma de Córdoba : Vientre fecundo de la transformación universitaria. In E. Sader, H. Aboites, & P. Gentili (Eds.), *La Reforma Universitaria Desafíos y perspectivas noventa años después* (Vol. Actualidad y Perspectivas de la Reforma Universitaria, pp. 16–19). Buenos Aires: Consejo Latinamericano de Ciencias Sociales CLASCO.

UDEP. (2010). Universidad de Piura e Instituto Confucio iniciarán clases de chino mandarín en abril. *Centro de Noticas*. http://www.dircom.udep.edu.pe/index.php?t=2009/febrero/352_02. Accessed 1 Nov 2010.

Wang, G. (1993). Greater China and the Chinese overseas. *The China Quarterly Greater China, 136*(Special Issue), 926–948.

Wang, G. (2001). *Don't leave home: Migration and the Chinese*. Singapore: Times Academic Press.

Wharton Health Economics. (2011). Health diplomacy: In Africa, China's soft power provides a healing touch. *Arabic Knowledge@Wharton*, p. 5. http://knowledge.wharton.upenn.edu/arabic/article.cfm?articleid=2748. Accessed 24 Nov 2011.

Yang, R. (2010). Soft power and higher education: An examination of China's Confucius Institutes. *Globalisation, Societies and Education, 8*(2), 235–245.

Yen, C.-H. (1985). *Coolies and mandarins: China's protection of overseas Chinese during the late Ch'ing period (1851–1911)*. Singapore: Singapore University Press.

Zhang, X., & Liu, R. (2006). Study foreign policy on language dissemination to encourage the spread of Chinese [研究国外语言推广政策,做好汉语的对外传播]. *Applied Linguistics [语言文字应用]*, *1*(1), 39–47.

Zhang, X., & Liu, R. (2008). *Overview of language promotion policy from world's major countries [世界主要国家语言推广政策概览]*. Beijing: Foreign Language Teaching and Research Press.

Zucha, K. M. (2007). *Incomplete developmental counterinsurgency: The case of the Shining Path of Peru*. Theses and Dissertations-Political Science. http://ecommons.txstate.edu/polstad/1, Texas. Accessed 10 Jan 2015.

Chapter 6
The Academic Profession in Latin America: Between a Corporatist and a Professional Ethos

Elizabeth Balbachevsky

6.1 Introduction

As happens in other regions, discussions regarding the academic profession in Latin America are only possible if one is satisfied with a high level of abstraction. In fact, the region comprises 20 countries, all with their own academic systems, and some with a very long history. The oldest universities in the region date back to the colonial era and were created by the Catholic Church and the Spanish Crown following the medieval model of universities. Higher education in the region varies by, for example, size, balance of private and public sector, and the degree of institutional differentiation. All these dimensions have strong impacts on the way the academic profession is shaped in each country.

Nevertheless, in more general terms, higher education in the region has some relevant traits and has experienced similar dynamics in recent years. In all Latin American countries, modern higher education (post-medieval universities) emerged in the early nineteenth century, as part of the struggle to establish modern states after their independence. At that time, new universities were created, and older ones were reformed to fulfill a new role in the effort to overcome the colonial heritage of the region and to support the birth of modern nation states. Their assigned role was to form a new professional elite trained to the best technical and legal profiles available at the time. To achieve this aim, all countries adopted the Napoleonic model, where universities are licensed by the state to teach and certify established professions. Under this model, by completing the first level of university education (the undergraduate level under other systems), the student has access to the most

E. Balbachevsky (✉)
Department of Political Science, University of São Paulo, São Paulo, Brazil

Research Center on Public Policy, University of São Paulo, São Paulo, Brazil
e-mail: balbasky@usp.br

H. Horta et al. (eds.), *Trends and Challenges in Science and Higher Education*, Knowledge Studies in Higher Education 3,
DOI 10.1007/978-3-319-20964-7_6

relevant professional degree for her career. This is the degree responsible for assigning a clear and long-lasting professional identity, which also serves as a license, granting access to various protected niches in the labor market. The Napoleonic model also entails strict control and supervision by the state over the universities, as all institutions are expected to offer the same kind of instruction, following uniform, nationwide standards established for each recognized profession. Regarding this tradition, it comes as no surprise that academic life inside Latin American universities tends to gravitate around the all-important first-level university degree. In Latin America, graduate education was a late addition to the original institutional fabric. In most countries, some form of graduate instruction (i.e., master's or, more rarely, doctoral degrees) was introduced by the end of the second half of the twentieth century. Even if some countries have developed actions to organize and evaluate graduate education, it still has a tenuous existence in most universities, surviving with weak support, and little institutional and societal recognition.

The Napoleonic heritage has also made the Latin American university a teaching-centered institution. While research has developed in the last quarter of the twentieth century, the institutional locus for research and its connections with teaching in most Latin American universities is shaky, to say the least. In most countries, research tends to be insulated in small, protected environments, such as research centers and laboratories. These microenvironments usually sustain weak links with the hub of the university and, in some cases, develop inner rules for career and workload that may even go against the university's general rules.

The general trends outlined above help to describe the roots of some of the traits peculiar to academic employment in Latin America. First, in a teaching-centered university system for first-level professional degrees, it is easy to understand how part-time commitment to the academic life is so widespread. Historically, the traditional professor in a Latin American university was (and still is) the distinguished professional that also teaches, and sometimes even holds a chair, in a part-time arrangement. For these professionals, a position at a university is regarded as both a sign of prestige and a duty, taking pride in preparing the next generation in their profession. While the expanding costs of the recent growth in public systems are partially responsible for the permanence and expansion of part-time contracts in many countries, the roots of this kind of contract are historical in these universities.

Furthermore, the novelty of graduate education in the continent also means that academic staff, on average, are poorly qualified. Even today in many countries, the majority of academics only hold a bachelor degree; this is the case in Mexico, Cuba, Chile, and Argentina. In some countries there are a larger number of academics holding a master's degree; however, Brazil is the only country where the proportion of academics holding a doctoral degree exceeds 15 % (Brunner and Ferrada-Hurtado 2011). What is seldom acknowledged in the literature is the impact this situation has on the institutionalization of universities. Not only is there a lack of qualified personnel, the alternatives for training are scarce. Thus, while some young academics will have access to graduate education once employed, others will spend their entire professional career without sound academic qualifications. In this scenario, even if the academic degree (and not performance) becomes the most relevant credential for ascending the academic ladder, the opposite will also be true. Because many academics move through their professional career without further academic qualifications, there is also the pressure to mitigate the exigency of a graduate degree for promotion. The typical answer to this pressure is to create organizational bypaths, allowing academics to circumvent the requirement of a degree for promotion. In the worst-case scenario, this situation also creates a pressure to lower the knowledge and performance content of graduate education, which creates a system of hollow credentials. As institutional careers become disconnected from academic hierarchies, the former also lose their discriminant capacity and meritocratic role to "accrue to an individual or to a group of academics for meritorious or exemplary performance" (Moore 1992). In many Latin American systems, the role of recognizing exemplary performance is played by external entities, such as research scholarships controlled by science foundations, or the prestige derived from an individual's links to prestigious research groups, laboratories, or research centers.

Since the 1960s, two trends have affected the old Latin American university model. The first was the first wave of the massification of access. From the mid-1960s onwards, with industrialization and the growth of cities, an increasing number of people in Latin America gained entry into university. Although some of these were young people from the enlarged middle-class sectors, the majority were (and still are) older people, women, and poorer people that finished secondary education late in life and demanded access to a university education. In some countries, this new clientele was absorbed by traditional public universities, which grew to become mega-universities enrolling hundreds of thousands of students at the bachelor level. In other countries, the new demands for access were (and still are) absorbed by an enlarged private sector, comprising mostly low-cost teaching schools and staffed by part-time, low-qualified professors. These schools offer evening programs, and cater to students that could not meet the public sector's entrance requirements.

The demand for access to higher education experienced a new boon in the late-1990s and the beginning of the new century. The quick expansion of both public and private sectors led to the hiring of a large number of new academic staff. Contrary to the old professional professor profile described above, the new staff have shaky academic and professional credentials and are, at the same time, dependent on the salaries paid by the universities and unable to establish a personal reputation as an academic. In most Latin American countries, academics with this profile tend to be organized into strong unions, pressing for better salaries, employment conditions, and career perspectives.

The second trend responsible for change comes from the pressures posed by the new generation of PhD holders in the region. The first generation of these scholars was trained abroad, and returned in the 1960s. These academics looked to transform traditional Latin American universities into new institutions, organized in disciplinary lines via departments and institutes, committed to research and graduate education with full-time contracts and lower teaching loads that would enable an academic career dedicated to research and scholarly performance. Since the 1960s, the growth in the number of academics with this profile is one of the most important inner sources of change of Latin American higher education (Bernasconi 2007).

Even though these scholars share some aspects with the non-qualified lectures organized inside academic unions, the university ideal that inspired these two groups could not be more different. For the former group, university is a place that supports science, and merit is the best way to organize hierarchies within and across institutions. For the latter, university is a tuition-free public institution, supported exclusively by public funds and where academics share a similar flat career structure in which seniority, not merit, should be the main criteria for promotion.

6.2 Academics in Latin America: A Layered Profession

As I have argued elsewhere (Balbachevsky 2015), the academic profession in Latin America does not enjoy a unified profile where all members share some basic mode of socialization (doctoral training) and at the very least some core values. The opposite is true, and one could say (borrowing the expression from Olle Edqvist 2003) that the academic profession in Latin America is a layered profession, where different professional profiles are superimposed, each of them with roots in a different phase of history. Furthermore, these profiles all have their own methods to replicate themselves, and, in doing so, they perpetuate the fragmented profile of the academic profession in Latin America.

To identify the most relevant traits of these different profiles, I propose a typology that takes into account two different aspects of the realities of academic life in the region:

- The degree of congruence between academic rank and academic credentials
- The degree of engagement with research

Figure 6.1 below provides a simplified scheme for this typology.

Thus, considering the two dimensions proposed above, it is possible to identify four different academic profiles inside Latin American universities. First, there is the *old academic oligarchy* composed of academics with shaky academic credentials, but they have reached positions of authority inside the institution. Some correspond to the traditional professor in Latin America: a professional distinguished in his

		Degree of commitment to research	
		High	Low
Degree of congruence between academic credentials and career	High	Academic elite	Lecturers
	Low	Young scholars	Institutional oligarchies

Fig. 6.1 A typology of the Latin American academic profession

profession and occupying the higher ranks inside a professional school. Others are academics with no particular professional identity, but with extensive experience of the bureaucratic rules by which the university is organized.

Second, there is the *academic elite*, composed of scholars with good academic credentials, well positioned in the institution's ranks, and sharing a strong commitment to research and knowledge creation. Their daily institutional experience revolves around their research center or laboratory where they concentrate their research activity. For them, autonomy is a relevant issue. Only strong, autonomous research units are able to resist interference from the more or less politicized environment that is traditional in Latin American universities (Schwartzman 1993). In some measure, the existence of these semi-autonomous microenvironments cushions the research-oriented academics and their teams from the dysfunction present in the university. As an example, in 2007 when studying the roots of success of some research centers in Latin American universities, one interviewer reported that, "Yes, I have to speak very well of the institute … if you noticed, it was even difficult for me to find problems within UNAM (National University of Mexico). Apart from the structural inconveniences, it was difficult to define the problematic areas. Largely it is because I am at the Institute of Biotechnology" (Balbachevsky 2008, p. 38). On the other hand, their dependence on external support to sustain their research unit, combined with their experience with designing and implementing projects, reinforces a strong entrepreneurial ethos (Etzkowitz et al. 2000). Inside this group, it is possible further differentiate a small group of *internationalized academic elite*, sustaining strong links with the international community.

Third, there are the *young scholars*, academics with good academic credentials and strongly committed to research and knowledge creation but positioned in the lower ranks inside the universities. For diverse reasons, in their cases, research and commitment to academic life have not been translated into institutional recognition. They share the core values expressed by the academic elite, but are more vulnerable to pressures coming from the unions and the radical views sustained by political movements from the larger society. If they are employed in research-oriented universities, their career prospects are good. However, if they are employed in teaching-oriented institutions, then they may face a difficult situation with poor career prospects and even some open resistance to their engagement in research. In this situation, research becomes an individual endeavor, limited by a lack of funds and institutional support, and resulting in products of low impact and recognition. The best (but less accessible) alternative open to young scholars employed in teaching-centered institutions is to attach their projects to a collective research effort organized inside a major national research network. However, the opportunities for this kind of networking are rare and largely depend on the connections young scholars forged in their doctoral years.

Finally, for the majority of *lecturers* in Latin America, the lack of academic credentials and close to nonexistent commitment with research produce a profile more similar with a secondary school teacher than the one typically identified in the international literature as an academic scholar. They are almost entirely disconnected from their national and international peer communities. Hence, their professional identity is neither defined by their professional degree, as is the case with academics in the professional oligarchy, nor by their individual achievements as independent scholars. Instead, their identity is locally rooted, based on their institutional affiliation and the small group of colleagues with whom they share daily work experiences. In a sense, academics belonging to this group tend to sustain a "semi-professional identity" (Etzioni 1969) because they tend to emphasize intrinsic rewards like personal satisfaction (e.g., being a good teacher) as opposed to extrinsic ones (e.g., peer recognition or professional status). This fact explains why this group so fiercely opposes any attempt to introduce intra- and inter-institutional differentiation based on merit and/or prestige. For them, the only acceptable grounds for differentiation are those produced by externalities, in principle accessible to everyone, like seniority.

It is very difficult to find data to estimate how these different groups are distributed inside different national systems or how they differentiate in their work, beliefs, and values. There are little comparable data for all countries, and data that do exist are seldom detailed enough to identify the members of the different academic profiles described above. Nevertheless, in 2007, as part of the international network called the Changing Academic Profession,[1] a comparative survey was carried out in the three largest academic systems in Latin America: Argentina, Brazil, and Mexico. The data collected by this survey are detailed enough for an exercise to better identify these different profiles. Thus, in Sect. 6.3 I use the data collected under the Changing Academic Profession (CAP) research network to search for tentative answers to these questions. Only academics working in the public sector were considered in the analysis, which means that, in the case of Brazil, the sample was reduced, with the large subsample of interviewers from private sector omitted from the analysis.

6.3 Academic Profiles in Latin America

As observed above, Latin American universities have always been plagued by shortcomings derived from the small pool of academic competencies from where to recruit faculty.[2] The massification of access to higher education and the consequent growth of the number and size of institutions have only aggravated this problem. While all countries in the region have tried to circumvent the shortcomings of this situation by sending scholars abroad and supporting domestic efforts to build a graduate layer, these alternatives have never been enough, and most importantly, were never meet with universal acceptance by all academics. While some academics were attracted by the new prospects offered by graduate education, this profile was never dominant. For most academics, the easier path was (and still is) to weaken

[1] For more detail see http://www.uni-kassel.de/einrichtungen/en/incher/research/research-area-change-of-knowledge/the-changing-academic-profession-cap.html.

[2] This section is partially based on a previous analysis published in Balbachevsky (2015).

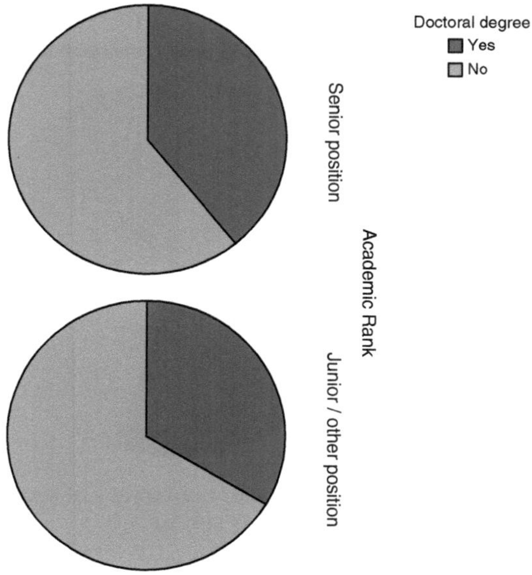

Fig. 6.2 Latin American academic profession: congruence between academic rank and academic credentials (Source: Balbachevsky (2015))

the relationship between academic credentials and the hierarchy inside the institutions. The negative side effect of this process is, of course, that not all academics with doctorates are able to find a senior position within a university, mostly because these positions are already filled by academics with lower academic credentials. Figure 6.2 shows the conditional distribution of academics credentials among academics in senior positions (equivalent to associate and full professors) and junior positions.

Figure 6.2 shows that in Latin America, although it is not the norm for senior academics to hold a doctorate degree, the proportion of lower-ranked academics with doctorates is not negligible. In fact, the 2007 CAP Survey shows that 61 % of the academics in senior positions have no doctoral degree, as do 33 % of those in junior positions.

When analyzing the indicators related to the research performance of academics among Argentina, Brazil, and Mexico, there is a very interesting pattern of association between publishing, securing external support to research activities, and sustaining an active network of peers from abroad. Considering these three indicators, the CAP data show that most Latin American academics are not involved in research-related activities. The majority reported no publications, have no access to external funds for research, and do not sustain active collaborations with international peers. However, another group reported some publishing activity, but no other research-related relevant activities. A minority reported success in publishing, access to research funds, but no external collaboration. Finally, some Latin American academics engaged in all three dimensions: they publish, have access to external resources for research, and hold active collaboration with peers from abroad. Others

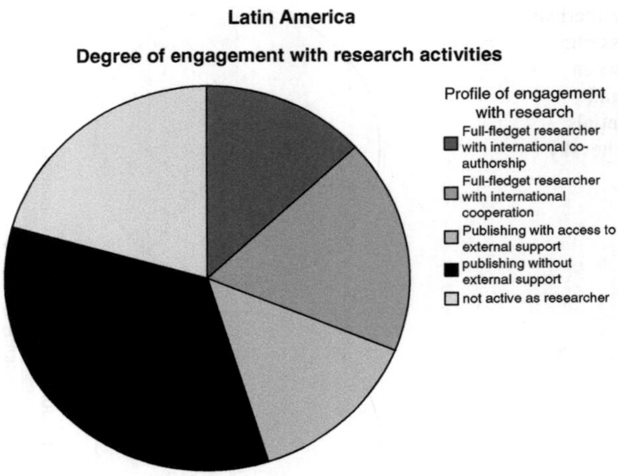

Fig. 6.3 Distribution of Latin American academics by different profiles of engagement with research (Source: Balbachevsky (2015))

collaborate and publish in co-authorship with colleagues from abroad, while some reported collaboration but no publishing with overseas colleagues. The pattern of association among these answers is shown in the scale[3] depicted in Fig. 6.3.

Thus, while 13 % of the sample are full-fledged researchers with international collaboration and co-authorship, 18 % also have an active profile as a researcher with international collaboration. A total of 14 % reported publishing with access to external funding, but without access to international networks, and 34 % reported some publishing activity but no access to external funding or any contact with international networks. Finally, 21 % of the sample are not active as researchers.

Looking at the information presented in Figs. 6.2 and 6.3, it is possible to estimate the presence of the Latin American academic profession in the different academic profiles identified by our typology. As shown in Table 6.1, the typology outlined above uncovers relevant differences in the way these academics organize their daily life.

Table 6.1 shows that young scholars and the international elite share the same strong intensity of commitment to research: both spend an average of 17 h per week on research. In contrast, compared with young scholars, members of the international elite reported a larger amount of time devoted to administrative tasks (6 h against 3.8 h per week), which relate to their position as research leaders. The domestic elite reported less time spent on research (12 h per week) and more time devoted to teaching and teaching-related activities (18.7 h per week). Members of the old oligarchy reported the largest amount of time spent teaching: on average, academics with this profile spend more than half of their working time (22.5 h of 41 h per week) on teaching-related activities. This group also reported the greatest

[3] The scale follows a Guttman scale. The coefficient of reproducibility of this scale is 0.91.

Table 6.1 Academic Profession in Latin America: commitment to research, teaching, and administration (in term classes) by diverse academic profiles

Academic profile		Hours per week: teaching	Hours per week: research	Hours per week: administration	Total hours per week: academic activities
International elite	Mean	15.8	17.1	6.2	44.1
	Std. deviation	9.1	11.0	6.9	15.1
Domestic elite	Mean	18.7	12.0	6.8	43.7
	Std. deviation	10.3	9.1	8.7	16.6
Academic oligarchy	Mean	22.5	5.0	8.0	41.0
	Std. deviation	11.9	6.7	10.4	17.8
Young scholars	Mean	13.5	17.8	3.8	39.6
	Std. deviation	7.7	11.3	5.9	14.7
Lecturers	Mean	15.9	11.0	3.4	35.0
	Std. deviation	10.1	9.6	6.6	16.6

Source: Balbachevsky (2015)
Eta2: teaching: 0.09; research: 0.24; administration: 0.04; total: 0.03

involvement with administrative tasks, an average of 8 h per week. This may be related to their particular source of prestige: the control of the bureaucratic channels of governance inside the university. Finally, lecturers, as expected, reported less time devoted to research-related activities, and more time devoted to teaching. They were also less involved with administrative tasks. Lecturers enjoyed the shortest academic week: on average, the academic activities (teaching, services, research and administrative responsibilities) of lecturers total 35 h per week, compared with 44 h for the academic elite and domestic elite, 41 h for the academic oligarchy, and 40 h for young scholars.

Similarly, academics with different profiles tend to have different evaluations regarding their core tasks, teaching, and research. When asked how they balance their preferences between teaching and research, the pattern of distribution for the responses closely follow the profiles described above, as one can see in Table 6.2.

In total, 83.2 % of the international elite reported a focused interest in research. Research is also the preferred activity by young scholars (76.8 %) and the domestic elite (54.5 %). However, teaching was the most valued activity by the old oligarchy (77.6 %) and lecturers (61.5 %). The data in Table 6.2 show that inside Latin American universities, preferences for teaching and research cut across institutional hierarchies. This profile helps to explain the mixed feeling with which these academics received the various implemented policies, which tended to emphasize the US model of university, with academics with a high profile of academic degrees and

Table 6.2 Latin America: distribution of preferences between research and teaching among different academic profiles

Regarding your own preferences, do your interests primarily lie in teaching or research?	Academic profile					Total
	International elite	Domestic elite	Academic oligarchy	Young scholars	Lecturers	
Primarily in teaching	0.9 %	6.5 %	29.4 %	1.1 %	13.6 %	13.9 %
In both, but leaning towards teaching	16.0 %	39.0 %	48.2 %	22.0 %	47.9 %	36.7 %
In both, but leaning towards research	67.2 %	48.3 %	20.4 %	63.4 %	33.6 %	41.9 %
Primarily in research	16.0 %	6.2 %	2.0 %	13.4 %	4.9 %	7.4 %
Total (100 %)	(701)	(600)	(1173)	(350)	(447)	(3271)

Source: Changing Academic Profession international files; countries: Argentina, Brazil, and Mexico
Chi square test: 939.5; df: 12; sig. 0.000

dedicated to both teaching and research (Bernasconi 2007; Leal and Marquina 2014; Gallaz-Fuentes et al. 2014).

In the last few decades, Latin American universities have been altered by the same winds of change present in other regions. Globalization and massification are global trends also present in Latin America. Furthermore, the regulatory frameworks under which these universities operate have changed. Since the 1990s, public universities have had to deal with greater interference from central government operating with a common agenda of reforms, which usually include the enforcement of quality assurance programs, expansion of undergraduate enrollments, and science policies enforcing competition and social and economic relevance.

As one would expect, the way academics perceive these changes and how they react to them vary according to how they experience academic life. Regarding the overall environment of the institutions, the issues that produced the greatest contrast between different academic profiles are presented in Tables 6.3 and 6.4 below.

When asked if they perceive a strong emphasis on the institution's mission, academics in junior positions tend to show more distrust than those in higher ranks. However, among higher-ranking academics, those categorized as belonging to the academic oligarchy seem more confident with the way their institutions are conducted. In total, 18 % of the members of the academic oligarchy strongly disagree with the statement of there being "a cumbersome administrative process", and only 19 % totally agree, compared with 8 % and 28 %, respectively, for the other groups.

When asked if they are kept informed about what is going on in their universities, higher-ranking academics tend to give more positive answers, and the lower ranking more negative evaluations (Table 6.4). Lower-ranking academics also tend to be more critical when evaluating how their university's administration deals with academic freedom. Finally, young scholars and the international elite are more skeptical regarding the provision of greater opportunities for student participation in the university's governance.

Table 6.3 Average level of satisfaction with the institution's environment; scale from 1 to 5, where 1 represents that the respondent agrees that the trait is very much present in the university and 5, the trait is not present

Academic profile		A strong emphasis on the institution's mission	A cumbersome administrative process	A supportive attitude of administrative staff towards teaching activities
International elite	Mean	2.4	2.5	3.1
	Std. deviation	1.3	1.28	1.2
Domestic elite	Mean	2.3	2.5	3.0
	Std. deviation	1.1	1.3	1.2
Academic oligarchy	Mean	2.2	3.0	2.8
	Std. deviation	1.2	1.3	1.3
Young scholars	Mean	2.7	2.3	3.1
	Std. deviation	1.3	1.1	1.2
Lecturers	Mean	2.6	2.5	3.0
	Std. deviation	1.3	1.2	1.2

Source: Changing Academic Profession international files; countries: Argentina, Brazil, and Mexico
ANOVA test indicates differences significant for $\alpha < 0.000$

Table 6.4 Average level of satisfaction with university governance; scale from 1 to 5, where 1 represents that the respondent strongly agrees with the statement, and 5 represents that the respondent strongly disagrees

Academic profile		I am kept informed about what is going on at this institution	The administration supports academic freedom	Students should have a stronger voice in determining policy that affects them
International elite	Mean	3.0	2.1	3.1
	Std. deviation	1.2	1.1	1.2
Domestic elite	Mean	2.8	2.0	2.9
	Std. deviation	1.3	1.1	1.2
Academic oligarchy	Mean	3.0	1.9	2.8
	Std. deviation	1.3	1.1	1.2
Young scholars	Mean	2.4	2.3	3.2
	Std. deviation	1.1	1.1	1.2
Teachers	Mean	2.6	2.5	2.9
	Std. deviation	1.2	1.1	1.2

Source: Changing Academic Profession international files; countries: Argentina. Brazil and Mexico
ANOVA test indicates differences significant for $\alpha < 0.000$

Table 6.5 Academic profession in Latin America: the presence of different professional profiles

		Country			
		Argentina	Brazil	Mexico	Total
Academic profile	International elite	9.7 %	41.8 %	20.2 %	21.2 %
	Domestic elite	13.9 %	20.6 %	19.4 %	18.3 %
	Academic oligarchy	7.6 %	5.1 %	57.6 %	36.5 %
	Young scholars	33.5 %	12.1 %	0.4 %	10.5 %
	Lecturers	35.3 %	20.4 %	2.4 %	13.5 %
Total (100 %)		(825)	(553)	(1954)	(3332)

Source: Changing Academic Profession international files; countries: Argentina, Brazil, and Mexico

Finally, Table 6.5 shows how the profiles of public sector academic personnel vary among the three Latin American countries. This table provides a glimpse of the large differences that separate one country from another. In fact, the data enable us to contrast the situation of the academic profession in the three largest academic systems in Latin America. Starting with Argentina, the data from the survey provides a cue of the precarious situation of the majority of academics in that country: 68 % of the respondents reported holding a junior position, which, in the Argentinian context, means that they are teaching assistants (Leal and Marquina 2014). Even when holding a PhD, many of these academics experience academic life from a very tenuous position, working part-time with unstable contracts.[4] In contrast, the reforms adopted by the Argentinian government since the 1990s have had some impact, pushing for more homogeneous and research-active hierarchies inside Argentinian universities. As shown in Table 6.5, 31.2 % of the sample are at the top of the country's academic hierarchies, that is, they are professors. Among them, only 24.4 % are not active in research and do not hold a PhD.

Graduate education and, in particular, doctoral training are institutionalized to a greater degree in Brazil. Thus, the larger proportion of Brazilian academics with an active research profile is of no surprise. Even considering that the design of the Brazilian sample overrepresented research-oriented universities, the proportion of academics with an active research profile is higher than in the other two countries. The figures for Brazilian graduate education are impressive: in 2013, more than 113,000 students were enrolled at the master's level and another 87,000 were enrolled in doctoral programs. Furthermore, Brazilian universities graduated more than 45,000 master's students and approximately 15,000 doctoral students in that year (GEOCAPES 2014). These figures make the Brazilian graduate system the largest in Latin America and one of the larger within the emerging countries.

[4] Argentina has a very peculiar situation, where all academic positions are held for a fixed period of time (5–7 years for professors and 3 years for assistants). Nevertheless, there is a large difference between those approved in a concourse and those working without the security of this temporary stability (Fanelli 2003; Leal and Marquina 2014).

Brazilian graduate education is also impressive for other reasons. Since the mid-1970s, the *Fundação Coordenação de Aperfeiçoamento de Pessoal de Nível Superior*, the Ministry of Education's agency in charge of graduate education, has been operating a sophisticated evaluation system based on peer-review. The assessment scheme successfully connects performance with support, creating a system that reinforces the best programs, while imposing a threshold of performance that limits growth without quality. With such an impressive profile, it is not difficult to understand the large proportion of research-oriented profiles among high-ranking academics in Brazilian public universities. Nevertheless, the data also show that Brazilian public universities still hire a large number of junior faculty without doctoral degrees. More importantly, as our previous analysis shows, academics with this profile tend to have little engagement in research-related activities. In fact, as explained elsewhere (Balbachevsky and Schwartzman 2007), the growth and consolidation of the graduate system in Brazil has created an informal divide among public universities. While some universities evolved to become real research universities with a large proportion of students enrolled at the graduate level and with almost all academic staff holding doctoral degrees, most of the public sector institutions have remained as teaching-oriented universities, where academic authority still lies mostly in the hands of the old oligarchy (Schwartzman and Balbachevsky 2014, p. 227).

Finally, Mexico is where the divisions are most expressive. The Mexican sample focused on full-time academics, and thus over-represents academics in higher positions in the institutional hierarchy. Even so, 57.6 % of high-ranking Mexican academics have a profile that can be termed as old oligarchy, marginally committed to research and largely focused on teaching at the undergraduate level. One should remember that Mexico has well-developed merit-pay programs that target high-performance academics. The payments arising from these programs represent a high proportion of the average academic income in the public sector. The data collected in the CAP survey provide relevant information to understand the main source of dissatisfaction among Mexican academics. While the public authorities try to enforce a uniform policy focused on the profile of the top research-oriented academics, most Mexican academics neither valorize this profile nor have the skills to reach this high level of research performance (Gil-Anton et al. 1994; Gallaz-Fuentes et al. 2014).

6.4 Conclusion

This chapter attempts to describe the diversity among the academic profession in Latin America, looking at various differences among the large number of countries in the region. While it is possible to highlight some similarities and more general convergent dynamics, many differences remained untouched by the analyses developed here.

Our study centered on proposing and developing a typology of academic profiles present in almost all academic systems in the region; these help to understand the main source of tensions inside Latin American universities. As proposed earlier in this chapter, it is possible to describe the Latin American academic profession as a layered profession, where old profiles, compatible with different university models, are placed alongside new profiles generated by recent changes in higher education dynamics. Essentially, the different professional profiles (with their diverse commitments regarding the core academic responsibility and values) share the same institution and are ruled by the same institutional rules.

This situation is an important clue to understand the strength and longevity of the old-fashioned formulas regarding university governance in Latin America. It is probable that weak central administration and strong autonomy by the faculties and institutes are necessary to accommodate the tensions arising from these differences. For those academics with a stronger commitment to research, the politicization of many universities could be a source of strain rather the fact that research-oriented microenvironment are kept in isolation from the university-wide environment. For these academics, it is the institute or center that constitutes the basic institutional reference, accessible for collegiate participation (Balbachevsky 2008). In contrast, the policies adopted by many countries that place a premium on the research-intensive academic profile clearly marginalize a relevant portion of the academics who have no qualifications, expertize, or resources to perform as world-class researchers.

As pointed out at the beginning of this chapter, Latin American academic tradition is mainly oriented toward teaching. Regarding the old institutional fabric of Latin American universities, teaching at the first university level (the bachelor degree) is the most relevant role performed by academics, and is possibly the role that receives the most recognition in society. Accordingly, a large proportion of academics in Latin America, regardless their institutional rank, tend to view themselves primarily as teachers, and not as researchers. In contrast, as opportunities for doctoral training expand, the number of academics that valorize research over teaching has also grown in the region. These academics feel justified in giving priority to research over teaching, as is clear in many higher education policies in the region. The tensions produced by these divisions are far from being solved in the first decades of the twenty-first century.

Acknowledgment The author would like to acknowledge the financial support given by FAPESP, Fundação de Amparo à Pesquisa do Estado de São Paulo (2006/03329-0).

References

Balbachevsky, E. (2008). Incentives and obstacles to academic entrepreneurship. In S. Schwartzman (Ed.), *University and development in Latin America: Successful experiences of research centers* (pp. 23–42). Rotterdam: Sense Publishers.

Balbachevsky, E. (2015). The Latin American University model and the challenges posed by the reforms: The perspectives from the academics. In W. Cummings, Teichler, U. (Eds.), *The relevance of academic work in a comparative perspective*. (pp. 239–252), Dordrecht: Springer.

Balbachevsky, E., & Schwartzman, S. (2007). Brazil; a typology of the academic profession and the impact of recent government and institutional policies. In W. Locke & Ü. Teichler (Eds.),

The changing conditions for academic work and career in selected countries. Kassel: INCHER-University of Kassel.

Bernasconi, A. (2007). Is there a Latin American model of the university? *Comparative Education Review, 52*(1), 27–52.

Brunner, J., & Ferrada-Hurtado, R. (2011). *Educación Superior em Iberoamérica: Informe 2011*. Santiago: CINDA.

Etzioni, A. (1969). *The semi-professions and their organization; teachers, nurses, social workers*. New York: Free Press.

Etzkowitz, H., Webster, A., Gebhardt, C., & Terra, B. R. C. (2000). The future of the university and the university of the future: Evolution of ivory tower to entrepreneurial paradigm. *Research Policy, 29*(2), 313–330.

Fanelli, A. M. G. (2003). *Academic employment structures in higher education: The Argentine case and the academic profession in Latin America*. Unesco, ILO: CEART/8/2003/HEES-2.

Gallaz-Fuentes, J., Martínez-Stack, J., Estévez-Nénninger, E. H., Padilla-González, L. H., Gil-Antón, M., Sevilla-Garcia, J. J., et al. (2014). The divergent worlds of teaching and research among Mexican faculty: Tendencies and implications. In J. C. Shin, A. Arimoto, W. Cummings, & Ü. Teichler (Eds.), *Teaching and research in contemporary higher education: Systems, activities and rewards* (pp. 199–220). Dordrecht: Springer.

GEOCAPES. (2014). Sistema de informações georeferenciadas. Brazil, CAPES, Ministry of Education. http://geocapes.capes.gov.br/geocapes2/. Accessed 21 Nov 2014.

Gil-Anton, M., Kuri, R. G., Franco, L. P., López, N. R., & Alvarado, M. A. C. (1994). *Los rasgos de la diversidad: um estudio sobre los académico mexicanos*. Azcapotzalco: Universidade Autónoma Metropolitana.

Leal, M., & Marquina, M. (2014). Current challenges facing the academic profession in Argentina: Tensions between teaching and research. In J. C. Shin, A. Arimoto, W. Cummings, & Ü. Teichler (Eds.), *Teaching and research in contemporary higher education: Systems, activities and rewards* (pp. 237–254). Dordrecht: Springer.

Moore, K. M. (1992). Faculty reward and incentives. In B. Clark & G. Neave (Eds.), *The encyclopedia of higher education*. Oxford: Pergamon Press.

Schwartzman, S. (1993). Policies for higher education in Latin America; The context. *Higher Education, 25*(1), 9–20.

Schwartzman, S., & Balbachevsky, E. (2014). Research and teaching in a diverse institutional environment: Converging values and diverging practices in Brazil. In J. C. Shin, A. Arimoto, W. Cummings, & Ü. Teichler (Eds.), *Teaching and research in contemporary higher education: Systems, activities and rewards* (pp. 221–235). Dordrecht: Springer.

Part II
Policies and Instruments to Foster Scientific Capacity

Chapter 7
On the Processes of Technical Change and Development in Latin America: A Proposed Framework of Analysis

Leandro Lepratte

7.1 Introduction

In recent years, the relationship between innovation, technological change, and development in contemporary capitalism has claimed importance in various fields within the social sciences. Many approaches recognize the importance of knowledge, information, higher education, and science and technology (S&T) as key factors for economic growth and development (Amsden 2004; Cimoli et al. 2010; Dabat and Rodriguez Vargas 2009; Dutrénit and Teubal 2011; Freeman 1991; Lastres et al. 2003, 2005; Lundvall 1988; Moncayo 2004; Ocampo 2005; Rodríguez Vargas 2009; Vessuri 2007). Even within the proposals of certain neoclassical scholars[1] related to the endogenous economic growth approach, human capital, as a broad category that encompasses these factors, has a leading role to define the purpose of progress in contemporary societies. As a result, a number of approaches have characterized these times as a *postindustrial society* or a *knowledge society*. Furthermore, this has led to a proliferation of studies on the *learning economy* (Lundvall and Johnson 1994), the *network* or *informational society* (Castell 1999), the *knowledge-based society* (OECD 1996), *cognitive capitalism* (Boutang 2004), and *global informational capitalism* (Dabat and Rivera 2004), among others.

Two of the most important academic fields emerging from the study of these phenomena regarding the *knowledge society* are innovation studies and science and technology studies (STS) (Fagerberg et al. 2012a).[2] In the case of innovation stud-

[1] They acknowledge the works by Smith and Marshall as precedents regarding the role of knowledge, learning and experience as wealth generators.

[2] These authors include studies on entrepreneurism as a third emerging field within knowledge society studies.

L. Lepratte (✉)
National Technological University FRCU, Concepción del Uruguay, Entre Ríos, Argentina
e-mail: leprattel@frcu.utn.edu.ar

H. Horta et al. (eds.), *Trends and Challenges in Science and Higher Education*, Knowledge Studies in Higher Education 3,
DOI 10.1007/978-3-319-20964-7_7

ies, the economics of innovation and technological change[3] (EITC) has been given significant weight within the field (Fagerberg and Verspagen 2009; Fagerberg et al. 2012b). Regarding STS, the history and philosophy of science and technology, on the one hand, and the sociology or social studies of science and technology on the other (Martin 2012), have been assigned the same level of importance. Both academic trajectories had their initial contribution in developed countries and, later, were incorporated into Latin America.

The trajectory of innovation studies (including EITC) and STS have been characterized by particular dynamics in Latin America (Kreimer and Thomas 2004; Arellano et al. 2012). The field of study science, technology, and society (CTS)[4] was generated by scientists and engineers rather than by researchers from the social sciences. The beginnings of Latin American CTS studies in the 1950s (Kreimer 2007) expressed the character of a movement and political praxis, rather than that of an academic field (Vaccarezza 2004).[5] However, between 1980 and 1995, CTS gained greater importance in Latin American academic *institutionalization*, causing a proliferation of empirical research and theoretical frameworks using approaches generated outside the region, such as the actor–network theory (ANT) and social construction of technology (SCOT), from the relativist–constructivist legacy in STS.[6] This same phenomenon of institutionalization also applies to science and technology policies (Arellano and Kreimer 2011; Casas et al. 2013), particularly the inclusion of instruments to promote innovation, using exogenous models from developed countries.[7] This originated a particular tradition of studies in the region that connect the role of science and technology with "macro" policies (Casas and Luna 1997; Villavicencio 2010). At the end of the twentieth century, the field of CTS consolidated in Latin America, with academic presence and dissemination, and the emergence of research groups (see Table 7.1).

What is relevant in this outlook (from 1995 to the present) is that for some scholars EITC has played an important role among the contributions of the tradition of the STS field in Latin America (Kreimer and Thomas 2004; Arellano and Kreimer 2011). However, within the trajectory of evolutionary and neo-Schumpeterian

[3] See Yoguel et al. (2013) for a broad description of the development of the innovation economics and technological change trajectory. They also include descriptive elements of its academic trajectory in Latin America.

[4] See Kreimer and Thomas (2004) for a reflection on these dynamics in Latin America. Arellano (2012) discusses the reflective nature of this work and proposes alternative interpretations on the evolution of this field.

[5] The following authors criticize the linear model (Dagnino et al. 1996) and use categories such as national project, social demand, and technological styles: Marcel Roche (Venezuela); Amilcar Herrera, Jorge Sábato, and Oscar Varsavsky (Argentina); José Leites Lopes (Brazil); Miguel Wionczek (Mexico); Francisco Sagasti (Peru); and Maximo Halty Carrère (Uruguay) (Arellano et al. 2012).

[6] Arellano and Kreimer (2011) point out this uncritical use of certain approaches in the region, where only some adaptation occurred.

[7] Dagnino and Thomas (2000) warn about this when referring to mainstream studies about the relation between industry and university and its application to Latin American contexts.

Table 7.1 Science and technology studies and neo-Schumpeterian evolutionary economics in Latin America (1950 to 2000s)

Period	Field orientation
1950–1980	Beginning of Latin American STS studies
1980–1995	Institutionalization; uncritical incorporation of approaches from developed countries (ANT, Triple Helix, SCOT, etc.)
1995–2000	Consolidation of training and research fields: social studies in technology, history of science, anthropology of science, and neo-Schumpeterian evolutionary economics
2000s	Differentiation of trajectories between EITC and EST Orientation towards problems of social utility and sustainability (social technologies, social inclusion technologies) Orientation towards problems of economic growth and development (micro–macro)

Source: Kreimer and Thomas (2004), Arellano and Kreimer (2011), Arellano et al. (2012), Yoguel et al. (2013), Silva and Teixeira (2009)

contributions in the region, there have been some specific processes of differentiation, legitimation, and institutionalization that do not recognize full participation in this field. As Yoguel et al. (2013) describe in the Latin American context, neo-Schumpeterian evolutionary economics significantly expanded between the 1990s and 2000s, together with the emergence and spread of a number of other ideas of heterodox economic thought; this included criticism regarding mainstream and neoliberal economic policies in the region. Recent related theories stem from the revival of development theory and Latin American structuralism. Thus, there is a tendency to align CTS issues with the production of knowledge, social utility of technology, social technologies, and sustainability processes. Meanwhile, in EITC, the importance lies mainly with the processes of structural change, specialization profile-determining factors, and ultimately the relationship with development problems in a predominantly economic sense (Casas et al. 2013).

Another particular element of the dynamics of both academic trajectories in the region is that their main studies focus on those phenomena related to the knowledge society (Arellano et al. 2012; Fagerberg et al. 2012b), being useful to evidence the global nature of innovation and technological change processes; however, the contributions regarding certain particularities in Latin America have appeared progressively. Some of the issues that have gained importance in the region are the recognition of local trajectories, the specific learning processes at an organizational and artifactual level, and the impulsion of capacities in an endogenous sense, among others: "The paradox is that this led to pay attention to local knowledge and situated practices, knowing that the reflection is worldwide" (Arellano et al. 2012, p. 24).

Together with this paradox exists a supplementary question exceeding the particularity of these trajectories in Latin America. In developed countries[8] and in

[8] Mackenzie's (1992) seminal study articulates EICT and STS as regards to technological change problems. He goes back to the evolutionist research of Nelson and Winter, and to ethnographic approaches of marketing construction and technological change ("ethnoaccountancy"). Another

our region, there have been scant interactions between social studies of technology and neo-Schumpeterian evolutionary economics, in the context of converging efforts from theoretical and methodological perspectives (Kreimer and Thomas 2004; Arellano et al. 2012). Thus, the early parts of this chapter focus on the paradox of understanding the phenomena related to the category of knowledge society in Latin America and, in turn, the almost nonexistent interaction between convergent trajectories in terms of theory and research (Swedberg 2014).

Despite the divergence and inconsistency, these trajectories share certain converging points of view in the identification of problems. For example, Latin American productive and innovative patterns are dependent on those from developed countries, scarce and unequal innovation abilities of people and organizations in the region, repeated reactions to technological change, difficulties in establishing incentives for technological and social innovation, and the predominance of institutional frameworks averse to structural and technological changes. Some agreements are also seen in the need to explain these problems from the specificity of the continent, either in the generation of empirical evidence, or in the frameworks that guide science and technology policies (Rivera Rios et al. 2009; Arocena and Sutz 2003; Dagnino and Thomas 2000). Moreover, there is an important point of convergence when considering Latin American limitations in innovation and technological change processes to be closely related to the permanent problems of economic

precedent line is the quasi neo-evolutionist line, the so-called Twente School. Such authors state that the evolutionist theory and SCOT contribute to a constructivist assessment of technologies (Van den Belt and Rip 1987; Schot and Rip 1997). Members of this school use a multidimensional perspective to reach a sociological understanding of the evolution processes of variation, selection, and retention in the evolutionist economics tradition. They consider that a social study of technology and innovation economics offers significant contributions to understanding the nonlinear nature of technological development, their dependence of different circumstances and contingencies, as well as the important role of those actors involved in the processes. They also criticize the permanent intents to the generalizations of some technology studies that pay no attention to particular aspects of "real world complexities" (Rip 1995, p. 418). In this converging sense, they highlight the divergent trajectories of technology developments, considering Freeman's contributions about the situated nature of technology construction and the interaction between artifacts and institutional infrastructures.

One of the most important contributions to converging efforts worldwide is the research by Bruun and Hukkien (2003). They articulate ANT, SCOT, and evolutionist economics, postulating the possibility of creating a framework that gathers common analytical elements from each theory. Another significant contribution is that of Geels (2005, 2007, 2010) articulating evolutionist economics, social studies on technology, and technological innovation management in relation to sociotechnical systems. Analytical and empirical data show possibilities of convergence via multidimensional models and inter-ontology crossover analyses.

In Latin America, the postulations of social technologies (Dagnino 2010) manifest a conceptual analytical framework supported by approaches of appropriates technologies, ANT, innovation economics, critic theory of technology, sociotechnical analysis, and the Latin American tradition of S&T policies analysis. Following this line, other authors (Thomas 2011) have recently postulated ideas of social inclusion technologies with converging contributions similar to those of social technologies. They are seen as "ways of designing, developing, implementing and managing technologies oriented to solve social and environmental problems, promoting social and economic dynamics of social inclusion and sustainable development" (Thomas 2011, p. 27).

development, social exclusion, and sustainability (Arocena and Sutz 2003; Albuquerque 2007; Ballasts and Arroio Cassiolato 2005; Rivera Rios et al. 2009; Robert and Yoguel 2010; Perez 2010; Dagnino and Thomas 2000; Figueiredo 2004; Dutrenit and Katz 2005; Dabat and Rivera Rios 2004; Arellano et al. 2012; Casas et al. 2013).

This overview of contributions and trajectories poses a key challenge for Latin America; that is, to propose (from the perspectives of social studies of science and technology and the economics of innovation and technological change) alternative ways of theoretical and political dialog.

From this starting point, we formulate the following questions: is it possible to propose some kind of convergence in certain contributions from social studies of science and technology and the economics of innovation and technological change to study problems regarding innovation, technological, and structural change in Latin America? If so, what are the chances of convergence? Do they depend on ontological, epistemological, and theoretical assumptions? In which way do the possibilities of convergence enable us to generate a research program for Latin America? How could this program have implications in the field of science, technology, and innovation policies to interpellate recurrent exogenous models applied in the region? Based on these core issues, this chapter presents results and progress by exploring the possibility of convergence between two trajectories, based on the tradition of Latin American thought in CTS. A theoretical and political effort is assumed to have implications in solving the *problems of development* in the region. From EITC, certain recent contributions were selected from scholars in neo-Schumpeterian evolutionary economics oriented to complex systems (Schumpeter 1947; Nelson and Winter 1982; Dopfer 1986, 2001, 2011; Dopfer and Potts 2008; Arthur 1990; Foster 2005; Foster and Metcalfe 2001, 2009; Antonelli 2007, 2011; Bloch and Metcalfe 2011; Von Hippel 1988; Saviotti and Pyka 2008; Saviotti 2011; Consoli and Patrucco 2011; Lane 2011; Lane and Maxfield 1997), as were those engaged in recent theorizing efforts in Latin American from a neo-Schumpeterian evolutionary economics perspective (Robert and Yoguel 2011; Dutrenit et al. 1996). A number of approaches were also selected from the social studies of technology, including ANT (Callon 1987, 1989, 1992, 1995, 1998, 2001, 2006, 2008; Latour 1999, 2007, 2008; Law 1987, 2008), SCOT (Bijker et al. 1987; Bijker 1993, 1995, 2010; Pinch and Bijker 1984, 1987; Pinch 1996, 2008, 2010), and contributions from Latin America regarding sociotechnical analysis (Dagnino 2010; Thomas 2008, 2010).

The main aim of this theorizing exercise is to establish ontological, epistemological, and theoretical interpretations of a convergent framework, based on certain contributions from social studies of technology (EST) and neo-Schumpeterian evolutionary economics oriented to complex systems (EEC),[9] to address those issues related to processes of innovation, technological change, and structural change from

[9]The nomenclature of EST and EEC needs to be understood as theoretical and conceptual contributions derived from an analytical exercise and its findings. Meanwhile, the denominations STS and EICT are used to identify broader disciplinary trajectories where EST and EEC operate, respectively.

a Latin American perspective. This objective has a number of specific achievements that represent an analytical theorizing exercise with five interdependent moments.

The first analytical moment considers the background of preliminary convergences in the social studies of science and technology and EITC; it also identifies certain general assumptions regarding the selection of theoretical contributions of these trajectories (m1). The second explores the possibilities of ontological convergence between EEC and EST, analyzing epistemological and theoretical implications, and framing general analytical dimensions (m2). The third moment presents a converging research agenda, setting dimensions that constitute specific analytical research lines (m3). The fourth is oriented to outline a framework that will generate empirical-based studies around the unit of analysis, sociotechnical systems of production and innovation (m4).

The fifth analytical moment, which is not involved in the process of theorizing, raises a meta-theoretical proposal on science and technology articulatory policies, able to be oriented and connected to the recurring problems of Latin American development (m5).

This chapter synthesizes some of the results of these analytic moments and discusses preliminary conclusions from the fifth moment.

7.2 Toward a Convergence Research Agenda

The contributions of EEC share basic assumptions with evolutionary economics where innovation and technological change are considered the engines of growth and transformation of capitalist economies (Antonelli 2011, p. 6). Theorists of this trajectory focus on generating explanatory models capable of understanding the structure and dynamic characteristics of economic systems and how they can be represented analytically (Foster 2005). Furthermore, complex systems are seen to have a number of properties (Foster 2005): they are dissipative structures representing totalities in themselves that are part of some systems and oppose others in different modes of interaction (including predation and conflict). These connections enable the emergence of organized complexities with higher levels of aggregation, which have some degree of structural irreversibility and undergo evolutionary processes that must be understood in historical terms (emerging phases, growth, seasonality, and structural transitions). These systems can be represented not only by relationships between micro- and macro-economic components, but also by meso factors (Dopfer 2011).

EST start from a methodological perspective based on the metaphor of the "seamless web," which seeks to break the technological and social determinism in addressing problems of STS (Hughes 1983, 1986; Bijker et al. 1987). According to these perspectives, the development of technologies should not be explained as a linear development of technical knowledge influenced by external social factors, as it is a framework where heterogeneous events (artifacts, institutions, rules, knowledge) and various actors (engineers, entrepreneurs, politicians, users) are integrated in complex and nonlinear ways (Thomas et al. 2008, p. 66).

According to Bijker (1995), the *sociotechnical* involves a theoretical–methodological stance as well as an analysis of complex units that requires the understanding that the technical is socially constructed and the social is technologically constructed (Thomas 2008). The networks of modern society are not divided analytically into economic, scientific, political, and technological parts. Hence, a priori distinctions about the technology and society relationship are not accepted (nor about the political, economic, social, etc.); instead they are integrated from a symmetrical perspective of the relationship between them (Latour 2008; Law 2008). The principle of symmetry used by the authors of EST follows the tradition of Bloor's strong program (1984) and Collins' relativistic program (1983), and positions itself against any social and technological determinism. The concept does not accept a priori distinctions and recognizes the ongoing co-construction between society and technology. Furthermore, it understands that the human, non-human, and non-natural cannot be regarded as incommensurable entities (Latour 2005).

The sociotechnical appears as a possibility to "describe processes of technological change and innovation through dynamic conceptualizations described in terms of relationships, processes and trajectories" (Thomas 2011) because it offers a more complex analysis in comparison with those studies focused on isolated subjects, unique artifacts, original situations, or factors of universal existence.

What are the chances that both trajectories (including selected recent contributions) achieve some sort of *extended convergence*? By extended convergence, we mean a type of convergence that, starting from an ontological perspective and trying to overcome the classical paradoxes of the social sciences, achieves at a theoretical and methodological level some kind of framework allowing, in our case, the study of such phenomena related to innovation, technological change, and structural change.

The theoretical contributions of EEC and EST, as we have seen in the introduction of this chapter, may pose some potentialities for extended convergence from the perspective of *inter-ontology crossovers* (Geels 2010; Gioia and Pitre 1990). Several authors have explored the possibility of articulation between different paradigms of social sciences (Burrell and Morgan 1994; Frost 1980; MacKenzie et al. 2007). Geels (2010), in search of theorizing multilevel approaches, proposes the use of an inter-ontology crossover approach to study sociotechnical problems. That study describes four types of meta-theoretical positions: full integration, incommensurability, eclecticism, and inter-ontology crossover. The inter-ontology crossover perspective recognizes that even if there are problems of incommensurability between theories, some strategies could be developed to make an approach via either general ontological assumptions or conceptual elements. The approach differs from those positions considering possibilities of full integration, and those who raise the incommensurability of theories. Furthermore, it departs from eclectic theoretical perspectives with ontologies that claim cases are difficult to combine.

What does it mean to build a converged framework from an inter-ontology crossover perspective? The challenge of this question lies in exploring the possibilities of *extended convergences*. The starting point is from an understanding or unraveling of the arguments that make possible the approach of both trajectories, particularly from identifying convergences based on assumptions about social questions that are not fully commensurable at the ontological level. It also proposes an analysis of the rupture of social science paradoxes, based on key ideas about the social, and its operating logic, historicity, conception of change, and the relationship and understanding of organizations and technology.

From the ontological analysis of selected approaches of EEC and contributions from EST comes an inter-ontological framework sustained by crossover convergence possibilities: *sociotechnical* and *complexity*, supported by a series of assumptions about the social reality. This is evidenced by three patterns as outlined below. The first is nonergodic dynamic relationships, of the complex systemic type, with an emphasis on meso issues and assumptions of the "seamless web" and "symmetry." These involve modes of the relationship between society and technology with different means of aggregation, structure, and logic operation that are predominantly endogenous (regarding methods to operate social relationships), especially by highlighting the importance of the construction of the interactions as well as constitutive of sociotechnical realities. The second pattern is that of fully nondeterministic temporality (on social processes and their historicity), which recognizes the effects of irreversibility but not in terms of behavior constraints, leaving space for unexpected responses, ruptures, and creative expressions. Finally, there are constitutive and constructive dynamics (order–disorder); both trajectories make an effort to conceive change under assumptions of sociotechnical complexity, actor–organization–network–technology (artifacts) (as units of analysis in the social relational) that demonstrate a constitutive level of understanding and interpretation of social relationships in terms of organizations–technologies.

The inter-ontology crossover does not postulate an ontology that sets unique limits, surpassing the particular ontologies of the contributions in each trajectory. On the contrary, it constitutes a ground that recognizes the incommensurability of approaches and operates under the principle that there are general assumptions with possibilities of extended convergence to enrich the theoretical and methodological production in social sciences, beyond its traditional paradoxes (micro–macro, structure–action, etc.) (Geels 2009).

The epistemological position that gives place to the theoretical and methodological strategy that emerged from the inter-ontology crossover approach is proposed by the good theory approach (Di Maggio 1995). The "good theory" aims at medium-range theoretical approaches that combine at least two of the following criteria for knowledge production: generality and scope, simplicity and parsimony, and accuracy and specificity (Di Maggio 1995; Geels 2007). These criteria arise from the analysis of Di Maggio (1995) who argues that social science theories can be grouped into three main types: regularity theories, critical theories, and narrative theories.

Without pretending to be a rigid analytic framework, the possibilities of inter-ontological convergence using at least two of these theory types may result in a series of movements in the good theory style. This may lead to convergence possibilities between different theoretical contributions of the framework, reassessing the strong core of relevant research.[10] By strong core we mean the theorizing and

[10] This implies that there are different levels of theory production and therefore ontological differences are recognized, differences that (from an inter-ontology crossover perspective) pretend to invigorate explanatory descriptive and interpretative capacities of the referential converging frameworks.

Table 7.2 Theorizing space proposed under "crossover inter-ontology" and the "good theory"

Theorizing space	SCOT (a)	ANT (b)	Evolutionary economic EEC (c)	*Selected study objects*
Generalized (G)	Technological cultures (a/N/struc)	Stabilizations (b/C/struc)	Macro models (*G/c/struc*)[a]	*Structural change*
Critical theory (C)	Technological frames (a/N/ctech)	Techno-economic and sociopolitical networks (*C/b/ctech*)[a]	Meso models (c/G/ctech)	*Technological change*
Narrative (N)	Social construction of artifacts (*N/a/innov*)[a]	Creating networks (b/C/innov)	Micro models (c/G/innov)	*Innovation*

Source: Di Maggio (1995), Geels (2010)
[a]Strong core research program

research modalities in each of the selected theoretical perspectives. In this way, some *research movements* can operate using the most important methodological perspective from the selected trajectories (core methodological and theorizing approach). Thus, they could lead to different movements in *converging research lines* (see Table 7.2).

This research proposal relates to the following selected Latin American development problems: innovation process, technological change, and structural change.[11] Each features provides some connection with the types of theories and theoretical contributions analyzed. It is necessary to consider that the strong core of each theory can provide generalizations, critical analyzes, and interpretations in accordance with the assumptions of theory typologies.

Thus, qualitative research can grasp the phenomena related to innovative processes using the strong core of the narrative methodologies of the SCOT approaches. Furthermore, the processes of technological change can be described and critically analyzed by means of the strong core of the forming and dismantling processes of heterogeneous networks of ANT. Finally, structural change processes can be connected with development problems in the region, and can benefit from the strong core of EEC, by its conceptual and methodological elements based on assumptions of complex macro factors (with micro and meso implications) (see Table 7.2).

The inter-ontology crossover epistemological perspective and the good theory approach make it possible to consider converging lines and research programs, assuming the postulates and overall dimensions raised at m1 and m2.

These research programs between EEC and EST can make significant contributions from a number of general theoretical convergences between general theoretical dimensions and multidimensional analytical connections. This results by forming

[11]We will refer back to these phenomena in analytical m4 and also in the meta-theoretical questions posed in m5. Both moments exceed the scope of the present article.

converging research cores about problems such as innovation processes, technological change, and structural change in Latin America.

A number of analytical dimensions matching the analytical convergence dimension founded by Bruun and Hukkinen (2003) can be derived from these research programs (constructed from the convergence opportunities given by the horizontal and vertical movements of theorizing and research). The contribution of Bruun and Hukkinen (2003) has been one of the most important attempts to link convergent contributions between EITC and EST (see Table 7.3). Based on the contributions of these authors we can set a number of *specific analytical dimensions* that connect to the converging research core proposed here, which in turn can give place to specific lines of research under our proposed converged framework. The dimensions described by Bruun and Hukkinen (2003) have been updated as part of this analytical theorizing exercise. Those authors describe a number of dimensions that can lead to convergence:

(i) An explanation of contextual stabilities and contingencies in networks;
(ii) A description of agency rooted in the social and heterogeneous networks;
(iii) Analysis of the orientations of action and divergent interpretations of convergent; and
(iv) An explanation of organizational, cognitive, and creative learning processes in instances of social interaction.

In turn, the concepts from the contributions of Latin American sociotechnical analysis have been added, as outlined above. They are included because we consider them mid-range concepts in the process of theory formation. These concepts are of central importance because they have been "tested" in the context of the sociotechnical perspective and research units of analysis regarding the economic, social, and technological dynamics of the region.

The analysis of the contingency and stabilization networks is connected to our convergent framework with the horizontal movements of generalization-type theories (1) (a/N/struc)<(b/C/struc)<(G/c/struc), which emphasize issues and phenomena related to processes of structural change (see Table 7.3, dimension d-i).

The description dimension of the agencies rooted in social and heterogeneous networks (Table 7.3, dimension d-ii) is related to the horizontal movements of critical-type theories (2) (a/N/CTech)<(C/b/CTech)>(c/G/CTech), which emphasize problems related to the processes of technological change.

Table 7.3 Converging research programs and development problems

Strong research core programs (inter-ontology crossover)	Research lines	Development problems
1. Generalized EEC (macro-economic focus)	Structural change (d-i)	*Sustainability*
2. Critical theory ANT (networks)	Technological change (d-ii)	*Social exclusion*
3. Narrative SCOT (artifacts and organizations)	Innovation processes (d-iii) (d-iv)	*Economic development*

The specific dimensions of intentional action orientations and divergent interpretations of convergent (see Table 7.3, dimension d-iii) and organizational learning, and creativity, and the cognitive dimension (see Table 7.3-dimension d-iv) are related to the horizontal movements of narrative-type theories (3) (N/a/innov) > (b/C/innov) > (g/C/innov), with cores focusing on issues related to innovation processes. Each of these specific dimensions constitutes a converging research core, and operates under the principles of the mid-range theory (Merton 1968).

The dimensions of contingency and stabilization networks may pose potential lines of research linking the study of sociotechnical groupings and technological frameworks of production and innovation systems in Latin America with the potentialities they offer to structural change processes in the region. The central question is in what way can the structural components of these systems generate feedback effects on other levels (micro and meso), regarding processes of change or resistance from their operating rules? This research core should problematize the structural components of the traditional productive specialization of Latin America, particularly the role of commodities and agribusiness. Furthermore, it may analyze the potential structural change that could occur from new sectors or productive dynamics, such as those related to technological activities (e.g., ICT, nanotechnology, and biotechnology). Change stemming from productive and innovative proposals in cultural industries, social economics, social technologies, or technologies for social inclusion may also be analyzed. Moreover, this research core should consider the role of sociotechnical alliances in production and innovation systems ("mature sectors" and stabilized areas) within the possibilities of change or resistance to sustainable development models, analyzing the lock-in effects driven by them (Windrum 1999). Here, the debate about the meaning of the transition to sustainability in the Latin American context is opened with criteria from the region rather than with parameters or future horizons adopted in developed countries. From this point of view, it becomes important to identify the role of certain free riders in the possibility of structural change towards sustainability in the region.

This dimension must be connected with principles of theorizing based on generality, being a research core focused on explaining macro-type problems regarding the behavior of production and innovation systems. Furthermore, the focus is on alternative changes to their own dynamics or in comparison with other modes of sociotechnical configurations and assemblies. This specific dimension could well constitute the core of studies on sociotechnical configurations and structural change in production and innovation systems in the countries and regions of Latin America.[12]

[12]This core implies a criticism of research that forces macro descriptions of innovation and production system functioning by means of stylizing links, cooperation modes, and institutional identity construction regarding models that are not from the continent. It also problematizes those neo development theories that interpret macro effects from an ingenuous holistic perspective (uni-dimensional and over-determined in economic terms (CEPAL); in technological terms, the technological paradigm approach; and/or in institutional terms, neo-institutional approaches), ignoring meso and macro components of complex system dynamics and their scope from the point of view of cultural (economical) policies (Dopfer and Potts 2008), as well as the diversity of socio-technical phenomena in seamless web and radical symmetry perspectives.

The descriptive dimension of agencies rooted in social and heterogeneous networks, based on theoretical principles of the critical tradition, focus on the study of phenomena related to relevant social groups in productive and innovative processes. This dimension also looks at the groups' forward and backward features in Gramscian terms (Laclau and Mouffe 1987; Rivera Ríos 2010), and on how the dynamics of heterogeneous networks of agents are multidimensional spaces—devices that either allow or block processes of technological change in a local–global relational sense. The central issue here is to describe the effects of the asymmetries, dependencies, and gaps generated by the dynamics of techno-economic and sociopolitical networks of production and innovation systems that operate at a local–global multi-scale. This is especially true for the role of scientific and technological poles (created in developed countries) and their relation to production poles in sectorial dynamics of Latin America. It is also important to focus the critical analysis on the implications of the technological change activated by these global networks, from the point of view of the problem of social inclusion. This core can address the historical dependence and lock-in effects of the sociotechnical dynamics of production and innovation systems in Latin American territories, from both techno-economic and sociopolitical perspectives. Furthermore, latent forces (updates or macro rules at a meso level) should be identified, as these contribute to the long-term historical processes that preserve backwardness traps in the region. In accordance with the critical tradition of social sciences, this core should expose how certain types of S&T policies in Latin America (under cluster-style stylized models and/or global value chains, other types of micro technology-based clusters promoting virtuous integration, or make use of opportunities for new techno-economic paradigms) act performatively in the dynamics of the region's production network to help strengthen the adverse effects of technological change, or those that do not promote endogenous change processes in the systems. We call this core a study of the sociotechnical dynamics of production and innovation systems and technological change in Latin America.

Finally, from the perspective of the interpretative tradition in social sciences, a core should be designed to address the study of intentional action and divergent interpretations converged in production and innovation systems, linking them to aspects of organizational learning and creativity (the cognitive dimension). This research program could focus on analyzing the sociotechnical trajectories of agent–artifact spaces in production and innovation systems, their specificities, and particular cognitive phenomena (imitation, transfer, reverse engineering, translation, transduction, problem-solution relations, performance). Furthermore, the program would focus on the generation of organizational abilities, to provide creative and adaptive responses in the context of heterogeneous networks and sociotechnical configurations with typical regional features. The core would be called a sociotechnical study of history and innovation processes, aimed at solving the region's economic development problems.

7.3 Multidimensionality of Development, S&T Policies, and Higher Education

How can these core research programs be connected to the region's S&T policies? And what role in particular might the actions of higher education institutions resignify in their academic research policies?

In Latin America, the first science and technology policies stemming from the big ideas of science were implemented in the 1960s. However, it was only in the 1990s that a process of *forced internationalization* (Arellano et al. 2012) was implemented, providing the opportunity to enter into large networks and mega science projects. Meanwhile, the integration of Latin American researchers within international research does not imply opportunities to negotiate because scientific elites are typically concentrated in developed countries.

From the perspective of the dependency theory, the founders of Latin American thought in science, technology, and society first approached this issue in the 1950s. They proposed, in terms of political praxis, an autonomous regional development program for science and technology in Latin America. Such is the case of Sabato's framework, widely known in the field of science and technology policies in Latin America, based on Sabato's triangle: Government–Businesses–S&T Institutions (Sabato and Botana 1968). This concept claimed that Latin American development should integrate science and technology with industry and higher education policies. In the 1980s, the prevailing ideas focused on technological learning, and there appeared early studies on science and technology policies about the industry–university phenomenon. Studies also emerged on the regulatory nature of technological cooperation approaches, converging, in an uncritical extent, with approaches such as Triple Helix, techno-scientific networks, and more recently on the incorporation of the concept of innovation systems (national, regional, local, and sectorial). Innovation systems have become, in some Latin American countries, regulatory frameworks operating from science and technology policies, with the central locus for innovation being at the enterprise level (Dutrénit et al. 2010; Villavicencio 2000). However, this situation does not recognize the specificities of the local and national dynamics of the relationship between government, enterprises, and institutions of science and technology (Sutz 2000). Thus, it is combined with science and technology policies that are delineated from international organizations and approaches designed for developed countries.

In this context, higher education institutions play a significant role in promoting science and technology activities in Latin America, because the majority of funding for such activities comes from the government's science and technology sector, and universities, while business enterprises (public and private) only contribute 30 % (RICyT 2013). In developed countries, the reverse is largely true.

However, some universities in Latin America, and in particular regarding activities that promote science and technology, operate in "glocanal" context (global, national, and local) (Marginson 2004). This issue represents the permanent tension between global sociotechnical configurations and articulations and local sociotechnical trajectories in our analysis of the convergence of sociotechnical production and innovation systems (Lepratte et al. 2015; Lepratte 2014).

Despite the protagonist role of scientific and technological activities, and some efforts in recent years to encourage investment in R&D by some countries in the region, the contributions of EST and EICT in Latin America demonstrate the difficulties in political science, technology, and innovation in the region with respect to connecting and making progress to solve development problems.

Development problems refer to three interrelated phenomena: economic development, social exclusion, and sustainability problems in the region. Here, the concept of economic development is used in the Schumpeterian sense of growth and transformation of an economy, through a process of creative destruction and the disruption of sociotechnical trajectories, highlighting also the contributions of EEC regarding the creative character and hence the cognitive development of these processes. Furthermore, economic development represents the ability to imagine intentional "possible worlds" (Foster and Metcalfe 2009) that are created and carried by organizations and artifacts (Dopfer 2011) and can achieve recognition of economic creativity (Metcalfe 2010). In our case, the possibility of creative responses to problems arising from the specialization profiles of the region (marked by the continued dependence on primary production and natural resources) is connected with the need to illuminate such development problems from a narrative position, via studies on innovation processes (see Table 7.3).

The other typical development problem in Latin America corresponds to the processes of social exclusion and poverty affecting large populations (González and Martner 2010; ECLAC 2013). In our view, this problem should be understood in the context of the processes generated from technological change in the dynamics of techno-economic and sociopolitical heterogeneous networks in the region. These can enable or hinder the integration of certain collective actors into the sociotechnical dynamics of convergence and irreversibility, in a permanent state of tension between global–national–local. Hence, certain actor–network relations can impose their interpretations as robust and durable normalizations with significant micro-political and semiotic effects in terms of the transfer of power. The problem of social exclusion relates to the study of technological change from a critical position (see Table 7.3).

Sustainability issues are understood here from a double perspective. As an intrinsic feature of the recurring problem of the impossibility of economic and political structural changes in Latin America, the semiotic weight of certain stabilized networks in the region, and therefore technological cultures that have been established as macrostructures over time, cause path dependence effects that are heavily deterministic. There is also the difficulty of having a "breeding ground" for new ideas that could subvert rules, in line with new global challenges and debates about the future direction of the economization processes of society. Thus, this issue is connected with the study cores on structural change and generalizations from EEC models (see Table 7.3).

Under a multidimensional conception, convergent research programs propose to provide empirical evidence, reflections, and discussions on development understood as a complex process of breaking trends toward techno-economic concentration, sociopolitical stabilization, and sociotechnical production and innovation systems. These are supported by lock-in effects and regressive articulations that do not allow transitions[13] to economic development, social inclusion, and structural change processes aimed at sustainability.

The multidimensionality of the development problems described above, identified with a number of problems that must be taken into account when considering science and technology policies in articulatory terms, is not intended to operate as an a priori category, or as a benchmark of the behavior of sociotechnical production and innovation systems. Rather, it constitutes an experimental space, where convergent research programs are challenged by the problems of development, and the need to translate their results and ongoing processes of theorizing, research, and discussion from the viewpoint of a sociopolitical praxis.

This problematized issue from the convergent framework should be approached from a multidimensionality perspective: narrative, critical, and explanatory, as stated above. It also must examine the impulse of political praxis implicit in the tradition of science, technology, and society of Latin American thought, recognizing that the institutions of science and technology, and especially university institutions, are invigorating examples of "glocanal" relations. In this sense, their research centers will have to propose articulatory policies, actions under a heterogeneous identity.[14]

Converging research efforts should occur within spaces of emerging articulatory political praxis (Laclau and Mouffe 1987; Mouffe 1999). These emerging spaces recognize the plurality of theoretical and political positions (identities), based on the possibility of integration (uncertainty, tension, and transitory) between preliminary approaches with potentialities in ontological, epistemological, and theoretical convergences. In turn, these efforts should promote opportunities for sociotechnical configurations capable of being development-oriented, as explained earlier.

In the narrative dimension, from the perspective of political praxis, it is possible to analyze possibilities to promote economic development by means of certain enterprises, sectors, or regions, either mature or new. Furthermore, the discussion about the nature of capitalist economic growth and alternatives (e.g., social economy) must be considered. The narrative dimension should indicate to policymakers, with higher degrees of specificity, the particular adaptations of sociotechnical space agents and sociotechnical trajectories, with artifacts that have become or could enable the emergence of new sociotechnical articulations. The identification of players acting against the rules (or potential ones) is crucial here to break the ongoing cycle of the typical adaptive responses of Latin American countries.

[13] In recent years, studies on transition have gained significance in developed countries (Geels 2002, 2005, Geels et al. 2011). Their approaches are based on inter-ontology crossover criteria, an interesting perspective to be resignified in Latin America. Our own perspective goes in that direction when analyzing structural change processes with their limitations and sustainability problems.

[14] The concept of articulation is taken from Laclau and Mouffe (1987); it gathers contributions from Thomas' sociotechnical alliances, Gramsci's historical bloc, and the technological framework of SCOT. Ultimately, it can be defined as a concept close to Haraway's articulation (1999).

Doing so will enable significant change processes in productive and innovative sociotechnical configurations in Latin America.

In relation to the previous dimension (the critical perspective) from a political point of view, it has to produce new ways of forming techno-economic and sociopolitical networks. This perspective is also concerned with the ways they can break lock-in effects, reinforce convergence, and ensure pockets of stabilization, which pose human–artifacts relations oriented by ideas and rules generated in global networks that promote lifelong dependency and social exclusion. The challenge here is to explicitly promote, via political science, technology, and innovation, a plurality of spaces of the emergence for new sociotechnical progressive articulations, democratizing their formulation, implementation, and monitoring processes. The effect of technological change is also discussed under assumptions that problematize the current rules and institutionalized articulations in science and technology, as well as the standardized cluster-style analysis and value chains that operate performatively over decision making in production policies. Such an analysis ignores situations of inequality, exclusion, and concentration that occur in productive and innovative networks at a global scale.

Finally, at a level that searches generalizable and explanatory elements, it is necessary to establish models of standard analysis of sociotechnical configurations for our production and innovation systems and the orientations in favor (or not) of structural change. In this respect, the problems of defining sustainable policies in scientific, technological, and industrial fields gain absolute relevance for Latin America, in terms of ensuring transitions to development in a multidimensional sense.

Transition Studies have become important in developed countries (Van de Poel 2003; Perez 2002; Geels 2002, 2005, 2011; Rotmans et al. 2001; Elzen and Wieczorek 2005) and should be reinterpreted in the context of the region, in light of particular and problematized matters of sustainability. In developed countries, transition studies focus on issues related to environmental sustainability, while in Latin America such studies should focus on the possibilities of sustainable structural changes operating at the level of sociotechnical configurations, introducing discussions about the complexity of sociotechnical horizons. These horizons will enable policymakers to map the emergence of possible spaces for technological innovation, and provide a heuristic ability to analyze their implications from the point of view of economic development, social inclusion, and sustainability of change in a structural sense. They would also provide a stable institutional framework that enables the transition process to be sustained and monitored, under metaphors of the possibility to build new heterogeneous networks, and a renewed narrative of necessary funds and resources (Smith and Raven 2012).

Thus, a converging framework between EST and EEC enables the development of a research program converged with articulatory spaces with emerging political ideas and actions directed to discuss and propose the transition to development. It is concerned with certain questions: How can a transition to sociotechnical trajectories be achieved, to trajectories that generate agent–artifact spaces promoting creative actions? In what way can the transition from the productive and innovative networks of the region to sociotechnical dynamics be achieved progressively, ensuring social inclusion? Finally, how can new identities in the science, technology, and innovation role be established in Latin America to operate as new horizons for sociotechnical configurations that contribute to sustainable structural change?

References

Albuquerque, E. M. (2007). Inadequacy of technology and innovation systems at the periphery. *Cambridge Journal of Economics, 31*, 669–690.

Amsden, A. (2004). Import substitution in high-tech industries: Prebisch lives in Asia! *CEPAL Review, 82*.

Antonelli, C. (2007). The system dynamics of collective knowledge: From gradualism and saltationism to punctuated change. *Journal of Economic Behavior & Organization, 62*, 215–236.

Antonelli, C. (2011). *Handbook on the economic complexity of technological change*. Cheltenham/Northampton: Edward Elgar.

Arellano, A., & Kreimer, P. (2011). *Estudio social de la ciencia y la tecnología desde América Latina*. [Social studies of science and technology in Latin America]. Bogotá: Siglo del Hombre.

Arellano, A., Arvanitis, R., & Vinck, D. (2012). Global connectivity and circulation of knowledge. Aspects of Anthropology of knowledge in Latin America. *Revue d'Anthropologie des Connaissances, 6*(2), 1–28. doi:10.3917/rac.016.i.

Arocena, R., & Sutz, J. (2003). Knowledge, innovation and learning: Systems and policies in the north and in the south. In J. Cassiolato, H. Lastres, & M. Maciel (Eds.), *Systems of innovation and development* (pp. 291–310). Cheltenham: Edward Elgar Publishing.

Arthur, B. (1990). Positive feedbacks in the economy. *Scientific American, 262*, 92–99.

Bijker, W. E. (1993). Do not despair: There is life after constructivism. *Science, Technology and Human Values, 18*(1), 113–138.

Bijker, W. E. (1995). *Of bicycles, bakelites, and bulbs: Toward a theory of sociotechnical change*. Cambridge: Londres, MIT Press.

Bijker, W. E. (2010). How is technology made?—That is the question! *Cambridge Journal of Economics, 34*(1), 63–76.

Bijker, W. E., Hughes, T. P., & Pinch, T. J. (1987). *The social construction of technological systems: New directions in the sociology and history of technology*. Cambridge: The MIT Press.

Bloch, H., & Metcalfe, J. (2011). Complexity in the theory of the developing firm. In C. Antonelli (Ed.), *Handbook on the economic complexity of technological change* (Chap. 3). Cheltenham/Northampton: Edward Elgar.

Bloor, D. (1984). The strengths of the strong programme. In J. R. D. Brown (Ed.), *Scientific rationality: The sociological turn*. Dordrecht: Riedel Publishing Company.

Boutang, Y., et al. (2004). *Capitalismo Cognitivo, Propiedad Intelectual y Creación Colectiva [Cognitive capitalism, intellectual property, collective creation]*. Madrid: Editorial Traficantes de sueños.

Bruun, H., & Hukkinen, J. (2003). Crossing boundaries: An integrative framework for studying technological change. *Social Studies of Science, 33*(1), 95–116.

Burrell, G., & Morgan, G. (1994). *Sociological paradigms and organizational analysis*. London: Heinemann.

Callon, M. (1987). Society in the making: The study of technology as a tool for sociological analysis. In W. Bijker et al. (Eds.), *Social construction of technological systems*. Cambridge: Cambridge University Press.

Callon, M. (1989). *La Science et ses réseaux: gènese et circulation des faits scientifiques*. Paris: Découverte.

Callon, M. (1992). The dynamics of techno-economic networks. In R. Coombs, P. Saviotti, & V. Walsh (Eds.), *Technological changes and company strategies: Economical and sociological perspectives*. London: Harcourt Brace Jovanovich Publishers.

Callon, M. (1995). Algunos elementos para una sociología de la traducción: la domesticación de las vieiras y los pescadores de la bahía de St. Brieuc. [Some elements of a sociology of translation: domestication of the scallops and the fishermen of St Brieuc Bay]. En: J. Manuel Iranzo, J. Rubén Blanco, T. González de la Fe, C. Torres y Alberto Cotillo, Comps, *Sociología de la ciencia y la tecnología* [Sociology of science and technology]. Madrid: CSIC.

Callon, M. (1998). El proceso de construcción de la sociedad. El estudio de la tecnología como herramienta para el análisis sociológico [Society in the making. The study of technology as a tool for the sociological analysis]. In M. Doménech & F. Tirado (Eds.), *Sociología simétrica [Symmetric Sociology]* (pp. 143–170). Barcelona: Gedisa.

Callon, M. (2001). Redes tecnoeconómicas e irreversibilidad [Techno economic networks and irreversibility], Redes – Revista de estudios sobre ciencia y tecnología. *Junio, 8*(17), 85–126.

Callon, M. (2006). Luchas y negociaciones para definir qué es y que no es problemático. La socio-lógica de la traducción [Struggles and negotiations to define what is problematic and what is not]. *REDES – Revista de estudios sociales de la ciencia y tecnología [Journal of Social Studies of Science]*, 12(23).

Callon, M. (2008). La dinámica de las redes tecno-económicas [Dynamics of techno-economic networks]. In H. Thomas & A. Buch (Eds.), *Actos, actores y artefactos: Sociología de la tecnología [Acts, actors and artifacts: sociology of technology]*. Bernal: Universidad Nacional de Quilmes.

Casas, R., Corona, J., & Rivera, R. (2014). Políticas de ciencia, tecnología e innovación en América Latina: entre la competitividad y la inclusión social [S&T policies in Latin America: between the competitiveness and social inclusion]. In: P. Kreimer, A. Arellano, H. Vessuri, L. Velho, (Eds.), *Perspectivas Latinoamericanas en el Estudio Social de la Ciencia, la Tecnología y el Conocimiento [Latin American perspectives in social studies of science, technology and knowledge]*. Red Cyted, FCCyT, Siglo XXI.

Casas, R., & Luna, M. (1997). *Gobierno, Academia y empresas en México: Hacia una nueva configuración de relaciones [Government, academy and industries in Mexico: Towards a configuration of relationships]*. Cuauhtémoc: Plaza y Valdés Editores/UNAM.

Casas, R., Corona, J., & Rivera, R. (2013). *Políticas de ciencia, tecnología e innovación en América Latina: entre la competitividad y la inclusión social [S&T policies in Latin America: between competitiveness and social inclusion]*. Brazil: LALICS.

Castell, M. (1999). *La era de la información: Economía, sociedad y cultura [The information age: economy, culture and society]*. México: Siglo Veintiuno.

Cimoli, M., Porcile, G., & Rovira, S. (2010). Structural change and the bop constraint: Why did Latin America fail to converge? *Cambridge Journal of Economics, 34*(2), 389–411.

Collins, H. (1983). An empirical relativist programme in the sociology of scientific knowledge. In K. D. Knorr-Cetina & M. Multar (Eds.), *Science observed: Perspectives on the social study of science* (pp. 83–113). London: Sage.

Consoli, D., & Patrucco, P. (2011). Complexity and the coordination of technological knowledge: The case of innovation platforms. In: C. Antonelli, *Handbook on the economic complexity of technological change*. Cheltenham/Northampton: Edward Elgar.

Dabat, A., & Rivera Ríos, M. A. (2004). Nuevo ciclo industrial mundial e inserción internacional de países en desarrollo [New world industrial cycle and international insertion of developing countries]. In A. Dabat et al. (Eds.), *Globalización y cambio tecnológico [Globalization and technological change]* (p. 75). México: Juan Pablos.

Dabat, A., & Rodriguez Vargas, J. (2009). *Globalización, conocimiento y desarrollo. La nueva economía global del conocimiento. Estructura y problemas [Globalization, knowledge and development. The new global economy of knowledge]*. Mexico: UNAM. Tomo 1.

Dagnino, R. (2010). *Tecnología social. Ferramenta para construir outra sociedade [Social technology. Tools to construct other societies]*. Campinas: SP, Komedi.

Dagnino, R., & Thomas, H. (2000). Elementos para una renovación explicativa-normativa de las políticas de innovación latinoamericanas [Elements for a renewal explanatory-normative for innovation policies in Latin America]. *Revista Espacios, 21*(2).

Dagnino, R., Thomas H., & Davyt, A. (1996). El pensamiento latinoamericano en ciencia, tecnología, y sociedad en Latinoamérica: Una interpretación política de su trayectoria [Latin American thought on science, technology and society: A political interpretation of this trajectory] *REDES, 3*(7), 13–51.

Di Maggio, P. J. (1995). Comments on "What theory is not". *Administrative Science Quarterly, 40*(3), 391–397.

Dopfer, K. (1986). The histonomic approach to economics: Beyond pure theory and pure experience. *Journal of Economic Issues, 20*(4), 989–1010.

Dopfer, K. (2001). History-friendly theories in economics: Reconciling universality and context in evolutionary analysis. In J. Foster & S. Metcalfe (Eds.), *Frontiers and evolutionary economics: Competition, self-organization and innovation policy* (p. 160). Cheltenham: Edward Elgar.

Dopfer, K. (2011). Mesoeconomics: A unified approach to systems complexity and evolution. In C. Antonelli (Ed.), *Handbook on the economic complexity of technological change (Chap. 13)*. Cheltenham/Northampton: Edward Elgar.

Dopfer, K., & Potts, J. (2008). *A cultural science (Kulturewissenschaft) manifesto*. Short paper prepared for FEAST, QUT.

Dutrenit, G., & Katz, J. (2005). Innovation, growth and development in Latin-America: Stylized facts and a policy agenda. *Innovation: Management, Policy & Practice, 7*, 105–130.

Dutrénit, G., & Teubal, M. (2011). Coevolution, emergence and economic development: Some lessons from the Israeli and Mexican experience. In C. Antonelly (Ed.), *Handbook on the economic complexity of technological change* (pp. 451–491). Cheltenham: Edward Elgar.

Dutrénit, G., Arvanitis, R., Capdevielle, M., Cruz, M. A., Del Valle, S., Rios, L., et al. (1996). La vinculación universidad-empresa en un macroproyecto de polímeros [University–industry relationships in a macro-project of polymers]. *Comercio Exterior, 46*(10), 808–816. http://ladb.unm.edu/econ/content/comext/1996/october/vinculacion.html

Dutrénit, G., Capdevielle, M., Corona Alcantar, J. M., Puchet Anyul, M., Santiago, F., & Vera-Cruz, A. O. (2010). *El sistema nacional de innovación mexicano: estructuras, políticas, desempeño y desafíos [The national system of innovation in Mexico: structures, performance and challenges]*. México: UAM/Textual.

ECLAC. (2013). *Panorama Social de América Latina 2012 [Social report of Latin America 2012]*. Naciones Unidas Edición.

Elzen, B., & Wieczorek, A. (2005). Transitions towards sustainability through system innovation. *Technological Forecasting and Social Change (special issue), 72*(6).

Fagerberg, J., & Verspagen, B. (2009). Innovation studies—The emerging structure of a new scientific field. *Research Policy, 38*(2), 218–233.

Fagerberg, J., Fosaas, M., & Sapprasert, K. (2012a). Innovation: Exploring the knowledge base. *Research Policy, 41*(7), 1132–1153.

Fagerberg, J., Landström, H., & Martin, B. R. (2012b). Exploring the emerging knowledge base of 'the knowledge society'. *Research Policy, 41*(7), 1121–1131.

Figueiredo, P. (2004). Aprendizagem Tecnológica e Inovação Industrial em Economias Emergentes: uma Breve Contribuição para o Desenho e Implementação de Estudos Empíricos e Estratégias no Brasil [Technological learning and industry innovation in emergent economies: A brief contribution to the design and implementation of empirical studies and strategies in Brazil]. *Revista Brasileira de Innovación [Brazilian Journal of Innovation], 3*(2), 323–361.

Foster, J. (2005). From simplistic to complex systems in economics. *Cambridge Journal of Economics, 29*, 873–892.

Foster, J., & Metcalfe, S. (2001). *Frontiers and evolutionary economics: Competition, self-organization and innovation policy*. Cheltenham: Edward Elgar.

Foster, J., & Metcalfe, S. (2009). Evolution and economic complexity: An overview. *Economics of Innovation and New Technology, 18*(7), 607–610.

Freeman, C. (1991). Network of innovators: A synthesis of research issues. *Research Policy, 20*(5), 499–514.

Frost, P. (1980). Toward a radical framework for practicing organization science. *Academy of Management Review, 5*(4), 501–507.

Geels, F. W. (2002). Technological transitions as evolutionary reconfiguration processes: A multi-level perspective and a case-study. *Research Policy, 31*, 1257–1274.
Geels, F. W. (2005). Processes and patterns in transitions and system innovations: Refining the co-evolutionary multi-level perspective. *Technological Forecasting and Social Change, 72*, 681–696.
Geels, F. W. (2007). Feelings of discontent and the promise of middle range theory for STS: Examples from technology dynamics. *Science, Technology & Human Values, 32*(6), 627–651.
Geels, F. W. (2009). Foundational ontologies and multi-paradigm analysis, applied to the socio-technical transition from mixed farming to intensive pig husbandry (1930–1980). *Technology Analysis & Strategic Management, 21*(7), 805–832.
Geels, F. W. (2010). Ontologies, socio-technical transitions (to sustainability), and the multi-level perspective. *Research Policy, 39*(4), 495–510.
Geels, F. W., Kemp, R., Dudley, G., & Lyons, G. (Eds.). (2011). *Automobility in transition? A socio-technical analysis of sustainable transport*. New York: Routledge.
Gioia, D. A., & Pitre, E. (1990). Multiparadigm perspectives on theory building. *Academy of Management Review, 15*, 584–602.
González, I., & Martner, R. (2010). Del síndrome del casillero vacío al desarrollo inclusivo: buscando los determinantes de la distribución del ingreso en América Latina. Documento presentado en el vigesimosegundo Seminario regional de política fiscal [The empty box syndrome inclusive development: Seeking the determinants of income distribution in Latin America. Paper presented at the twenty-regional seminar on fiscal policy]. Santiago: Comisión Económica para América Latina y el Caribe (CEPAL).
Hughes, T. P. (1983). *Networks of power, electrification in western society, 1880–1930*. Baltimore: Johns Hopkins University Press.
Hughes, T. P. (1986). The seamless web: Technology, science, etcetera, etcetera. *Social Studies of Science, 16*, 192–281.
Iizuka, M., & Katz, J. M. (2010). Natural resource industries, 'tragedy of the commons' and the case of Chilean salmon farming. UNU-MERIT, Maastricht Economic and Social Research and Training Centre on Innovation and Technology.
Kreimer, P. (2007). Estudios sociales de la ciencia y la tecnología, ¿para qué? y ¿quién?. *Revista REDES, 13*(26), 54–64.
Kreimer, P., & Thomas, H. (2004). Un poco de reflexividad o ¿de dónde venimos? Estudios sociales de le ciencia y la tecnología en América Latina [A bit of reflexivity or where do we come from? Social Studies of Science and Technology will in Latin America]. In P. Kreimer et al. (Eds.), *Producción y uso social de conocimientos, estudios de sociología de la ciencia en América Latina [Production and social use of knowledge, research on the Latin American sociology of science]* (pp. 11–90). Bernal: Universidad Nacional de Quilmes.
Laclau, E., & Mouffe, C. (1987). *Hegemonía y estrategia socialista [Hegemony and socialist strategy]*. Madrid: Editorial Siglo XXI.
Lane, D. (2011). Complexity and innovation dynamics. In C. Antonelli (Ed.), *Handbook on the economic complexity of technological change*. Cheltenham/Northampton: Edward Elgar.
Lane, D., & Maxfield, R. (1997). Foresight complexity and strategy. In B. Arthur, S. Durlauf, & D. Lane (Eds.), *The economy as an evolving complex system II*. Redwood City: Addison Wesley.
Lastres, H., Cassiolato, J. E., & Maciel, M. L. (2003). Systems of innovation for 21 development in the knowledge era: An introduction. In J. E. Cassiolato, H. M. Lastres, & M. L. Maciel (Eds.), *Systems of innovation and development – Evidence from Brazil*. Cheltenham/Northampton: Edward Elgar.
Lastres, M., Cassiolato, J., & Arroio, A. (2005). *Conhecimento, sistemas de inovação e desenvolvimento/Organização [Knowledge, innovation systems and development]*. Rio de Janeiro: Editora UFRJ; Contraponto.
Latour, B. (1999). Give me a laboratory and I will raise the world. In M. Biagioli (Ed.), *The science studies reader*. New York: Routledge.

Latour, B. (2007). *Nunca fuimos modernos: ensayo de antropología simétrica [We have never been modern: essays in symmetric anthropology]*. Buenos Aires: Siglo XXI.
Latour, B. (2008). *Reensamblar lo social: una introducción a la teoría del actor red [Reassembling the social. An introduction to actor – network theory]*. Buenos Aires: Manantial.
Law, J. (1987). Technology and heterogeneous engineering: The case of Portuguese expansion. In W. T. Bijker, T. Hughes, & T. Pinch (Eds.), *The social construction of technical systems: New directions in the sociology and history of technology* (pp. 111–134). Cambridge: MIT Press.
Law, J. (2008). Actor network theory and material semiotics. In B. S. Turner (Ed.), *The new Blackwell companion to social theory*. Oxford: Wiley-Blackwell.
Lepratte, L. (2014). Complexity, transition and development. *Journal of Technology Management & Innovation, 9*(4), 85–97.
Lepratte, L., Blanc, R., Pietroboni, R., & Hegglin, D. (2015). Socio-technical system innovation. An analysis of the dynamics of the poultry meat production sector in Argentina. *Ibero American Journal of Science, Technology and Society, 28*(10), 57–82.
Lundvall, B. (1988). Innovation as an interactive process: From user-producer interaction to the national system of innovation. In G. Dosi et al. (Eds.), *Technical change and economic theory* (pp. 349–369). London: Frances Pinter.
Lundvall, B.-Å., & Johnson, B. (1994). The learning economy. *Journal of Industry Studies, 1*(2), 23–42.
MacKenzie, D. (1992). Economic and sociological explanation of technical change. In R. Coombs, P. Saviotti, & V. Walsh (Eds.), *Technical change and company strategies: Economic and sociological perspectives* (pp. 25–48). London: Academic.
MacKenzie, D., Muniesa, F., & Siu, L. (Eds.). (2007). *Do economists make markets? On the performativity of economics*. Princeton: Princeton University Press.
Marginson, S. (2004). Competition and markets in higher education: A ‘glonacal’ analysis. *Policy Futures in Education, 2*(2), 175–244.
Martin, B. R. (2012). The evolution of science policy and innovation studies. *Research Policy, 41*(7), 1219–1239.
Merton, R. K. (1968). *Social theory and social structure* (3rd ed.). Glencoe: Free Press. Previous editions in 1949 and 1957.
Metcalfe, S. (2010). Dancing in the dark, la disputa por el concepto de competencia. *Desarrollo Económico, Revista de Ciencias Sociales, 50*(197), 59–79.
Moncayo, E. (2004). El debate sobre la convergencia económica internacional a interregional: enfoques teóricos y evidencia empírica [The debate about international and interregional economic convergence: theoretical approaches and empirical evidence] [versión electrónica]. *Eure, 30*(90), 7–26.
Mouffe, C. (1999). *El retorno de lo político*. Barcelona: Paidos.
Nelson, R. R., & Winter, S. (1982). *An evolutionary theory of economic change*. Cambridge: The Belknap Press.
Ocampo, J. A. (2005). RaÃºl Prebisch and the development agenda at the dawn of the twenty-first century. *CEPAL Review, 75*, 23–37.
OECD. (1996). *The knowledge-based economy*. Paris: OECD.
Perez, C. (2002). *Technological revolutions and financial capital: The dynamics of bubbles and golden ages*. London: Elgar.
Pérez, C. (2010). Technological dynamism and social inclusion in Latin America: A resource-based production development strategy. *CEPAL Review, 100*, 123–145.
Pinch, T. (1996). The social construction of technology: A review. In R. Fox (Ed.), *Technological change: Methods and themes in the history of technology* (pp. 17–36). Amsterdam: Harwood.
Pinch, T. (2008). Technology and institutions: Living in a material world. *Theory and Society, 37*, 461–483.
Pinch, T. (2010). On making infrastructure visible: Putting the non-humans to rights. *Cambridge Journal of Economics, 34*(1), 77–89.

Pinch, T., & Bijker, W. (1984). The social construction of facts and artifacts: Or how the sociology of science and the sociology of technology might benefit each other. *Social Studies of Science, 14*, 399–441.

Pinch, T. J., & Bijker, W. (1987). The social construction of facts and artifacts: Or how the sociology of science and the sociology of technology might benefit each other. In W. E. Bijker, T. P. Hughes, & T. J. Pinch (Eds.), *The social construction of technological systems: New directions in the sociology and history of technology* (pp. 17–50). Cambridge: The MIT Press.

RICYT. (2013). *El Estado de la Ciencia. Principales Indicadores de Ciencia y Tecnología Iberoamericanos/Interamericanos [The state of science. Main indicators of science and technology in Ibero-America]*. Buenos Aires: REDES Center.

Rip, A. (1995). Introduction of new technology: Making use of recent insights from sociology and economics of technology. *Technology Analysis & Strategic Management, 7*(4), 417–432.

Rivera Ríos, M. (2010), *Desarrollo económico y cambio institucional. Una aproximación al estudio del atraso económico y el desarrollo tardío desde la perspectiva sistémica [Economic development and institutional change. An approach to the study of economic backwardness and late development from the systemic perspective]*. México: UNAM.

Rivera Ríos, M., Robert, V., & Yoguel, G. (2009). Cambio tecnológico, complejidad e instituciones: Una aproximación desde la estructura industrial e institucional de Argentina y México [Technological change, complexity and institutions: An approach from the industrial structure and institutions in Argentina and Mexico]. *Revista Problemas del Desarrollo, 40*(157), 75–109.

Robert, V., & Yoguel, G. (2010). La dinámica compleja de la innovación y el desarrollo económico [The complex dynamics of innovation and economic development]. *Desarrollo Economico [Economic Development], 50*(1), 423–453.

Robert, V., & Yoguel, G. (2011). The complex dynamics of economic development. In C. Antonelli (Ed.), *Handbook on the economic complexity of technological change (Chap. 17)*. Cheltenham/Northampton: Edward Elgar.

Rodríguez Vargas, J. (2009). El nuevo capitalismo en la literatura económica y el debate actual [The new capitalism in economic literature and the contemporary debate]. In A. Dabat, & J. Rodriguez Vargas (Eds.), *Globalización, conocimiento y desarrollo. La nueva economía global del conocimiento. Estructura y problemas*. UNAM, México. Tomo 1.

Rotmans, J., Kemp, R., & van Asselt, M. B. A. (2001). More evolution than revolution: Transition management in public policy. *Foresight, 3*(1), 15–31.

Sábato, J., & Botana, N. (1968). La Ciencia y la Tecnología en el desarrollo futuro de América Latina [Science and technology in the future development of Latin America]. *Revista de Integración*, 3, noviembre. Buenos Aires.

Saviotti, P. (2011). Knowledge, complexity and networks. In C. Antonelli (Ed.), *Handbook on the economic complexity of technological change (Chap. 6)*. Cheltenham/Northampton: Edward Elgar.

Saviotti, P., & Pyka, A. (2008). Micro and macro dynamics: Industry life cycles, inter-sector coordination and aggregate growth. *Journal of Evolutionary Economics, 18*(2), 167–182.

Schot, J., & Rip, A. (1997). The past and future of constructive technology assessment. *Technological Forecasting and Social Change, 54*, 251–268.

Schumpeter, J. (1947). The creative response in economic history. *Journal of Economic History, 7*(2), 149–159.

Silva, S., & Teixeira, A. (2009). On the divergence of evolutionary research paths in the past 50 years: A comprehensive bibliometric account. *Journal of Evolutionary Economic, 19*, 605–642.

Smith, A., & Raven, R. (2012). What is protective space? Reconsidering niches in transitions to sustainability. *Research Policy, 41*(6), 1025–1036.

Sutz, J. (2000). The university–industry–government relations in Latin America. *Research Policy, 29*, 279–290.

Swedberg, R. (2014). *Theorizing in social science: The context of discovery*. Stanford: Stanford University Press.

Thomas, H. (2008). Estructuras cerradas vs. Procesos dinámicos: trayectorias y estilos de innovación y cambio tecnológico [Closed structures and dynamics processes: trajectories and innovation and technological change styles]. In H. Thomas & A. Buch (Eds.), *Actos, actores y artefactos. Sociología de la tecnología [Acts, actors and artifacts: sociology of technology]* (Chap. 6). Bernal: Universidad Nacional de Quilmes.

Thomas, H. (2010). Los estudios sociales de la tecnología en América Latina [The social studies of technology in Latin America], Íconos. *Revista de Ciencias Sociales [Social Sciences Journal], 37*, 35–53.

Thomas, H. (2011). Tecnologías para la inclusión social en América Latina: de las tecnologías apropiadas a los sistemas tecnológicos sociales. Problemas conceptuales y soluciones estratégicas [Technologies for social inclusion in Latin America: from the appropriate technologies to the sociotechnical systems]. In H. Thomas, G. Santos, & M. Fressoli (Eds.), *Tecnología, desarrollo y democracia. Nueve estudios sobre dinámicas socio-técnicas de exclusión/inclusión social [Technology, development and democracy. Nine studies on the social inclusion/exclusion of sociotechnical dynamics]*. (Chap. 1). Buenos Aries: MINCyT.

Thomas, H., Fressoli, M., & Lalouf, A. (2008). Estudios sociales de la tecnología: ¿hay vida después del constructivismo? [Social studies of technology: is there life after the constructivism?]. *REDES, 14*(27), 59–76.

Vaccarezza, L. (2004). El campo CTS en América Latina y el uso social de su producción. *Revista Iberoamericana Ciencia, Tecnología y Sociedad, 1*(2), 211–218.

Van de Belt, H., & Rip, A. (1987). The Nelson–Winter–Dosi model and the synthetic dye chemistry. In W. E. Bijker, T. P. Hughes, & T. Pinch (Eds.), *The social construction of technological systems: New directions in the sociology and history of technology* (pp. 159–190). Cambridge: MIT Press.

Van de Poel, I. R. (2003). The transformation of technological regimes. *Research Policy, 32*(1), 49–68.

Vessuri, H. (2007). *O inventamos, o erramos: La ciencia como idea-fuerza en América Latina [Invent or fail: science as a task force in Latin America]*. Bernal: Universidad Nacional de Quilmes Editorial.

Villavicencio, D. (2000). La innovación en la empresa como espacio de análisis sociológico [Innovation in the firm as a field of sociological analysis]. *Sociología del Trabajo, 40*, 59–78.

Villavicencio, D. (2010). Cambios en la concepción de las políticas de CT+I [Changes in S&T policies ideas]. *Ide@s CONCYTEG, 5*(56).

Von Hippel, E. (1988). *The sources of innovation*. New York: Oxford University Press.

Windrum, P. (1999). *Unlocking a lock-in: Towards a model of technological succession*. Research Memoranda 010. Maastricht: MERIT, Maastricht Economic Research Institute on Innovation and Technology.

Yoguel, G., Barletta, F., & Pereira, M. (2013). De Schumpeter a los postschumpeterianos: viejas y nuevas dimensiones analíticas [From Schumpeter to the Neo-Schumpeterians: old and new analytical dimensions]. *Problemas del Desarrollo [Problem of development], 44*(174), 35–59.

Chapter 8
Research Assessment of Public Research Centers in Mexico: An Instrument of Science Policy

Salvador Estrada and Joao Aguirre

8.1 Introduction and Objectives

Public research centers (PRCs) are typically considered an instrument of public policies on science, technology and innovation (ST&I). These organizations have been valued as a means to create science and technology (S&T) capabilities, as well as a mechanism to ignite development, particularly at regional and local levels. This process leads to the emergence of key actors who can execute relevant policies, as they are the recipients of subsidies for research and development (R&D), S&T education and training, as well as those aimed at S&T services.

Evaluating publicly funded S&T activities has been a long-standing practice at the international level. We raise the relevance and strategic aspects of this intervention to make real the promises of the benefit of S&T development, particularly in relation to progress and social wellbeing (e.g., the promotion and maintenance of a country's competitive and comparative advantages). With a scarcity of funds and increased competition for access to projects, there is the aim to allocate funds to projects that have the greatest impact on society. Thus, PRCs enter as an instrument to ignite regional and industrial development. However, there is a lack of consensus on the ways to evaluate their impact. Although this situation is not particular to the administration of S&T, public management has aimed to use methodologies that not

S. Estrada (✉)
Departamento de Finanzas y Administración, División de Ciencias Sociales y Administrativas, Universidad de Guanajuato,
Campus Celaya-Salvatierra Sede, Celaya, Guanajuato, Mexico
e-mail: salvador.estrada@gmail.com

J. Aguirre
Instituto Tecnológico Metropolitano, Medellin, Colombia
e-mail: ing.joao@gmail.com

H. Horta et al. (eds.), *Trends and Challenges in Science and Higher Education*, Knowledge Studies in Higher Education 3,
DOI 10.1007/978-3-319-20964-7_8

only monitor expenses but also provide information on results and contributions to the resolution of problems or make the most of available opportunities.

This chapter is born from a meta-analysis project to assess the main instruments of the 2008–2012 Mexican Special Program for Science, Technology and Innovation (*Programa Especial de Ciencia, Tecnología e Innovación*; PECiTI). The objective of this program was to evaluate the development of ST&I capacities and policies, as well as their effects on the recomposition of national ST&I efforts and their impacts on socioeconomic development. In the reformulation, there has been an examination of the main performance evaluation instruments designed by public administration to determine awareness in public policy regarding PRCs in terms of failures, areas of improvement, and limitations.

8.1.1 Rationale of the Intervention

PRCs are a government-supported infrastructure; the services they offer provide collective benefits that are indivisible and whose short-term utilities cannot offset expenses due to unpredictable risks related to technical, financial, or commercial decisions concerning their production.

Education services to provide S&T abilities cannot be restricted to clearly outlined uses, nor are they offered extensively as they are reserved for a highly specific demand segment. Such services also face a high level of uncertainty regarding labor markets and a lack of information about costs and opportunities. On the other hand, research entails high execution and evaluation costs, its infrastructure, installations and equipment cannot be limited to certain uses, and it requires a mastery of technical knowledge that sometimes cannot be separated from administrative aspects. Furthermore, the transfer of results requires an intensive learning process in both knowledge and various skills from engineering and design to administration, finance, and marketing.

Given the limited capacity of technology absorption by companies—resulting from insufficient information, aversion to investment risks, choice of incorrect alternatives, the drive of conflicting interests, conservatism, and difficulties inherent in the learning process—it is desirable to have local R&D capabilities that enable access to information, support, and technical advice in an increasingly global environment. Moreover, some state activities, in addition to providing collective goods and services, occur in markets where there are no other competitors, in such a way that to minimize costs and increase efficiency, there is a need for information and technical support to exploit economies of scale.

Thus, research centers can be considered an instrument of public policy as they enable improvements in the content and quality of higher education, and provide highly specific skills and infrastructure. Furthermore, PRCs offer information and technical support to various organizations, businesses and industries, link to other countries, provide quality public services, and efficiently operate public companies.

8.1.2 Assessment of PRCs

There have been numerous studies covering a wide range of perspectives on public finance systems. Many studies focus on the design of public policy and the impact on the mix of instruments (Guston 2000; Larédo and Mustar 2001); others involve comparative studies between countries on the policies surrounding research (Elzinga and Jamison 1995; Senker et al. 1999; Lemola 2002), while some are motivated by the role and organization of research councils (Braun 1998; Van der Meulen 2003; Slipersæter et al. 2007) and the financing structure of public projects (Lepori et al. 2007).

The impact of finance mechanisms oriented by priority has been addressed from a demand perspective, as have the answers provided by scientists (Laudel 2006), higher education institutions (HEIs) (Jongbloed 2007, 2008), and research organizations (Sanz Menéndez and Cruz-Castro 2003).

Diverse bodies of literature document the importance of ST&I activities for the economic and social development of a country (Nelson 1991; Pavitt 1998; Lundvall et al. 2002). However, this view has been questioned as social science studies on the matter have shown (via case studies) the singular ways in which relationships between social and scientific contexts develop. Furthermore, they highlight a lack of understanding regarding how social aspects intervene in the production of knowledge and its joint use for social structure (Zabala 2004).

The establishment of research institutes has been regarded as an instrument of public policy to create regional advantage (Cooke and Leydesdorff 2006). The reasoning is that these structures enable one to localize economic benefits and stimulate links with industry (Debackere and Veugelers 2005). In the case of the United States and the European Union, evidence shows the importance of these contributions to economic development (Feller 2004) and to achieve strategic national objectives (Graversen et al. 2005). This influence, it seems, is limited by economic, organizational, and disciplinary contexts, the orientation of research, mechanisms of transfer, and the absorption capacity of local businesses (Bilboa-Osario and Rodríguez-Pose 2004; Zahra and George 2002; Roper and Love 2006; Anselin et al. 2000; Fischer and Varga 2003; Hewitt-Dundas and Roper 2011).

Past publications show extensive evidence concerning the economic benefits of public investment in R&D. However, there is no consensus on evaluation methods (Georghiou and Roesner 2000). Among them, econometrics and financial methods stand out, as do those that measure product impacts (e.g., bibliometrics, patent counts, and partial convergence indicators) (Moore et al. 2007).

In the early 1990s, the US government established an administrative mechanism to make decisions for financing programs based on an analysis of the results of past programs. As part of the Government Performance and Results Act, scientific agencies were required pay attention to the assessment of impacts, given that one of the requirements was to present strategic plans that would include the establishment of a mission, results-based goals, and defined plans for performance evaluation that included measurable indicators. These plans and measurements are used in performance-based budget reviews and play a leading role in the recommendations of the Office of Management and Budgeting on budget allocations. Additionally, the Program Assessment Rating Tool was introduced this century; it summarizes the analysis of the factors that impact program performance in such a way that comparisons over time and similar programs can be conducted (United States Government Accountability Office 2005; as cited in Corley 2007).

8.1.3 Historical and Contextual Background

While there was no consolidated framework for steering and funding national S&T policies until the creation of the National Council on Science and Technology (*Consejo Nacional de Ciencia y Tecnología*; CONACYT) in 1970, policies appeared early on the Mexican government's development agenda and date back to the 1930s.

These developments were followed by an initial discourse for capacity building aimed at the exploration and exploitation of natural resources for industrial development. While it was assumed that social and economic issues on the public agenda should be served, scientists were still seeking funding for basic research and the training of human resources. Several research institutions and universities were founded, or re-founded, during this period with the mission to undertake public service, that is, support for an administrative sector (i.e., agriculture, health, or energy) and, in the case of universities, technical and professional education.

It is important to note the creation of the National Polytechnic Institute (*Instituto Politécnico Nacional*; IPN) and the reopening of several state universities during this period. At the National Autonomous University of Mexico (*Universidad Nacional Autónoma de México*; UNAM), institutes for mathematics, physics, and chemistry were founded. Spanish exiles came to Mexico and began several research initiatives in chemistry, biology, health, and social sciences via an organization called *Casa de España* (Spain's House; this then became the pubic research center known as *Colegio de México*, focused on social science and humanities). It is also important to highlight the founding of the Healthcare and Tropical Diseases Institute (*Instituto de Enfermedades Tropicales*), which was the first formal research institute in the field of health in Mexico (Retana-Guiascón 2009).

During the 1940s, various institutes were created with the mission to provide research and other scientific services, including the National Institute for Agricultural Research, the National Cardiology Institute, the National Hospital of Nutrition Diseases, and the National Laboratories for Industrial Development (Unger 1995, cited in Gutiérrez 2004). Thanks to the establishment of Mexico's Children Hospital and the Mexican Social Security Institute, hospital facilities began to combine healthcare with teaching and research (SSA 2002; SSA 2001). For regional development, a number of national commissions were established to foster several crops (corn, coffee, olive) to ensure multiregional coverage. These commissions included representatives from the federal government (finance, agriculture, and economy), international agencies (i.e., the Rockefeller Foundation), and scientific institutions, to work on the development of improved varieties with higher productivity. This development was oriented to providing better food sources for Mexicans and had a direct impact on infrastructure (e.g., developing experimental agricultural stations and laboratories, like the International Maize and Wheat Improvement Center) and the creation of new education programs (Guerrero 1989).

Twenty years later, concerns regarding the organization of the academic profession, interdisciplinary research, and the offer of industrial services, led to the strengthening of basic research as well as the development of applied research in the energy and farming fields. The most distinguished research institutes established in that period were the Center for Research and Advanced Studies of the National Polytechnic Institute (*Centro de Investigación y de Estudios Avanzados*; CINVESTAV). Also included were institutes focused on materials, instruments, and applied mathematics at UNAM, and across various administrative sectors including the Mexican Petroleum Institute (IMP), the Nuclear Research Institute (under the Energy Secretariat), and the National Agriculture and Livestock Research Institute (INIFAP, under the Agriculture Ministry) (Unger 1995, cited in Gutiérrez 2004).

The next decade was characterized by several changes in the public institutional framework, thanks to the creation of CONACYT, which provided the first national inventory of S&T activities. The state promoted these activities and focused them toward social and economic development by setting up several new research centers, mostly outside the metropolitan area of Mexico City to look after sectoral and/or regional demands. According to Campos and Jiménez (1994), 29 PRCs were founded between 1972 and 1980. These centers were related to para-state companies or regulators seeking to explore new areas (i.e., astrophysics, social anthropology, biotic resources, and applied chemistry) and decentralize research activities via the creation of research centers devoted to support regional small and medium-sized businesses (SMEs).

In 1979, the management of CONACYT was removed from the presidency and assigned to the planning and budget secretary to better coordinate planning efforts in terms of administrative execution. At that time, the majority of new research centers came under the secretary administration (Estrada 2012).

In the early 1980s, the country suffered an unprecedented crisis and researchers' salaries fell significantly, resulting in a mass exodus. A new instrument was proposed to retain them, called the National System of Researchers (*Sistema Nacional de Investigadores*; SNI), consisting of a salary supplement dependent on academic achievements and with peer-review bodies to provide an opinion on whether to award a tenure grant (Estrada 2012). This instrument, nowadays, is the most important incentive for academic research of scientific value, helping to blur applied research and link activities while receiving an important share of the fiscal budget for S&T.

The next generation of sectoral plans in the S&T field were drafted in the mid-1980s and early 1990s, showing a high techno-economic discourse and demonstrating a growing interest in linking academic endeavor with industrial applications as to the provision of technical services (Estrada 2012).

The economic crises helped to advance a new vision regarding research activity, which reassessed the legitimization to develop applied projects. University administrations established administrative units for the planning, promotion, operation, and regulation of services and consulting (De Gortari 1997). Meanwhile, research centers led by the energy sector (basically those linked to petroleum and electricity research), along with the National Laboratories of Industrial Development, united their managers to establish a private nonprofit organization called the Mexican Association of Directors of Applied Research and Technology Development (*Asociación Mexicana de Directivos de la Investigación Aplicada y el Desarrollo Tecnológico*; ADIAT) to increase collaboration with industry, develop technology knowhow management, and to improve communication with public administration (ADIAT 2009).

By the 1990s, several initiatives existed to foster industry–university linkages with international credits to fund novel programs (e.g., to create a consulting services market, to appoint industrial representatives or businesspeople to higher education consulting boards, and to establish private research centers, technology transfer offices, and academic spinoffs) with the requirements of established industry linkages to access funds, or performance-based assessment criteria based on resources obtained from private sources. Another key measure was to group research centers under the education arena and the programming and budgeting secretary creating the Network of SEP-CONACYT Centers, mainly due to the adhesion of CONACYT to the Education Secretariat (Estrada 2012).

Another mechanism to orchestrate the cooperation of various actors was the establishment of regional research systems where actions are suited to different potential users (state delegations of concerned ministries such as agriculture, health, ecology, and social development). These were anchored with resources supplemented by CONACYT, state governments, and other stakeholders, leading to the assignment of competition projects between universities and research centers localized in specific regions (Estrada 2012). A general assessment of these regional research systems detected positive impacts in terms of human resource formation, infrastructure, networking, and learning (Campos-Ríos and Sánchez-Daza 2008; Sampere 2003).

At the end of the 1990s, a new legal framework was drafted (the Federal Law on Scientific Research in 1999, which was replaced by the Science and Technology Law in 2002), proposing that funding for research must relate to the requirements of sectors within various federal government, states, and municipalities. The framework also represents an active and committed participation of stakeholders as it integrates a mechanism called the Science and Technology Forum, providing independent advice to the president and voice to the expressions of the different actors in the S&T community. Thus, there emerged a new institutional configuration defining the National System of Science, Technology, and Innovation in force today (Estrada 2012). A legal diagnosis showed that among the main problems, there was poor linkage between research and education, defective support for research groups at universities, unclear promotion rules for conducting research in para-state organisms, and the inefficient management of self-generated or third-party provided resources, for which a proposal to create new funds for research and development was established (Estrada 2012).

The current legal framework was reformed in 2009, with the state extending its role beyond that of strengthening and development to consolidation, and not just in scientific research and technological development, but also in innovation. Such action enlarges the spaces of interaction and widens the spread of actors involved in policy design and in the training of professionals in this area (i.e., researchers, engineers, designers, marketers, economists, financial advisers, and lawyers involved in the introduction of new products, processes, or services in the market) (Estrada 2012).

Thanks to innovation surveys, we know that those (few) companies with academic collaboration have a greater propensity to do so with research centers (mostly from the CONACYT network) rather than with universities. Thus, universities do not seem a relevant source for businesses, possibly because they are more guided by research curiosity. However, a case study (Estrada 2006) examining the results of CONACYT's joint fund with state governments (known as *FOMIX*) shows that state universities have a higher use of these funds. These results suggest that there may be a problem of predatory competition between research centers and universities; some universities (depending on the kind of request) obtain funding in areas in which they are not as strong as PRCs (Estrada and Pacheco-Vega 2009).

To explain the increasing competition among universities for these funds, it has been suggested that they have very little institutional funds to use for their own research agenda; therefore, each researcher exploits the funds known to them without a coordinated effort. There are positive elements to state funds, they establish agendas that allow universities to acquire relevancy and direct their efforts toward state demands; however, universities do lose the autonomy to propose their own agenda. It is also possible that regional systems of innovation gain variety and power by establishing various agendas (and each with its own complementary, or not, funds) and derive strength and richness by fully aligning research agendas. Further analysis is required to identify the situation regarding government agencies that provide funding and the research agendas of both universities and PRCs (Estrada and Pacheco-Vega 2009).

In comparing complementary and competitive activities among HEIs and public administration research centers (see Table 8.1), it is noted that the university sector serves bachelor-level education and public centers, especially those ascribed to CONACYT and complement this activity with a short supply; this activity is chiefly performed by public universities. Furthermore, the offering of postgraduate programs is mainly based in the education sector, but compared with bachelor-level education, this service is of major importance in public administration research centers. On this topic, UNAM is outstanding in terms of the number of researchers associated with SNI, papers in indexed peer-reviewed journals, and patents (patents is the only area in which UNAM does not hold top ranking, second only to IMP). The postgraduate degrees are also important at state universities, and research is

Table 8.1 Performance indicators for PRCs and HEIs

Organization	Indicators							
	Teaching staff	Enrollment	Bachelor programs	Graduate programs	SNI staff	ISI papers	Patents	
	Total	Total	Total	Total	Total	Total	Applied	Granted
	No. Ntl. %	No. Ntl. %	No. Ntl. %	No. Ntl. %	No. Ntl. %	No. Ntl. %	No. Ntl. %	No. Ntl. %
TOTAL	342,617 100	3,322,469 100	29,155 100	1,374 100	17,639 100	9,793 100	1,378 100	996 100
	FEDERAL UNIVERSITIES							
COLPOS	337 0.1	948 0.03	32 0.11	19 1.38	239 1.35	228 2.33	6 0.44	5 0.5
UNAM	27,414 **8**	204,581 **6.16**	333 1.14	135 **9.83**	3583 **20.31**	3142 **32.08**	163 **11.83**	149 **14.96**
UAM	5,641 1.65	51,935 1.56	142 0.49	63 4.59	906 5.14	582 5.94	88 6.39	64 6.43
IPN	10,479 3.06	100,452 3.02	224 0.77	73 5.31	779 4.42	787 8.04	60 4.35	29 2.91
CINVESTAV	588 0.17	2,312 0.07	58 0.2	56 4.08	727 4.12	847 8.65	106 7.69	66 6.63
	PUBLIC STATE UNIVERSITIES							
UDG	7,684 **2.24**	92,451 **2.78**	389 **1.33**	81 **5.9**	662 **3.75**	329 **3.36**	14 1.02	4 0.4
UANL	5,095 1.49	79,246 2.39	276 0.95	72 5.24	443 2.51	326 3.33	49 **3.56**	13 1.31
BUAP	3,657 1.07	54,434 1.64	189 0.65	35 2.55	384 2.18	283 2.89	13 0.94	0 0
UG	1,915 0.56	17,812 0.54	153 0.52	27 1.97	265 1.5	201 2.05	39 2.83	7 0.7
UABC	4,450 1.3	47,878 1.44	173 0.59	33 2.4	216 1.22	184 1.88	9 0.65	1 0.1
UADY	955 0.28	13,841 0.42	107 0.37	20 1.46	167 0.95	132 1.35	10 0.73	4 0.4
	SECTORAL CENTERS AND CONACYT SYSTEM							
INIFAP	0 0	0 0	0 0	0 0	188 **1.07**	133 1.36	3 0.22	0 0
IMP	0 0	0 0	0 0	3 0.22	169 0.96	149 1.52	391 **28.37**	477 **47.89**
IIE	0 0	0 0	0 0	0 0	48 0.27	26 0.27	64 4.64	58 5.82
INCMNSZ	0 0	0 0	0 0	0 0	133 0.75	158 1.61	3 0.22	3 0.3
INNNMV	0 0	0 0	0 0	0 0	75 0.43	83 0.85	6 0.44	5 0.5
CICESE	179 0.05	428 0.01	18 0.06	19 1.38	150 0.85	183 **1.87**	2 0.15	1 0.1
INAOE	110 0.03	409 0.01	8 0.03	8 0.58	112 0.63	141 1.44	6 0.44	4 0.4

(continued)

Table 8.1 (continued)

	Indicators							
	Teaching staff	Enrollment	Bachelor programs	Graduate programs	SNI staff	ISI papers	Patents	
	Total	Total	Total	Total	Total	Total	Applied	Granted
Organization	No. Ntl. %	No. Ntl. %	No. Ntl. %	No. Ntl. %	No. Ntl. %	No. Ntl. %	No. Ntl. %	No. Ntl. %
CIBNOR	91	146	2	2	111	120	1	0
	0.03	0	0.01	0.15	0.63	1.23	0.07	0
CIAD	138	198	3	3	108	136	6	5
	0.04	0.01	0.01	0.22	0.61	1.39	0.44	0.5
CICY	73	242	5	7	89	92	3	0
	0.02	0.01	0.02	0.51	0.5	0.94	0.22	0
IPICYT	0	0	0	13	63	92	5	0
	0	0	0	0.95	0.36	0.94	0.36	0
CIO	57	64	3	3	62	103	1	2
	0.02	0	0.01	0.22	0.35	1.05	0.07	0.2
CIMAT	69	148	9	4	61	57	0	0
	0.02	0	0.03	0.29	0.35	0.58	0	0
COMIMSA	48	81	5	4	56	91	16	10
	0.01	0	0.02	0.29	0.32	0.93	1.16	1
CIQA	49	48	4	4	51	40	15	4
	0.01	0	0.01	0.29	0.29	0.41	1.09	0.4
CIATEJ	24	67	2	1	46	31	7	3
	0.01	0	0.01	0.07	0.26	0.32	0.51	0.3
CIATEQ	0	0	0	0	19	44	1	0
	0	0	0	0	0.11	0.45	0.07	0
CIDESI	20	59	6	4	6	6	6	1
	0.01	0	0.02	0.29	0.03	0.06	0.44	0.1
CIATEC	0	0	0	1	4	1	16	3
	0	0	0	0.07	0.02	0.01	1.16	0.3

COLPOS: College of Postgraduates; UNAM: National Autonomous University of Mexico; UAM: Metropolitan Autonomous University; IPN: National Polytechnic Institute; CINVESTAV: Center for Research and Advanced Studies; UDG: University of Guadalajara; UANL: University of Nuevo León; BUAP: Autonomous University of Puebla; UG: University of Guanajuato; UABC: Autonomous University of Baja California; UADY: Autonomous University of Yucatán; INIFAP: National Agriculture and Livestock Research Institute; IMP: Mexican Petroleum Institute; IIE: Institute of Electrical Research; INNSZ: 'Salvador Zubirán' National Nutrition Institute; INNMV: 'Manuel Velasco' National Institute of Neurology and Neurosurgery; CICESE: Ensenada Center for Scientific Research and Higher Education; INAOE: National Institute of Astrophysics; Optics and Electronics; CIBNOR: Center for Biological Research of the Northwest; CIAD: Center for Food Research and Development; CICY: Scientific Research Center of Yucatán; IPICYT: 'Potosino' Institute of Scientific and Technological Research; CIO: Center for Research in Optics; CIMAT: Center for Research in Mathematics; COMIMSA: Mexican Corporation for Materials Research; CIQA: Center of Applied Chemistry; CIATEJ: Center for Research and Assistance in Technology and Design of the State of Jalisco; CIATEQ: Advanced Technology Center; CIDESI: Center for Engineering and Industrial Development; CIATEC: Center for Applied Innovation in Competitive Technologies.

Source: Authors' own data, based on Explorador de datos del Estudio Comparativo de Universidades Mexicanas (ExECUM).

increasingly important to this sector. It is common to rate single or double ratios between researchers and productivity (across all institutions). Regarding patenting contributions, both HEIs and PRCs have similar outputs, with the majority of patent holders coming from these entities.

These patent holders are basically comprised of inventors based in universities and PRCs, with the dependency ratio decreasing from approximately 30 (patents applied by foreigners in comparison with nationals) in 2000 to 12 in 2011; thus, in the share figures of all patents applied for in Mexico, national contributions range from 3 % to 8 % (CONACYT 2012). PRCs led by the IMP produce patents to be used by the petrochemical sector, especially PEMEX (Mexican Petroleums), to substitute expensive imported inputs as is the case of Mexican Social Security Institute (IMSS), producing patents to substitute expensive or orphan drugs demanded by "beneficiaries" and incidentally transferring to commercial laboratories to attack prevalent generalized diseases such as colitis. Agriculture research has generated few patents but has managed to develop improved seeds and vegetable varieties as well as diagnosis and treatment methods against plagues and cattle diseases.

Federal universities, headed by UNAM, have developed several organizational groups to cope with public and private demands by traditional mechanisms of linkage-based training, consulting, and service delivery. Between 1991 and 2009, UNAM applied for 1,108 patents and 816 were granted. Modest, but relatively important in the Mexican context, are the numbers obtained by other HEIs at both national and local levels; however, patenting activities are concentrated in larger institutions in Mexico City and the main industrial and central states, such as Nuevo Leon, Jalisco, Puebla, Guanajuato, or Coahuila. In terms of technology transfer activities, there has been little experience in the area in comparison with PCRs (Calderón-Martínez 2013).

8.2 Method

A descriptive analysis of various documents was undertaken to describe the current situation of PRCs in Mexico. Of particular interest are noncompetitive sources of funding based on fiscal resources and various government evaluation instruments. Inquiries were also made into various legislative and planning frameworks, especially concerning commitments in planning, management, and evaluation via both current legislation and the accountability process in programming and budgeting cycles (all public policies must report on the use of funds and results obtained). The main evaluation instruments used, and their results, are identified. With the goal of validating and complementing these government indicators, we turn to independent evaluations. An assessment of PCRs was conducted, focusing on their main weaknesses and possible areas of improvement. This accounting- and accountability-based focus benefits from the use of compound indicators that transcend the tally of resources and results. This approach also focuses on the search for causes and effects, although with a perceptual bias from civil service. While the organizational structure and culture are not considered, this focus presents the situation in aggregate form, with which we lose the interpretive context regarding the reality of the discipline, sector, or region of the relevant environment.

8.2.1 Population Framework

PRCs[1] are semi-public entities of the public administration or are private institutions of public interest, that in accordance with their articles of incorporation, have a primary objective to carrying out S&T research or innovation activities, and in a systematic way. They were established from public policy that looked to encourage and produce knowledge as well as to develop diverse S&T fields in the country and regions. PRCs are committed to respond to the environment and contribute to the development and progress of society. They count on a workforce with a high level of specialization in various knowledge areas (Table 8.2).

Table 8.2 PRCs: relevant population

PRC	Area/sector	Organizational structure
CONACYT system (10 % SNI)	S&T/social sciences and humanities	3 centers, 4 colleges, 1 institute
	S&T/exact and natural sciences	7 centers, 3 institutes
	S&T/technological and services development	6 centers, 2 trusts and one corporation
Sectoral centers (7 % SNI)	Energy	3 institutes
	Health	Various institutes and specialty hospitals
	Agriculture	2 institutes, 1 college and 1 university
	Environment and natural resources	1 institute
	Communications	1 institute and 1 agency
	Justice	1 institute
Education (67 % SNI)	Federal universities	71 campuses (UNAM), 19 centers (IPN), 10 branches (CINVESTAV), 5 branches (UAM)
	Public state universities	88 universities
	Other	1 center, 1 college, 1 faculty

Note: The first column shows the participation of national researchers by subsystem between parentheses.
UNAM: National Autonomous University of Mexico; UAM: Metropolitan Autonomous University; IPN: National Polytechnic Institute; CINVESTAV: Center for Research and Advanced Studies

[1] Art. II, Fr.7, BASES of organization and functioning of RENIECYT. http://www.conacyt.gob.mx/Acerca/Documentos%20Normatividad/BASES-organizacion-funcionamiento-RENIECYT_10sep08.pdf. Accessed 10 November 2012.

To ensure a comprehensive understanding of the research centers in Mexico, the federal government sectors (executive and legislative) should be considered, including social, private, and federal governments. However, there is no precise tally and thus, the potential population is insufficiently covered by the current public policy. All research centers that aim to receive funds—or that have indirectly participated in the public budget through competitive subsidies for the implementation of research, development, and innovation activities—are registered in the National Registry of Institutions and Businesses in Science and Technology (*Registro Nacional de Instituciones y Empresas Científicas y Tecnológicas*; RENIECYT). Of the full registry (consisting of 6,978 individuals and legal entities that develop S&T activities), research centers represent just 1.1 % of all agents (distributed as follows: 72.4 % federal; 13.1 % state; and 14.5 % without classification). For their part, PECiTI emphasizes that, "The private sector and the Federal Government operate different centers and institutes that carry out scientific research, technological development and innovation. The public sector directly finances 39 research centers that are coordinated by different State agencies and decentralized, nonspecialist, agencies."

The centers and institutions that the public sector coordinates via federal executive power are divided into two groups. One is functionally ascribed to CONACYT, which despite its disciplinary and organizational diversity that fragments it into three well-defined subgroups, portrays an institutional identity that is reaffirmed when submitted to the same environmental pressures (the vagaries of budgets, self-generation of resources, evaluations, audits, and other such norms). The second group is composed of disperse and heterogeneous sectoral centers that act in diverse fields such as energy, health, agriculture, environment, and natural resources. They respectively contribute 10 % and 7 % to the national capacity of research certified through membership in SNI.[2]

In addition, the budget branch of the education administrative sector ascribes to various federal universities, state public universities, and a single center, college, and faculty.[3] These are not considered part of the public administration, but are the main force behind research in Mexico; moreover, they have departments within their structures that act as research centers.

[2] SNI is constituted by regulations, administrative staff, a beneficiary list, peer-review units, and an annual budget. Academic bureaucracy designs an annual Request For Proposals where potential beneficiaries—according to their academic merits undergo a paired peer-review as well as an evaluation by consultative commissions—authenticate their reputation and given a positive response, the successful applicants may maintain their productivity through the provision of a full salary via a tax-free scholarship for a certain period of time.

[3] This block, ascribed to the education sector, considers itself independent of other public centres, given that they are not para-state entities. However, for the purposes of this explanation, it is maintained here as given that some HEIs include research centers in their structure, in such a way that these centers are controlled, administered, or directly associated with the education sector.

These centers, research departments, and public institutions of higher education are financed in different ways: direct allocations from the Federal Expense Budget, federal agencies, subsidies (direct or assigned during competitive processes), and self-generated income from services to public and private sectors.

8.3 Results and Discussion

8.3.1 Normative and Planning Framework

In the case of public centers, they were founded with the aim of supporting public services (e.g., sanitation and agricultural development), the activity of para-state businesses and their regulating organisms (regarding energy or water management), and to attend to new research areas besides decentralizing activity (like CONACYT centers). Later, they were assigned postgraduate-level education and to support the development of SMEs in the region.

The current Science and Technology Law (LCyT 2002) recognizes research centers as a collective actor, known as the National Network of Research Centers and Groups of the National System of Science and Technology. Moreover, it is empowered to the highest level, being named a member of the General Council of Scientific and Technological Research; the council is responsible for defining federal public expense criteria and assigning priorities, approving the consolidated budget, and monitoring and understanding its evaluation.

PECiTI is also attributed as an actor, involved in the system and required (as are others, including companies, HEIs, states, and municipalities) to contribute to transversally maintaining the theme of ST&I in all areas and to achieve greater social ownership of the national scientific and technological culture. It also participates in the advancement of activities (regional vocations via postgraduate degrees, research in academic institutions, and technological development and innovation in companies) by creating relations with all agents. Thus, it is hoped that the generation and application of S&T knowledge will increase.

To date, there exists no public policy regarding research centers. As an instrument, they are identified as a mechanism of support for SMEs, in that they achieve the technological development of products, processes, materials, and services of aggregate value. Research centers also engage in collaboration to take advantage of the relevant infrastructure, to create and strengthen links via collaborative networks, technological parks, and consortiums that promote joint projects. Furthermore, they keep international cooperation alive. Finally, such centers are considered among strategic objectives as a mechanism to evaluate the application of public resources, via results-based administration contracts.

Since 2002, a planning and evaluation framework has been in place for research centers, constituted by the Planning Law, Science and Technology Law, CONACYT

Organic Law, Para-state Entities Federal Law, National Development Plan, Special Program for Science, Technology and Innovation, and Operative Rules for substantive CONACYT projects. For example, the National Development Plan states that the Federal Public Administration must advise of its results and resources. The Science and Technology Law points out that the results of its activities will be evaluated and that support policies will be subject to the evaluation of results and impacts on satisfying the country's needs. PECiTI expresses that progress and accountability reports are essential for evaluation and follow-up; therefore, a planning system must be established to determine priorities, activities, and goals. An indicator system should also be established to give certainty to compliance and evaluation, in particular with regard to the objectives of the National Development Plan.

8.3.2 Application of Expenses and Results

It is necessary to emphasize that research centers can be evaluated in an aggregate way using budget programs, as they are executed by responsible parties (that are sometimes the same research centers or body that coordinate with these entities). The *Programmatic Structure of the Budget Expenses of the Federation*[4] distinguishes a group of programs that aim to provide the public service of research, technological development, and postgraduate degree formation, and can therefore provide important information on the collective performance of research centers. These programs are identified with the prefix E and refer to fiscal resources that the public administration directs towards the operation and maintenance of research centers.

In the budget, public centers represent the public provision of high-value services (PpE Modality), which are subsidized by programs subject to rules of operation (Pp S).[5] If the functional classification of budget programs is analyzed, it is possible to interpret that the centers fulfill an economic development function in the ST&I administrative sector and develop scientific research, technological research, and publications to improve the wellbeing of the population (Pp E001 Modality). Those that develop technology and innovations in addition to also publications (Pp E002), do so with the purpose of improving the competitiveness of the country. Through subsidies, both contribute to the strengthening of S&T capacity.

[4] See document: Secretary of Public Tax and Credits, Sub-Secretary of Expenses Programmatic Structure to Employ in the proposed Expense Budget of 2013. http://www.hacienda.gob.mx/EGRESOS/PEF/programacion/programacion2013/estructura_programatica_ppef_2013.pdf.

[5] The idea of this duality can be clarified by understanding the origin of the resources. According to Estrada González (2009), CONACYT centres of natural and exact sciences receive 25 % of their funding from fiscal resources (Pp E Modality) and 33 % from subsidies (Pp S and U Modalities); that is, through competitive funding. Technological development receives 12 % of its funding from fiscal resources and 17 % from subsidies.

Furthermore, the budget presents programmatic elements that allow for the evaluation and assignation of the exercise of expenses. Derived from the Federal Budgets and Tax Responsibility Law, the System of Performance Evaluation (*Sistema de Evaluación del Desempeño*; SED)[6] is as a set of methodical elements that allow for the realization of an objective performance valuation of budget programs, under the principles of verification of the level of compliance with goals and objectives, and based on strategic and management indicators. SED indicators form part of the budget program and for the process of elaboration of the relevant budget. SED implies a dynamic between the process of budgeting and planning activities, such as in the execution and evaluation of policy, programs, and public institutions. The two principal components of SED are the evaluation of public policy and budget programs, and management for the type of expense.

The budget incorporates elements of the program that are derived from the analysis of the logical framework[7] that shapes a matrix of result indicators (*Matriz de Indicadores de Resultados*; MIR), where direct impacts of program execution are stated; this is their 'end', the direct results for the beneficiary population, which is the named 'purpose' of the program (in addition to the public goods or services that are produced or delivered through the operation of the program), and the tasks and inputs required to provide a high standard are labeled 'components' and 'actions,' respectively. These four elements create a narrative summary of the program. For the assessment, monitoring, and verification of achievements, each element has indicators, objectives, and means of collection, in addition to the assumptions or circumstances that can affect their fulfillment. For example, the indicators of the E002 Program, refers to the lending of services of research centers dedicated to technological development.

Table 8.3 infers that centers seek to improve the competitiveness of the country via the products of their research, teaching, and transfer (e.g., projects, articles, thesis, services and intellectual property), especially those oriented towards socioeconomic development, maintaining standards for their projects, and arbitrated by academic practice (e.g., the use of peers and the market, obtaining self-financing resources, results in competitive funds, structure of the range of services, and the growth of the client base). This matter is only pertinent to centers that depend on the

[6] See http://shcp.gob.mx/EGRESOS/PEF/Paginas/sed.aspx.

[7] The focus of analysis on the logical framework is a way to structure the main elements of a project, highlighting the logical links between expected input, planned activities, and expected results. It was created by USAID in the late 1960s, and has been used since in projects to develop public investments. On March 31, 2007, the Secretary of Public Tax and Credits (*Secretaría de Hacienda y Crédito Público*), the Secretary of Public Function (*Secretaría de la Función Pública*), and the National Commission of Social Development Policy Assessment (*Comisión Nacional de Evaluación de la Política de Desarrollo Social*) published General Guidelines for the Evaluation of Federal Programs of the Public Federal Administration in the *Official Journal of the Federation*. The methodology of the logical framework was established with the objective to carry out the evaluation and monitoring of federal programs included in the Budget of Federal Expenses of each tax year, to promote result-based management, and to consolidate the system of performance evaluation.

Table 8.3 Matrix of results indicators by type of indicator, method of calculation and results (2009–2011)

Level (MRI)	Strategic indicators (effectiveness)		E002 Program 2009	2010	2011
END	Mexico's position regarding university–industry collaboration in the R&D variable of the Global Index of Competitiveness	Methodology of the World Economic Forum: the global index of competitiveness is composed of 113 variables, organized into 12 groups or pillars; two-thirds of these come from an opinion survey and remainder from public sources.		59	45
Purpose	Percentage of research projects supported and thesis completed, oriented towards economic development	(Number of research projects oriented towards socioeconomic development + number of postgraduate thesis completed oriented towards socioeconomic development/total number of research projects + total number of postgraduate thesis completed) × 100	57	56	66
Purpose	Percentage of research projects in knowledge transfer	(Knowledge transfer projects/total developed projects) × 100	62.5	64	67
Purpose	Percentage of intellectual property registered by center	(Number of patents licensed and/or models of utility, author rights transferred/total investigations realized by the center) × 100	30	10	20
Management indicators (effectiveness)					
Component	Percentage of projects that contribute to the solution of regional and sectoral demand	(Number of research projects approved in "*Fondos Mixtos*" program + number of research projects approved in "*Fondos Sectoriales*" program/number of research projects approved in the regional fund ("*Fondos Regionales* program") + total number of research projects) × 100	39	32	53.5
Component	End result efficiency	(Number of students graduated by cohort/number of students matriculated by cohort) × 100	51	67	65.5

Component	Ratio of related products generated by ascribed academic personnel	(Related products/total academic personnel)	4	5	16
Activity	Percentage of resources self-generated	(Amount of total budget of resources self-generated (own)/amount of total budget from fiscal resources) × 100	64	48	208
Activity	Ratio of users of services offered by research centers	Number of users of the services/total researchers	8	7	0.4
Activity	Rate of increase in clients	(Clients of the center in year n/clients of the center in year n-1) × 100	99.5	101	45.4
Management indicators (quality)					
Activity	Number of graduates in programs of the National Registry of Quality Postgraduate Degrees (*Padrón Nacional de Posgrados de Calidad*; PNPC), by researcher	Number of graduates in specialty programs of PNPC + number of graduates in masters programs of PNPC + number of graduates in doctorate programs of PNPC/total number of researchers	0.4	0.6	0.6
Activity	Index of quality of postgraduates of the center	Number of programs registered in the PNPC as a new creation + (2) × number of programs registered in the PNPC in consolidation + (3) × Number of consolidated programs registered in the PNPC + (4) × number of programs registered in the PNPC of international character/(4) × total number of postgraduate programs offered by institution	0.65	0.7	0.7
Management indicators (efficiency)					
Activity	Projects by researcher	(Total research and technology development projects in year n/Total researchers of the center in year n)	1.4	1.6	1.4

Source: Authors' own, based on http://www.conacyt.mx/index.php/el-conacyt/evaluacion-de-programas-conacyt. Accessed 17 January 2013

governmental administrative sector of S&T, although the indicators could also apply to centers in the agriculture, energy, education, and health sectors. Be it in terms of competitiveness or national development, sectoral research centers (or those tied to universities through the allocation of resources) aim for their human capital to orient their research and teaching products to improve the level of income of agricultural producers, the confidence and quality of energy inputs, the quality of higher education, and the health conditions of the population.

8.3.3 Results-Based Goals

In an assessment carried out towards the end of the last presidential term (2000–2006),[8] PRCs were shown to be organizations that contribute to the generation of knowledge, the formation of specialty human capital, and the resolution of problems. Their principle needs relate to the increased volume of finance obtained from competitive funds, the continuation and amplification of support in training human resources, increases in resources for infrastructure and equipment, the improvement of work conditions, evaluation systems, and incentives for researchers, and the realization of structural and functional reforms, such as procuring the decentralization of institutions and territory.

According to the document *General Information of the State of Science and Technology in Mexico* (CONACYT 2012), the performance of the health and energy sectors excelled during the reference period (2006–2012); both grew in financial volume, for research in particular. In the case of postgraduates, where the budget grew fastest, health and agro-fisheries were both standouts. Scientific production also increased, and at a similar rate to the growth of expenses. This behavior did not alter the profile of Mexican production, where agriculture and animal farming predominate, followed by medicine and physics. The most productive institutions (and of greater impact) are still those of higher education followed by those in the health sector (IMSS and the "Salvador Zubirán" National Institute of Nutrition) and energy sector (IMP). In total, 60 funding sources were available to strengthen Mexico's S&T capacity. These funds were operated by CONACYT, in coordination with secretaries of state ("*Fondos Sectoriales*" program), state governments, and federal entities ("*Fondos Mixtos*" and "*Fondos Regionales*" programs). Historic data show that the institutions with the greatest number of funded projects were federal education institutes (UNAM, CINVESTAV, IPN, and UAM), followed by the National Institute of Forest, Agricultural and Livestock Research (INIFAP), organizations in the health sector (IMSS and the National Institutes of Public Health), and public state universities. The production of invention patents shows that the main players in the field of technological development are transnational companies. National centers with the best patent performance are IMP followed by UNAM, UAM, and the Monterrey Technological Institute for Higher Education (a private university

[8] Sanz et al. (2007). *Evaluación PCT México*. International Panel. Draft paper.

Table 8.4 Variations in the performance of public centers (2006–2012)

Performance indicators	Expense per administrative sector (median annual growth rate)
Increase in expenses 6.6 %	Health 14.8 %
Increase in scholarships 47.9 %	Energy 10.8 %
Increase in personnel 5.2 %	CONACYT centers 9.4 %
Increase in number of projects −4.5 %	Education 1.1 %
Increase in infrastructure 10.5 %	Agriculture −1.9 %

Source: Authors' own data, based on CONACYT (2012) and Notacefp/007/2012, February 8, 2012. Center of Public Finance Studies

with an outstanding performance). Research institutions have a limited capacity for technology transfer and, consequently, require market intelligence activities, the development of extension services and commercialization, projects management, and intellectual property management (Table 8.4).

The centers are evaluated in an aggregate manner, looking at the strategic indicators reported by administrative sectors for the elaboration of the Federal Expenditures Budget (*Presupuesto de Egresos de la Federación*) and that are valued by the Secretary of Public Tax and Credits (*Secretaría de Hacienda y Crédito Público*; SHCP). The design and implementation of the SED indicators constitute one of the central elements of the management for results model. Thus, there is a profound transformation in the way in which the performance of the institution and its programs are assessed, considering elements of results-based decision making and accountability. Because of the diversity and complexity of the information generated through SED, the Synthetic Model of Performance Information (*Modelo Sintético de Información del Desempeño*; MSD) was created. This model is composed of the following variables[9]:

1. Budget Performance: considers the efficiency of the expenses of each budgeted program over the previous 4 years, establishing the difference between the authorized budget and the actual one, to determine underspending and overspending.
2. Matrix of Indicators for Results (MIR): Examines three variables related to MIR— (a) quality of the MIR; (b) annual relationship between achievement of goals and the difference in the exercised budget in the t-1/ approved in t; and (c) the advance in the results of the Pp compliant with the indicators of the levels of purpose and components.
3. Evaluation: considers the external evaluations most recently registered in the inventory of the Annual Evaluation Program. Regarding these evaluations, representative queries are selected according to the terms of reference for the type of assessment.
4. Program to Improve Management (*Programa de Mejora de la Gestión*): takes into account the results of the Index of Institutional Development (*Índice de*

[9] See SHCP, *Unidad de Evaluación del Desempeño* (2012). MSD Methodological Annex.

Desarrollo Institucional) of the Special Program of Management Improvement (*Programa Especial de Mejora de la Gestión*). The improvement projects are aligned with the corresponding Pp.

5. Aspects Susceptible to Improvement (*Aspectos Susceptibles de Mejora*): registers the advance of aspects susceptible to improvement in accordance with the "Mechanism for the follow-up of aspects susceptible to improvement derived from external information and evaluations to federal programs."

A global assessment of the performance of the relevant research, technological development, and postgraduate training departments during the present 6-year period could come from the behavior of synthetic indices. The system of PRCs, in which the management and provision of high value-added public services is referred to, has transitioned from that of a medium-low performance to a mid-level performance. Thus, Mexico has gone from a scarcity of information or evidence for the achievement of certain goals to the next level. Without doubt, there are many areas of improvement, but we have reached an early stage of the learning process, a stage that can strengthen and consolidate as continuity is maintained and the responsible departments undertake external evaluation and make a commitment to improve.

Following these criteria, the following entities have the worst relative budget performance: the relevant departments for spatial research, primary sector research projects, health in the IMSS, the Institute of Security and Social Services of the Workers of the State (ISSSTE), postgraduate programs in the health sector. In contrast, departments responsible for education programs dependent on the agricultural and educational sector as well as programs of technological R&D in energy, education, S&T and penal sciences, have executed their resources in a timely and appropriate manner.

With regard to the indicator of public management, various improvements have not resulted in the expected outcomes, as improvements to quality of services, institutional effectiveness, and operational costs have been mid-level at best. However, better results have been achieved in the health sector, particularly in departments of direct control (IMSS, ISSSTE). It seems that the IMP and the Mexican Space Agency have not attempted to introduce any improvements into their organizations.

Considering the extent of the commitment to improve, this indicator is dependent on external evaluation; only those institutions related to health and educational research present results, therefore they are the only ones that can capitalize. This is proven by the advanced level in committed activities by these research departments, despite the lag in the commitment to improve in postgraduate programs.

The programs that fund the operation of research centers have rarely been subject to external evaluations.[10] Only those departments responsible for the activities

[10] The SHCP matrix does not consider the PRC as a figure that forces it to subscribe to a results-based management agreement, which integrate assessment activities by various actors such as SHCP, SFP, CONACYT, the Federation Superior Audit, and external evaluation committees. Furthermore, self-evaluation reports are put forth for consideration by a public commissary; as para-state entities, their annual work plan is also audited by an internal control office.

of postgraduates and research in the health and education sectors have been evaluated, obtaining a higher grade in postgraduate activities in both instances and in health research. In general, it can be said that they tend to have a moderately good alignment, design, planning, focus, and orientation towards results.

Regarding S&T systems, executors of only a few centers have well-constituted index matrices. They have expended the budget according to their goals, in addition to their actions to obtain direct results in the target population. The following standouts in this sense, are the Mexican Space Agency, institutes of public health, and specialty hospitals.

With regard to the global indicator, the priority programs of education and health departments are the most rigorously evaluated, given that they are externally evaluated and that these criteria are dominant the SED. The remainder of these programs present improvements in global performance, albeit in the mid-level range. In this sense, less hierarchal programs in this classification are those departments from the S&T sector. To analyze the components, it would seem that the indices matrix is not consistent in its construction; however, the introduction of changes in management has not produced significant results either.

8.3.4 Performance

Another evaluation mechanism of PRCs is results-based management agreements, in which performance indicators and annual goals are defined to ensure compliance; agreements are valid for 5 years. In 2011, all CONACYT research centers were reported to have entered results-based management agreements, while in other sectors, namely agriculture and energy, such agreements were only held by INIFAP, the College of Postgraduates, the Institute of Electrical Research, and IMP. However, the National Institute of Public Health was recently approved as a PRC, categorized as such by the current Law of Science and Technology; thus, it will now have to enter to a results-based management agreement.

Within the framework of management agreements, CONACYT centers group their indicators according to the following activities: knowledge generation, training of human resources, and support for regional socioeconomic development and strengthening (for competitiveness). Each center has its own set of indicators, depending on the area in which it is set (e.g., natural and exact sciences, social sciences and humanities and technological development). However, the structure maintains a common pattern that could trigger a process of learning for all public centers.

The indicators of knowledge generation refer to production, dissemination, and transfer as a development of invention. Training indicators show the excellence of academic personnel, certification of postgraduates, personal participation in educational programs, enrollment, graduation and grades, participation and culmination of the thesis, end efficiency, and labor market insertion rates.

Support for development is estimated by the portfolio of projects—whether it attends to the request for proposals (RFP) of joint or sectoral funds,[11] which assumes a link between regional issues or with certain needs of the population—and the population attended or benefitted, the orientation of the thesis themes to socioeconomic development, agreements that impact public policy, the profile and orientation of the work of the researcher, as well as the number of organizations served.

Finally, those products and projects referred to as strengthening are those with links to industry, contracts, and those supporting a high number of companies supported and service users. This is in addition the (financial) sustainability index, which is usually calculated as a ratio between self-generated resources (which includes self-generated income for the sale of services, as well as competitive funds granted) and the total budget (which includes self-generated resources and those obtained through fiscal resources).

These agreements establish a commitment to evaluation on the part of the tax authority, public function, and the organization responsible at the sector level. They are strongly oriented towards overseeing the financial management of both the annual work programs and the medium-term strategic programs. Regarding performance indicators, variances are observed in the agreed goals; the organization under evaluation must justify these aims. The same occurs regarding the framework of results-based indicators. In the end, according to the level of compliance, they offer a qualitative addition regarding performance.

In addition to these indicators managed in the coordinated centers by CONACYT, public centers ascribed to other sectors present some differences with respect to the indicators included in results-based management agreements.

8.3.5 *Detected Problems*

The evaluations of the National Council of Evaluations (*Consejo Nacional de Evaluación*) and SED regarding the lending of advanced value-added services (scientific research, technological development, and postgraduate training) underscore the fact that many problems in the vertical logic of the matrix (i.e., the relations of causality between activities, products or components, purposes, and ends), cannot be sufficiently established. It is important to emphasize that this situation is common in the creation of public policy and, in particular, in S&T. It requires a coordinated effort and inter-institutional collaboration to study these causal relationships.

The evaluation of the impact of budget programs on some of the aims and purposes included in the matrix of indicators take time to carry out because their measurement would need to be adapted to longer periods lapses than usual (trimesters, semesters, or years). However, it is possible that they require the development of

[11] Trusts to promote S&T activities based on matching funds between CONACYT and state governments (mixed funds; *Fondos Mixtos*) and state agencies or federal public administration sectors (sectoral funds; *Fondos Sectoriales*), respectively.

tools and methodologies of complex analysis. The horizon can stretch so far ahead that any follow-up becomes impracticable. More importantly, it is of value to use mechanisms to assess additional benefits other than the efforts, processes, and results. For example, relationships that limit productivity depend on the discipline, center size, the number of projects that can be undertaken and students that can attend, and the publications that can be generated. The projects tend to present high heterogeneity in, for example, complexity, and the number and type of actors involved and reached. They can have a high level of internationalization and be a source of expert mobility, as well as exploiting highly specific physical infrastructures, with which they would qualitatively benefit students, researchers, groups, centers, and regions.

One of the most important aspects for the achievement of impacts is to rely on mechanisms that can mobilize research results. Specialized publications can limit the dissemination of cutting-edge methods or processes for prevention, diagnosis, treatment, and rehabilitation. It is desirable to bring these methods to clinical applications so that they can lend a practical hand via professional training. However, sometimes, motivation is not enough. In the case of public health research, González-Block (2009) proposed that the growing complex needs of the population require answers that consider systemic, program, organizational, and instrumental changes in addition to existing leadership, experience, and vision in the sector, together with specific social and economic situations. In this way, on occasion, the policies of the state and government are the principle vehicles to disseminate research results.

The expected outcome also depends on the precision with which demands are established. For example, centers try to burden the validation of demands to leading agencies. In this way, mixed and sectorial funds work to flag social, regional, and sector needs and problems (of specific populations) from the perspective of the centers. To what extent this expectation is achieved should be studied, given that some demands remain unaddressed or are inconsistent between one RFP and another. What can be interpreted, under the prior argument, is that a minority of these demands do not reflect the sensed problems, nor are they expressed with sufficient clarity to be undertaken.

It might be worth mentioning that in situations where the results are restricted to a new project, the participation of other specialists, the use of other services (such as the manufacture of prototypes or another technical service), or a highly specialized infrastructure including the development of methods or protocols (similar to what occurs in the field of food safety, vegetable sanitation or health) may be required to ensure an appropriate diffusion or dissemination process. Furthermore, an intermediary is required to reach the market potential.

To ensure that successful cases are replicable, it would be pertinent to create a forum for the exchange of experience and knowledge. Such forums could occur under the framework of research networks or even involve these mechanisms. This medium may also create a link between problems; for example, as in the area of health, research has been linked with the attention to health via various themed networks such as breast cancer, human papilloma virus, aging, neurodegenerative, and cardiovascular diseases.

The quality of the network is a function of the competencies of the people involved, in addition to the availability of modern infrastructure. The establishment of the mission should follow priority areas and themes, and exploit the strengths of the research, human resources, and technological on offer. It should include a game plan for finance from seed funds, concurrent and specific, to attend to areas not covered by other RFPs. The network should develop competencies that are congruent with the areas and themes defined, considered integral training programs. It should also encourage the appropriate participation of technical and support personnel. To ensure that the results can be transferred, it would be desirable to count on a network of extended services that give continuity to the generation and application of knowledge.

To illustrate these ideas, the INIFAP matrix of indicators could be a paradigm changer. That matrix emphasizes that to impact competitiveness, it is necessary for the institute to use productive processes, products, and developed services. Thus, it is essential to train professional service providers on the subject, together with technical service agencies; additionally, given the protection of property rights, to find commercialization with providers of inputs or representatives of agro-industries. Regarding aims or impact, long-term indicators of technology transfer, as well as net income from adopted technology (a benchmark set by 10 "successfully" transferred technologies) that is commercially available, while the purpose or direct results consider the technology adopted by a producer with respect to technology generated by the institute 3 years before. Concerning components or products, indicators relating to the training of technicians and professionals, technical publications, scientific publications, technology transferred, and technologies, are all validated in the field. Work and activities are measured by the skill level of personnel, dissemination of events, allocation to investments (salaries and wages, infrastructure, equipment, operational costs), maintenance, service provision, receiving donations, licensing of intellectual property, finance from competitive funds, and even savings.

To ensure that there is no delay in the development of support and stimulus for the dissemination of these initiatives, the following are also offered: finance for additional costs incurred in technology setup, assurance of risk of use, as well as appropriate use; that is, support for training and technical assistance. Moreover, given the economic, social, and cultural factors that influence the dissemination of information, it is necessary to have mechanisms of observation and documentation on the rate of adoption. Without doubt, some of the programs implemented since the inception of CONACYT have been valuable; however, they have been introduced in a temporary manner and do not have the required continuity.

Fifteen years ago, in parallel with the SED mechanism, it was established that public centers must assume both short- and long-term commitments to produce the desired results, provide criteria and indicators about their management and services, and reduce uncertainty surrounding project reports (finances and investment). Thus, the effects surrounding the planning capacity of the centers should be evaluated. Available evidence (from reports of various commissions and the conventions of management via results online) shows a heterogeneous learning pattern. The standout cases are the INIFAP and the Center of Investigation in Advanced Materials (CIMAV); the former shows integral management of the generation, application, and transfer of knowledge, and the latter shows solvency via diverse planning tools, such as a balanced scorecard for the implementation of their strategies. Solleiro and Escalante (2009) identify best practice regarding knowledge management, looking at a group of 14 university centers and a number of organizations with the CONACYT system (Advanced Technology Center, CIMAV, and Center for Research and Assistance in Technology and Design of the State of Jalisco).

8.3.6 External Evaluations

The findings and recommendations from academic studies on research centers (Díaz-Pérez 2011; Solleiro and Escalante 2009; Merritt 2006; Estrada-González 2009; Rubio 2009) present their organizational aspects, management, and performance; the majority of studies have focused on CONACYT centers. They identify design flaws in mission compliance and identify assorted challenges they face. For example, only a few S&T development centers are identified as having technology transfer in their mission. It seems that the norms and policies do not promote the generation of solutions, nor do they motivate technology and knowledge transfer. Thus, they will have to conduct a reference exercise to determine best practice to ensure compliance with missions.

With regard to the extent of linkage, this experience it is considered successful. In CONACYT centers (in the natural and exact sciences subsystems, as well as those of technological development), between one-half and two-thirds of all income comes from private sources. For example, technological development centers face a market where more than two-thirds of their clients are in the industrial sector and a similar number are SMEs that mostly seek consultancy, training, and routine and analytical services. Regarding research services, these are typically requested by businesses that lag behind others (i.e., that do not have formal engineering or research departments), as well as by those that have a formal structure in their R&D activities but demand complementary research. As expected, the latter find the centers' services (quality, value, relevance and opportunity) extremely useful, including their problem-solving capacity, equipment, intellectual capital (human capital, hiring, and information), and elements of static competitiveness (e.g., costs, location, and access).

However, not all linking activities are equally developed; for example, technology transfer. Only 0.3 % of income from services is derived from this activity (licensing of patents and utility models, sale of knowhow). The main obstacles identified are a lack of demand and commercial ability knowledge, small market, lack of public support programs, and barriers due to internal and external norms. Facing this situation, it is necessary to establish a specialized infrastructure that can offer transfer services to centers, such as the design of an intellectual property policy, consultancy and training, technology and market intelligence, negotiation of rights and licensing, R&D partner identification, commercialization, consulting for business plans for technology-based companies, and the coordination of actions with similar organizations. Within the matrix of results indicators for the budget program of technological development, knowledge transfer activities are indicated and assessed via projects that have the potential to be adopted through commercialization, collaboration, portfolio consolidation, promotion, and dissemination. However, when we aggregate the information, these essential activities are not capable of truly impacting on national and regional development and competitiveness.

8.4 Conclusions

Research centers are both a mechanism and a beneficiary of public policy. The numerous PRC evaluations mention impacts of different kinds (on the creation of ST&I capacities, on companies, return processes, and others). The main results of assessments from the last two 6-year terms are shown in multiple ways (e.g., Federal Budget of Expenses, the performance evaluation system, results-based administration agreements, the process for the betterment of public management, among others), all of which show a mid-level performance by our research centers. Regarding the federal public administration, it fulfills its planning and budgetary requirements, and its scope of goals, in a more or less satisfactory manner.

In recent years, it is evident that PRCs, as an instrument of ST&I policy, have received public resources for a wide range of purposes, from local problems, collaboration with regional entities, the shared use of equipment and infrastructure, personnel mobility, attention to human resource formation, and the search for external resources and operation under self-sufficiency or expansion of coverage. Based on this evidence, PRCs appear to have multiple goals.

The main problems stem from assuming what the impacts or direct results of PRC services may be, as well as the establishment of certain causes and effects of their activities on products and services, and strategic ends and purposes. Our analysis considers these planning exercises valuable because of the knowledge gained regarding how to measure public action, how to characterize public service and its results, as well as its potential impacts—where we perhaps need to create political, structural, and institutional conditions so that research results can have the expected social and economic impact.

Without doubt, investment in PRCs contributes to fill the knowledge and ability reservoir (i.e., knowledge capital). However, to have access to this kind of capital we need both society and businesses to improve their capacity and culture to assimilate and innovate in regard to demand. We also need centers to develop their capacity to relate and transfer knowledge such as the commercialization of research results. It is possible that even with these improvements, we will face articulation problems for which the involvement of intermediaries is necessary.

Regarding public policy recommendations, performance budgeting measures have been developed with a rationale to accomplish science policy (Salomon 1994), in the sense that they reflect the fostering of S&T and the exploitation of results for general policy objectives. Therefore, different administrative assessment mechanisms consider the productivity of a researcher in terms of S&T products (i.e., publications and citations, human resource formation, projects, and grants funds as intellectual property), as well as the management and policies of the center (e.g., timely spending of budget, enrollment, delivery of services, certification of educational programs and personnel, strategy, steering projects, industry and government contracts, population attended, and self-generation of resources).

Via training and education, PRCs contribute to the regional capacity to absorb technical knowledge, by dissemination activities that help improve the general understanding of S&T. Furthermore, PRCs produce intellectual property assets that support knowledge in a specific field of technology and accomplish generalized social diffusion through its publication in an industrial property gazette. Although information failures are addressed via these mechanisms, they are too weak to realistically challenge systemic failures such as population educational levels, the lack of private capital for technological development, general industrial strategies to compete with those non-technology-based, sparse government coordination, insufficient development of the market for industrial labor for scientists and technicians, and the lack of extended networks for supporting administrative and technical services to transfer knowledge through market or institutional procedures. Thus, the rationality to intervene still lags in the linear model of innovation, expecting that the diffusion and application of knowledge is an automatic process and resides in the good will of PRCs to steer research and adapt strategies to put knowledge on the market shelf. There is an expectation too that knowledge will limit the role of the state to arbitrate the accurate spending of resources, the efficient management of PRCs, and the competence of researchers to deliver products.

These policies will maintain S&T resources as development potential; however, a more aggressive intervention is required to exploit these resources and impact directly on regional development and population welfare. Two viable alternatives could establish the necessary environmental conditions to assure social diffusion and the economic use of knowledge. The first option is that the state can question a PRC's mission in the knowledge society and implement strategies to transform public research infrastructure into a service provider for the generation, application, and transfer of knowledge. The other alternative is to develop a platform of services to mobilize research results via networks of support services for knowledge transfer to industrial or societal applications.

Various methods to manage knowledge in PRCs and in public policy have been around for some time now in Mexico, countries belong to the Organisation for Economic Co-operation and Development, and indeed worldwide, representing nations with diverse levels of development. Such methods include public–private collaborations for financing, training or research activities, common infrastructures to exploit highly specialized facilities and equipment, emerging intermediaries to coordinate policies, agents and information flows, technology brokerage for knowledge transfer, be it independent, in collaboration, or even inside PRCs. We are eager to point out strategies to mobilize research results via available funding for follow-up projects (e.g., steering project to transfer project), professional training for updates in new methods and procedures, validation of technology through scaling up, demonstration facilities, training for specialized providers of support services, and the development of technology packaging knowhow. Regarding societal transfer, two aspects must be considered to develop protocol-drafting knowhow to put ensure new knowledge on health, energy, agriculture, and public services, and to employ several communication and exchange mechanisms to participate in public policy processes. Considering the whole cycle takes us from agenda formation (supplying diagnosis, information, discussion groups, formal debates, public hearings and think tanks) through to the policy-design stage (participating and assessing legislatures, councils, committees, boards and interorganizational networks to flag technical restraints of alternative actions and results), and from the implementation phase (offering technical assistance and training for managers and executives) to evaluation procedures (suggesting assessment criteria and methods).

Over the next few years we will need to strengthen the operation of new centers with further resources, as well as finding financial resources to create new ones. Mexico's expansion strategy should consider building new infrastructure, consolidating consortiums, developing local networks and public/private/social intermediaries, as well as private funding. We can consider the experience gained in the agriculture and health sectors, in addition to that of international foundations, as that of the joint work among countries, universities, and multinational businesses. Finally, there also exists the possibility of working on building virtual models (like those currently in place between Argentina and Brazil for nano- and biotechnology) and in taking advantage of installed capacity to incubate new HEI centers or businesses.

The social–budget analysis enables us to accept PRCs as an instrument to create S&T capabilities, but we must work on the mechanisms to ignite development based on those products obtained from their advanced services. PRCs are developing as a mechanism that enables the transfer of knowledge and competencies to respond to certain demand signals, but they are not yet able to deal with information flaws and the absorption capacity of agents in their relevant environment to directly influence the level of development.

References

ADIAT. (2009). *De la investigación aplicada a la innovación. Historia de la ADIAT* [*From applied research to innovation. History of ADIAT*]. México: Asociación Mexicana de Directivos de la Investigación Aplicada y el Desarrollo Tecnológico A.C. [Mexico City: Mexican Association of Directors of Applied Research and Technology Development AC].

Anselin, L., Varga, A., & Acs, Z. (2000). Geographic and sectoral characteristics of academic knowledge externalities. *Papers in Regional Science, 79*, 435–443.

Bilbao-Osorio, B., & Rodríguez-Pose, A. (2004). From R&D to innovation and economic growth in the EU. *Growth and Change, 35*, 434–455.

Braun, D. (1998). The role of funding agencies in the cognitive development of science. *Research Policy, 27*, 807–821.

Calderón-Martínez, M.G. (2013). *La gestión de patentes universitarias. El caso de la UNAM.* (Cuaderno de Trabajo; 11) [The management of university patents. The case of the UNAM. (Workbook 11)]. México: Dirección General de Evaluación Institucional-UNAM [Mexico City: UNAM General Directorate of Institutional Assessment].

Campos, M.A., & Jiménez, J. (1994). La actividad científico-tecnológica en México: un factor marginado en la sociedad mexicana [Scientific-technological activity in Mexico: a marginalized factor in Mexican society]. In R. Varela, & L. Mayer (Eds.), *Los grandes problemas de la ciencia y la tecnologia* [Major problems in science and technology] (pp. 171–186). México: UAM.

Campos-Ríos, G., & Sánchez-Daza, G. (2008). El desarrollo de la ciencia y la tecnología en el ámbito regional [The development of science and technology within a regional scope]. *Tecsistécatl. Revista Electrónica de Ciencias Sociales* [Electronic Journal of Social Sciences]. No. 3, http://www.eumed.net/rev/tecsistecatl/index.htm. Accessed 31 May 2014.

CONACYT. (2012). *Informe del estado de la ciencia y tecnología en México 2011* [*Mexico's 2011 report of the state of science and technology*]. México: Consejo Nacional de Ciencia y Tecnología [Mexico City: National Council for Science and Technology].

Cooke, P., & Leydesdorff, L. (2006). Regional development in the knowledge-based economy: The construction of advantage. *Journal of Technology Transfer, 31*, 5–15.

Corley, E. A. (2007). A use-and-transformation model for evaluating public R&D: Illustrations from polycystic ovarian syndrome (PCOS) research. *Evaluation and Program Planning, 30*, 21–35.

De Gortari, R. (1997). Los académicos: de la producción a la comercialización del conocimiento [Academic staff: From knowledge production to marketing]. In R. Casas & M. Luna (Eds.), *Gobierno, Academia y Empresas en México. Hacia una nueva configuración de relaciones* [Government, academy and business in Mexico. Towards a new array of relationships] (pp. 137–160). México: IIS-UNAM /Plaza y Valdés.

Debackere, K., & Veugelers, R. (2005). The role of academic technology transfer organizations in improving industry science links. *Research Policy, 34*(3), 321–342.

Díaz-Pérez, C. (2011). La calidad y certificación como estrategia de difusión en los centros de investigación y desarrollo tecnológico: repercusiones organizacionales y problemas pendientes [Quality and certification as dissemination strategies in research and technological development centers: Organizational outcomes and pending issues]. In C. Díaz-Pérez (Ed.), *El dilema de la innovación. Redes, clusters y capacidades tecnológicas* [The dilemma of innovation. Networks, clusters and technological capabilities] (pp.173–195). Zapopan: Centro Universitario de Ciencias Económico Administrativas. Universidad de Guadalajara [University Center for Economic and Administrative Sciences. University of Guadalajara].

Elzinga, A., & Jamison, A. (1995). Changing policy agendas in science and technology. In S. Jasanoff, G. E. Markle, J. C. Petersen, & T. J. Pinch (Eds.), *Handbook of Science and Technology Studies* (pp. 573–597). Thousand Oaks: Sage.

Estrada, S. (2006). Estudio 5. Valoración de los Fondos Mixtos [Study 5. Assessment of matching funds], en Dutrénit, G. (coord.). Diagnóstico de la Política Científica, Tecnológica y de Fomento a la Innovación en México (2000–2006) [Diagnosis of science, technology and innovation promotion policy in Mexico (2000–2006)] (pp. 201–224). Mexico City: Foro Consultivo Científico y Tecnológico, A.C. [Scientific and Technological Consultative Forum, A.C.] http://www.foroconsultivo.org.mx/libros_editados/diagnostico.pdf. Accessed 6 June 2014.

Estrada, S. (2012). *Las políticas de ciencia y tecnología en la agenda legislativa.* (Cuaderno de Trabajo; 17) [The politics of science and technology in the legislative agenda (Workbook 17)]. México: Dirección General de Estudios Legislativos: Política y Estado del Instituto Belisario Domínguez del Senado de la República. [Mexico City: General Directorate for Legislative Studies: Politics and State, Belisario Domínguez Senate Institute].

Estrada, S., & Pacheco-Vega, R. (2009). Sistemas y políticas de investigación, desarrollo e innovación. Algunas propuestas [Research, development and innovation systems and policy: Some proposals]. *Revista Espiral. Estudios sobre Estado y Sociedad* [Spiral Magazine. Studies on State and Society], 1 (44), 31–76.

Estrada-González, G. (2009). *Creación de OTT's como mecanismos de impulso de la transferencia de tecnología en los Centros de Investigación CONACYT* [Creating TTO's as mechanisms for boosting technology transfer in CONACYT Research Centers]. Tesis, Maestría en Política y Gestión del Cambio Tecnológico. Dirección Humberto Merritt Tapia y Adolfo Sánchez Aguilar [Thesis. Master of policy and management of technological change. Supervisors Humberto Merritt Tapia and Adolfo Sánchez Aguilar]. México: Centro de Investigaciones Administrativas, Económicas y Sociales. Instituto Politécnico Nacional [Mexico City: Management, Economics and Social Research Center. National Polytechnic Institute].

Feller, I. (2004). S&T-based development and university. *Economic Development Quarterly, 18*(2), 138–150.

Fischer, M. M., & Varga, A. (2003). Spatial knowledge spillovers and university research: Evidence from Austria. *The Annals of Regional Science, 37*(2), 303–322.

Georghiou, L., & Roesner, D. (2000). Evaluating technology programs: Tools and methods. *Research Policy, 29*, 657–678.

González-Block, M. A. (2009). Leadership, institution building and pay-back of health systems research in Mexico. *Health Research Policy and Systems*. doi:10.1186/1478-4505-7-22.

Graversen, E. K., Schmidt, E. K., & Langberg, K. (2005). Dynamic research environments: A development model. *The International Journal of Human Resource Management, 16*(8), 1498–1511.

Guerrero, O. (1989). *El Estado y la Administración Pública en México. Una Investigación sobre la Actividad del Estado Mexicano en Retrospección y Prospectiva* [The state and public administration in Mexico. Hindsight and foresight research on the activity of the Mexican State]. México: Instituto Nacional de Administración Pública [Mexico City: National Institute of Public Administration].

Guston, D. H. (2000). *Between politics and science. Assuring the integrity and productivity of research*. Cambridge: Cambridge University Press.

Gutiérrez Serrano, N.G. (2004). La Vinculación en el ámbito Científico-Tecnológico de México. Instituciones de Educación Superior en Interacción con Distintos Actores [Linkages in the science and technology arena of Mexico. Higher education institutions interactions with different actors]. *Revista Latinoamericana de Estudios Educativos* [Latin American Journal of Educational Studies], 34(2), 47–94.

Hewitt-Dundas, N., & Roper, S. (2011). Creating advantage in peripheral regions: The role of publicly funded R&D centres. *Research Policy, 40*, 832–841.

Jongbloed, B. (2007). *Reforming the Dutch academic research enterprise: universities' responses to project funding and performance monitoring*. Kassel: Consortium of Higher Education Researchers Conference.

Jongbloed, B. (2008). Creating public-private dynamics in higher education funding. A discussion of three options. In J. Enders & B. Jongbloed (Eds.), *Public-private dynamics in higher education funding* (pp. 113–138). Bielefeld: Transcript.

Larédo, P., & Mustar, P. (2001). *Research and Innovation policies in the new global economy. An international comparative analysis*. Cheltenham, Northampton: Edward Elgar.

Laudel, G. (2006). The art of getting funded: How scientists adapt to their funding conditions. *Science and Public Policy, 33*, 489–504.

LCyT. (2002). Ley de Ciencia y Tecnología. Diario Oficial de la Federación [Science and Technology Act. Federation's Official Journal], DOF 05-06-2002. México: Secretaría de Gobernación [Mexico City: Ministry of the Interior].

Lemola, T. (2002). Convergence of national science and technology policies: The case of Finland. *Research Policy, 31*, 1481–1490.

Lepori, B., van den Besselaar, P., Dinges, M., Potì, B., Reale, E., Slipersæter, S., et al. (2007). Comparing the evolution of national research policies: What patterns of change? *Science and Public Policy, 34*(6), 372–388.

Lundvall, B. A., Johnson, B., Andersen, E., & Dalum, B. (2002). National systems of production, innovation and competence building. *Research Policy, 31*(2), 213–231.

Merritt, H. (2006). La vinculación industria-centros tecnológicos de investigación y desarrollo: el caso de los centros CONACYT de México [Linking industry-research and development technological centers: the case of Mexico's CONACYT centers]. México: I Congreso Iberoamericano de Ciencia, Tecnología, Sociedad e Innovación [Mexico City: Iberoamerican Congress on Science, Technology, Society and Innovation].

Moore, M. C., Arent, D. J., & Norland, D. (2007). R&D advancement, technology diffusion, and impact on evaluation of public R&D. *Energy Policy, 35*, 1464–1473.

Nelson, R. (1991). Why do firms differ, and how does it matter? *Strategic Management Journal, Special Issue: Fundamental Research Issues in Strategy and Economics, 12*, 61–74.

Pavitt, K. (1998). The social shaping of the national science base. *Research Policy, 27*(8), 793–805.

Retana-Guiascón, O.G. (2009). La Institucionalización de la Investigación Científica en México. Breve Cronología [The institutionalization of scientific research in Mexico]. *Ciencias [Sciences]*, 94(2), 46–51.

Roper, S., & Love, J. H. (2006). Innovation and regional absorptive capacity: The labour market dimension. *Annals of Regional Science, 40*(2), 437–447.

Rubio Castillo, F.A. (2009) *La estructura organizacional en centros de investigación, desarrollo e innovación; una aproximación a la experiencia internacional* [The organizational structure in research, development and innovation centers; an approach to the international experience]. Tesis, Doctorado en Administración. Dirección Enrique Cabrero Mendoza. Santiago de Querétaro: Universidad Autónoma de Querétaro. Facultad de Contaduría y Administración [Thesis, PhD in management. Supervisor Enrique Cabrero Mendoza. Santiago de Querétaro: University of Queretaro. School of Accounting and Management].

Salomon, J.J. (1994). Modern science and technology. In J.J. Salomon, JF. Sagasti, C. Sachs-Jeantet (Eds.), *The uncertain quest: science, technology and development* (pp. 29–64). Tokyo: United Nations University Press. http://archive.unu.edu/unupress/unupbooks/uu09ue/uu09ue00.htm. Accessed 31 May 2014.

Sampere López, J.C, (2003). *Política regional de ciencia e innovación: El caso del Sistema de Investigación Miguel Hidalgo de CONACYT* [Regional policy on science and innovation: The case of CONACYT's Miguel Hidalgo Research System]. Tesis de Maestría en Estudios Regionales [Master thesis on regional studies]. México, DF: Instituto de Investigaciones Dr. José María Luís Mora [Dr. José María Luis Mora Research Institute].

Sanz Menéndez, L., & Cruz-Castro, L. (2003). Coping with environmental pressures: Public research organization responses to funding crises. *Research Policy, 32*(8), 1293–1308.

Senker, J., Balázs, K., Higgins, T., Larédo, P., Muñoz, E., Santesmases, M., et al. (1999). *European comparison of public research systems*. Final report. EC TSER project SOE1-CT98-1117. Brighton: SPRU, University of Sussex.

Slipersæter, S., Lepori, B., & Dinges, M. (2007). Between policy and science: Research councils' responsiveness in Austria, Norway and Switzerland. *Science and Public Policy, 34*(6), 401–415.

Solleiro, J.L., & Escalante, F. (2009). *Knowledge management and value creation in Latin American research centers*. Final report. IDRC project number 102497–003. México, Mexico City: CCADET-UNAM.

SSA. (2001). *Programa de acción: investigación en salud* [Action program: Health research]. México, DF: Secretaría de Salud [Ministry of Health].

SSA (2002). *Los institutos nacionales de salud de México* [The national institutes of health of Mexico]. México: Secretaría de Salud [Ministry of Health].

Unger, K. (1995). El desarrollo industrial y tecnológico mexicano: estado actual de la integración industrial y tecnológica [The Mexican industrial and technological development: Current state of the industrial and technological integration]. In P. Mulás del Pozo (Ed.), *Aspectos Tecnológicos de la Modernización Industrial de México [Technological aspects of the industrial modernization of Mexico]* (pp. 44–79). Mexico City: Academia.

Van der Meulen, B. (2003). New roles and strategies of a research council: Intermediation of the principal-agent relationship. *Science and Public Policy, 30*, 323–336.

Zabala, J.P. (2004). La Utilidad de los Conocimientos Científicos como Problemas Sociológicos [The advantage of scientific knowledge as sociological problems]. In P. Kreimer, H. Thomas, et al. *Producción y uso social de conocimientos. Estudios de sociología de la ciencia y la tecnología en América Latina* [Social production and use of knowledge. Studies in sociology of science and technology in Latin America] (pp.151–172). Buenos Aires: Universidad Nacional de Quilmes Editorial.

Zahra, S. A., & George, G. (2002). Absorptive capacity: A review, re-conceptualization, and extension. *Academy of Management Review, 27*, 185–203.

Chapter 9
Promoting Research Careers in the Business Sector: Assessing Public Policy Instruments in Brazil and Chile

Eliana A. Arancibia Gutiérrez and Fábio Couto e Silva Neto

9.1 Introduction

The promotion of technological innovation in the business sector in Latin America has been intensified in recent decades via the establishment of legal and institutional instruments, and many incentives to such end. The latest innovation policies have established indirect mechanisms, such as tax incentives,[1] to promote innovative activities in firms, while also promoting innovation programs that directly support firms' research and development (R&D) activities. The university–firm relation has also been an objective of several recent policies in the region.

In this context, one of the critical factors to foster the innovative capacities of industrial firms is the existence of human resources with the necessary skills to adapt, generate, and transform knowledge into new business opportunities. In Latin American countries, this critical mass is insufficient to foster the innovative capacities of firms. According to data from the Inter-American Science and Technology Indicators Network (RICyT) and the Brazilian Ministry of Science, Technology,

[1] In Brazil, the main laws marking this policy are the Innovation Law (10.973, 2004), which proposes the means to stimulate innovation, scientific, and technological research activities in a productive environment and the Wellness Law (11.986, 2005) that creates special tax incentive schemes for technological innovation, for example, the removal of certain tax requirements. Chile introduced the Tax Incentive Law for R&D (20.241, 2008), which allows private firms to reduce their tax obligations by up to 35 % of their R&D resources. This law was amended in 2013 to extend access to this benefit.

E.A. Arancibia Gutiérrez (✉)
National Authonomous University of Mexico, CEPHIS-UNAM, Merida, Mexico
e-mail: earanci@gmail.com; elianagutierrez@ige.unicamp.br

F. Couto e Silva Neto
Department of Science and Technology Policies, Unicamp and Renato Archer Information Technology Center (CTI Renato Archer/MCTI), Campinas, Brazil
e-mail: fabioi.chaves@gmail.com; fabioneto@ige.unicamp.br

H. Horta et al. (eds.), *Trends and Challenges in Science and Higher Education*, Knowledge Studies in Higher Education 3,
DOI 10.1007/978-3-319-20964-7_9

and Innovation (MCTI), 10–40 % of researchers in Latin American countries work in firms,[2] compared with 55–75 % in East Asian countries (OECD 2012).

Recently, several countries in Latin America sought to change this trend and designed specific programs to stimulate the hiring of researchers with postgraduate degrees by firms (e.g., Argentina, Brazil, Chile, Colombia, and Mexico). These programs were conceived within the legal framework of science and technology (S&T) policies, and are managed by national S&T agencies. In general, they offer subsidies to firms that incorporate researchers with master's and doctorate degrees that are capable of conducing innovation projects and to provide the creation of R&D departments within industrial firms.

This chapter analyzes these programs to attract researchers into the private business sector in Latin America, focusing on the Brazilian and Chilean experience because of their systematic character within the Latin American context. After this introduction, Sect. 9.2 describes the problem of the poor incorporation of researchers into the business sector of these countries, and how S&T policies have been implemented in Latin American to address this problem. Section 9.3 describes the main characteristics and operation mechanisms of these programs and Sect. 9.4 provides a comparative analysis between Brazilian and Chilean experiences, considering a set of indicators related to the demand for program resources, the industry, and the regional distribution of participating firms. The concluding remarks take into account the observed trends and a brief discussion on the main challenges these programs face.

9.2 Distribution of Researchers by Employment Sector: The Latin American Stylized Fact of the Problem

Several authors have highlighted that the innovative capacity of firms depends on, in a greater sense, the educational and professional level of its R&D personnel, as well as the number of employees dedicated to such activities. The reliance on a greater and well-trained number of employees with the tacit and technical skills to use, diffuse, and generate knowledge, has positive effects on the innovative performance of firms (Schoenecker et al. 1995; Souitaris 2002; Velho 2007). According to Velho (2007), there is a positive correlation between the number of postgraduate employees dedicated to industrial R&D activities and technological outputs, which is not limited to higher technology firms but spans the entire industrial sector. Velho

[2] The case of Mexico can be considered an outlier, given the increase in the rate of researchers employed by firms (from 19.7 % in 2000 to 45.3 % in 2005, and falling to 41.5 % in 2010). This rate surpassed even Spain (33.7 %), Portugal (22.9 %), and Brazil (25.9 %) (RICyT 2013). Dutrénit and Arza (2010, p. 111) stresses that despite such a phenomenon, this has not been reflected in an increase in innovative capacities in the industrial sector; the redistribution of researchers represents one of the most remarkable structural changes in the Mexican innovation system profile in recent years.

(2007) also states that for firms to innovate, firms must employee human resources with postgraduate degrees. Without them, firms limit their internal capacity to search for external innovative solutions to their problems, and will rarely be able to create knowledge-based innovations.

In Latin America, however, human resources with postgraduate degrees notably concentrate in the academic sector. This situation inhibits its potential contributions for innovative activities in other sectors, especially the business sector (see Fig. 9.1). Viotti (2010) stresses this situation, showing that 76 % of Brazilians with doctorate degrees obtained between 1996 and 2006 and employed in 2008 worked in universities, colleges, and research centers. In contrast, only 2.4 % worked in the business sector. Similarly, Luchillo (2009) analyzed the employment paths of former scholarship students (graduating between 1997 and 2006) from the Mexican National S&T Council (CONACyT). His study revealed that only 3.5 % of former doctorate scholarship recipients from national institutions, and 5.1 % of those from foreign institutions were employed in the business sector. Among master's degrees, this proportion was higher, at 22.2 %.[3]

The employment sector distribution of these researchers is even more pertinent if one considers the steady expansion of postgraduates in Latin America. Despite the country differences of this phenomenon, the number of postgraduate programs and

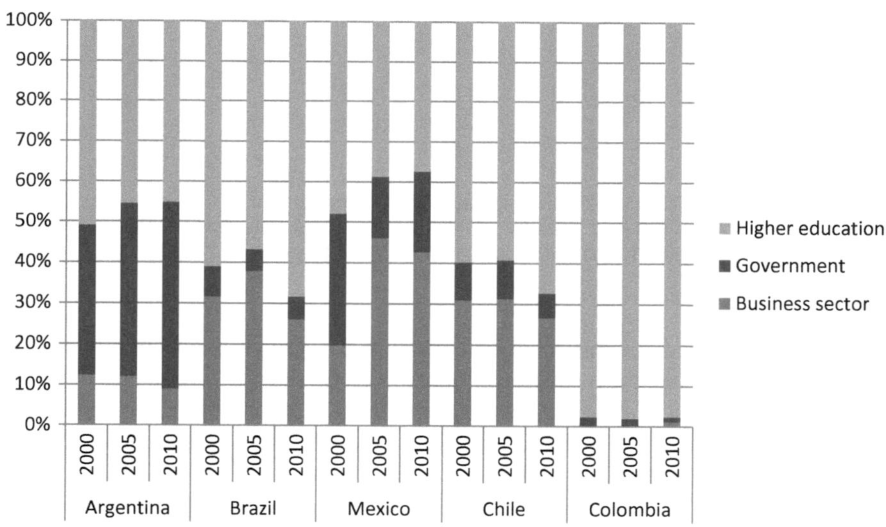

Fig. 9.1 Distribution of researchers by employment sector in selected Latin American Countries, selected years (%). (Source: Authors' elaboration based on data from RICyT (2013) and the Brazilian MCTI (2013)) Note: Chilean data for 2000 and 2005 are estimated according RICyT data for 2007–2010

[3] Luchillo (2009) points out that these data present a significant correlation between master's degree holders and their employment in the business sector in Mexico. This could explain the increase of the proportion of researchers employed in the business sector in this country during the last decade.

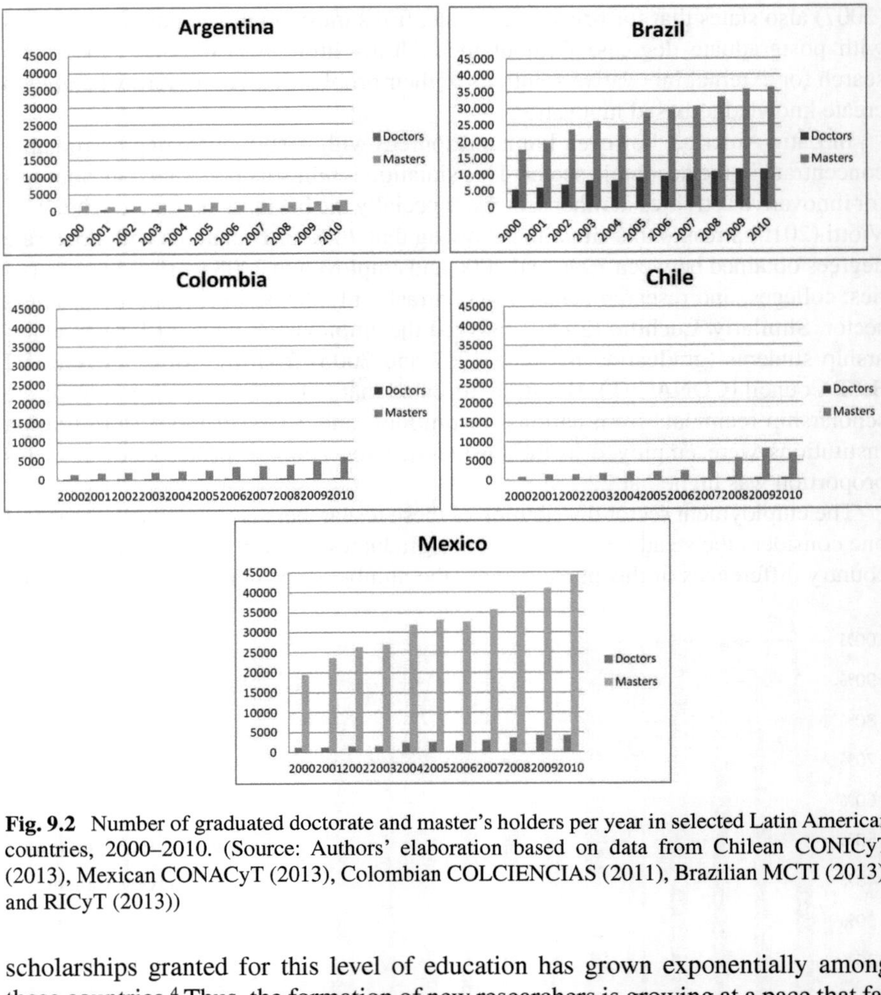

Fig. 9.2 Number of graduated doctorate and master's holders per year in selected Latin American countries, 2000–2010. (Source: Authors' elaboration based on data from Chilean CONICyT (2013), Mexican CONACyT (2013), Colombian COLCIENCIAS (2011), Brazilian MCTI (2013), and RICyT (2013))

scholarships granted for this level of education has grown exponentially among these countries.[4] Thus, the formation of new researchers is growing at a pace that far exceeds the creation of new jobs in the higher education and research sectors (see Fig. 9.2). If such trends persist, the public investment in postgraduate human resources will hardly produce greater innovative dynamics in the region.

[4]To illustrate this phenomenon, in 2000, 1440 postgraduate programs in Brazil (master's and doctorate) granted by CAPES (Brazilian Council For The Improvement of Third Level Education Programs). In 2010, this number increased to 2840. However, in 2000, CAPES granted 20,470 scholarships for master's and doctorate degrees; this number more than doubled in 2010 to 54,848 (CAPES 2013). In Chile in 2004, there were 394 master's degrees and 115 doctorate programs offered by the Universities of the Rectors' Council. By 2012, these numbers increased to 610 and 171, respectively (Chilean National Educational Council 2012). Similarly for this country, the number of doctorate scholarship recipients under CONICyT (national and foreign) increased from 429 in 2006 to 816 in 2012 (Estadísticas Becas, CONICyT 2013).

In an effort to solve this problem, several Latin American countries have designed programs to attract researchers into business-sector firms (see Frame 9.1). Argentina introduced the "Researchers in firms" program in 2003, and Mexico ran its "Scientific and technological inclusion of workers into Mexican social and productive sectors" program between 2005 and 2008. This program was replaced by the publication of state requests to attract master's and doctorate holders into industry, in association with regional state governments interested in the initiative. In 2011, Colombia formalized the "Placement of Colombian and foreign doctors in Colombian firms" program with two initial calls for applications in 2008 and 2010. Within this regional context, the Brazilian program "Human Resources in Strategic Areas: Researchers in firms" (RHAE) and the Chilean program "Attraction and Placement of Advanced Human Capital: researchers in the productive sector" (PAI) represent the most developed initiatives to address the problem. These programs are managed by the Brazilian National S&T Development Council (CNPq) and the Chilean National S&T Research Committee (CONICyT), respectively. The history and operational mechanisms of these programs are described in the following section.

Frame 9.1: Main Programs to Attract Researchers into the Business Sector in Latin America

Country	Year of operation	Program	Manager institution
Argentina	2003	Researcher in firm modality	National S&T Research Council (CONICeT)
Brazil	1987 (first version); 2007 (current version)	Human resources in strategic areas: researchers in firms	National S&T Development Council (CNPq)
Chile	2008	Attraction and placement of advanced human capital	National S&T Research Committee (CONICyT)
Colombia	2011 (two pilot calls in 2008 and 2010)	Placement of Colombian and foreign doctors in Colombian firms	Administrative Department of S&T and innovation (COLCIENCIAS)
Mexico	2005–2008 (State calls from 2011)	Scientific and technological inclusion of workers into Mexican social and productive sectors (IDEA) State calls for the inclusion of master's and doctorate holders in firms—regional state governments of: Nuevo León, Yucatán, Morelos, Hidalgo, among others	National S&T Council (CONACyT)

Source: Authors' elaboration based on Argentinean CONICeT (2013), Chilean CONICyT (2014), Mexican Conacyt (2013), Colombian COLCIENCIAS (2011), and Brazilian CGEE (2011).

9.3 Incentives to Attract Researchers into Firms: Brazilian and Chilean Experiences

9.3.1 Brazil: RHAE Program—Researchers in Firms

The RHAE program was the first program in Latin America with the objective to subsidize the training of human resources for R&D activities in strategic industries, as well as the placement of researchers into private firms. This initiative began in 1987, and currently exists as an MCT program, managed by CNPq. Since its creation, the program has been amended three times, the first in 1997, when the program was oriented towards the training of researchers for R&D activities in micro, small, and medium-sized enterprises (MSMEs). Later, in 2002, the program was changed again, this time to include the training of researchers for R&D in firms in strategic industries, in line with those included in national industrial, technological, and foreign trade policies (PITCE). The program was renamed "RHAE—Innovation" and four calls for applications were published before 2006.

In 2007, the agency overseeing the program again redesigned the program to encourage the hiring of master's and doctorate researchers by MSMEs. In this current version, the program aim is to encourage the appointment of highly skilled personnel to conduct R&D activities in firms, and to strengthen the linkages between industry and higher education institutes. Firms interested joining this program must present an R&D project intended to boost the firm's innovative activities. Furthermore, firms must also create a research team comprising researchers with master's and doctorate degrees, who can be accompanied by graduate and technological graduate students, visitor researchers, consultants, and other technical staff. To these personnel, the RHAE program offers scholarships of different values according to recipients' educational levels. The duration of these scholarships is between 2 and 3 years, according to the nature and maturity of the project.

Frame 9.2: Summary of "RHAE: Researchers into Firms" Calls for Applications Between 2007 and 2012

2007
Objectives: Support research and technological innovative activities via the placement of master's and doctorate holders in firms. The projects must comprise the priority industries from PITCE. **Priority industries**: Those assigned by PITCE, in addition to semiconductors, software, pharmaceuticals, capital goods, biotechnology, nanotechnology, alternative power sources, biofuel, nuclear energy, aeronautics, and aerospace. **Eligible firms**: SMEs, according to the following range: a) microfirms (up to 19 employees); b) small firms (between 20 and 99 employees); c) medium firms (between 100 and 499 employees). At least 30 % of the available resources must be directed towards firms located in north, northeast and center-west regions.

(continued)

Frame 9.2 (continued)

2007

Project requirements: Projects of technological development or product or process innovations regarding the priority industries mentioned previously. The project coordinator must be registered with the candidate firm. Maximum duration of the project is 30 months and with a maximum expenditure of USD 157,000. Firms must also have material conditions for the project execution, using own or third-party resources.
Benefits: The scholarships granted through industry funds cover a period of 24 months (doctorate holders: up to USD 2,356.00; master's holders: up to USD 1,728.00). The firms must contribute to 20 % of the project costs, which can be assigned to operational costs (including salaries) and equipment.
Evaluation criteria:

- Degree of innovation and technological impact of the project
- Relevance for the priority industries defined in the call for applications
- Adequacy of the cooperative arrangements of the project development
- Adequacy of the **counterpart** contribution in quantitative and qualitative terms
- Technical, economical, and market feasibility of the project
- Profiles of the team and requested scholarships

2008 and 2009

Objectives: As in 2007, but now aim to encourage R&D activities in the firms' production processes, according to the "Productive Development Plan" (PDP) and "ST&I Action Plan"
Priority Industries: In addition to the 2007 industries, includes industries to increase competitive edge (automotive, naval, textile, and furniture) and to consolidate and expand leadership (bioethanol, oil and gas, chemistry, and mining).
Eligible Firms: MSMEs, according to the following range: microfirms (gross revenue less than USD 126,650), small firms (gross revenue between USD 126,650, and USD 1.26 million), medium-sized firms (gross revenue over USD 1.26 million).
Requirements and benefits: As in 2007
Evaluation Criteria:

- Degree of innovation and technological impact
- Relevance to the priority industries defined in the call for applications
- Adequacy of the cooperative arrangements of the project development
- Technical, economical, and market feasibility
- Profile of the team and requested scholarships

2010

Objectives, priority industries, requirements, benefits, evaluation criteria: As in 2008–2009
Eligible firms: Medium-sized firms must now present a gross revenue between USD 1.26 million and USD 47.12 million. For the remaining MSMEs, the same criteria apply as in 2008 and 2009.

2012

Objectives, priority industries, requirements, benefits, evaluation criteria: As in 2008–2010
Eligible firms: The limits of required gross revenues increase for MSMEs—near 50 %. In this round, the participation of large firms was allowed, with minimum gross revenue of USD 47.12 million. A share of up to 20 % of the available resources can be directed to such firms.

Source: Authors' elaboration, based on RHAE CNPq call for applications between 2007 and 2012 (CNPq 2007, 2008b, 2009, 2010, 2012).

Thus, RHAE provides firms the advantage of relying on highly educated researchers, supported by the government via scholarships. The development or improvement of a product or process, and the possibility to absorb human resources to work on R&D projects summarizes the main objectives of this program (CNPq 2007, 2008b, 2009, 2010, 2012). As Frame 9.1 presents, the program has had successive modifications through its five calls for applications up to the publication of this book.

9.3.2 The Chilean Experience: Attraction and Placement of Advanced Human Capital into the Productive Sector Program

The PAI program was created in 2009 and is currently managed by CONICyT, working through three different modes: "Attraction of foreign researchers", "Placement of researchers into the academic sector" and "Placement of researchers into the productive sector". This section stresses the latter mode, oriented specifically to strengthen the R&D capacities of Chilean industries via the attraction of doctorate holders who can develop innovation projects. This mode, the "placement of researchers into the productive sector" works through annual calls for applications, in which Chilean firms or the Chilean subsidiaries of foreign firms and national research centers[5] are eligible to request financial resources from the program. These firms (or research centers) must develop an R&D project related to their business that presents a significant impact on the capacities of the Chilean productive sector. These projects must involve up to two researchers with a doctorate degree in such R&D activities. The program also offers a scheme to fund the salaries of researchers during the project term (80 % in the first year, 50 % in the second, and 30 % in the third), and a scholarship for a postgraduate student to develop his/her thesis (of related nature to the project). Furthermore, the program offers two additional grants to cover the project operational costs (up to USD 6000 per year) and to support researcher participation at a technical conference (in Chile or overseas) in line with the project objectives (USD 3600).

Frame 9.3 summarizes the path of the PAI program from 2009 to 2012, in which the main objectives of the four published calls for applications remained unchanged. However, there were several modifications intended to relax some requirements for the participation of firms and to renew the grant schemes, adjusting the benefits for participant firms and researchers. Thus, the two first calls for applications (2008 and 2009) focused on priority industries, a requirement that was omitted in subsequent rounds. From 2010, the program extended the possibility for recently established

[5]An established legal entity, at the time of its application, related and led by one or more firms, holdings, entrepreneurs, or a group of entrepreneurs (e.g., professional associations and cooperatives), with its main activities being in R&D and being able to effectively demonstrate technological transfer capacity to the productive sector.

firms (up to 2 years old) directly related to R&D activities to present their R&D projects. New spin-off firms from academic and technological institutes between 6 months and 2 years old are also eligible. The program also allows the possibility to include a scholarship for a postgraduate student with a thesis related to the project and for collaboration in research activities. Finally, the gross annual salary of the researcher was decreased in 2010, ranging between USD 36,000 and 45,600.

Frame 9.3: Summary of PAI Program Calls for Applications Between 2009 and 2012

2009

Objective: Finance the participation of researchers in R&D projects that generate impacts in Chilean priority industries.
Priority industries: Economic (mining, aquaculture, food, special interests tourism, and global services); Public interest platforms (Energy, biofuel, environment, ICT, and biotechnology); Social (education, health, dwelling, public safety and public policies).
Eligible firms: Chilean firms and foreign firms with headquarters in Chile and national technological centers that intend to develop R&D projects or continue an existent R&D line.
Project requirements: a) Seize an opportunity or solve a business or productive sector problem in any of the priority industries; b) Involve the participation of a doctorate holder or technologist with the skills required for project development; c) Provide a **placement** for the researcher before, during, and after the project duration; d) Grant researchers a gross annual salary of USD 42,000; e) Duration between 1 and 3 years.
Benefits: The salary is co-funded by a government subsidy that corresponds to up to 80 %, 50 %, and 30 % of the minimum salary required for the researcher during the first, second, and third year, respectively. This is independent of the duration of the project. The following grants are offered: up to USD 6000 annually for the operational costs of the project and up to USD 3600 to fund the researcher to attend a conference related to the project (in Chile or overseas).
Evaluation Criteria:
- Demonstration of the firm's (or technological center) commitment to project success (15 %)
- Demonstration of the innovative capacity of the R&D project (30 %)
- Previous references or recommendations relating to the researcher (30 %)
- Quality of the benefits obtained by the firm (or technological center) and researcher with the **project** (25 %)

2010

Same objectives and priority industries as in 2009.
Eligible firms: As in 2009, with the inclusion of recently established firms in the R&D business (1 to 2 years old) and spin-off firms (between 6 months and 2 years old).
Project Requirements: As in 2009 but also consider the possibility to include a master's or doctorate student working on a thesis related to the R&D project. The age limit for researchers is 35 years old. The firm must ensure that the researcher receives a gross annual salary between USD 36,000 and 45,600.
Benefits: An additional subsidy of up to USD 18,000 for the inclusion of a postgraduate student. The cofounding scheme remains, as well as the other subsidies.
Evaluation criteria: The weight of researcher references or recommendations is reduced to 25 %. A new element is appended: Prior technological transfer that ensures the deployment of the obtained R&D results (5 %).

(continued)

Frame 9.3 (continued)

2011

Objectives: As in 2009 and 2010, with an emphasis on the fostering of R&D activities to support the social and economic development of Chile. In addition, there are no longer priority industries.
Eligible firms: As in 2010.
Project requirements: The participant researchers must now have a doctorate degree. The projects minimum duration is extended to 2–3 years, and the age limit regarding researchers has been removed.
Benefits: The gross annual salary range is increased, and must be between USD 36,000 and 54,600. The subsidy for conference costs for researchers is increased to USD 4000, and this benefit is also granted to postgraduate students at USD 3000.
Evaluation criteria: As in 2010. A bonus is granted for those projects that include a postgraduate student.

2012

Same objectives as in 2011; no priority industries defined. However, a new feature is included: Consultancy for the installation of new R&D capacities oriented to MSMEs. This is an additional support to foster the presentation of projects, providing consultancy services for presentation preparation and further applications for funding.
Eligible firms: As in 2011.
Project requirements: As in 2011.
Benefits: Same as in 2011 for the case of researcher attraction projects. Regarding project preparation, a subsidy of 80 % of the 4-month consultancy period is granted.
Evaluation criteria: As in 2009, with equal weights for each criterion (25 %).

Source: Authors' elaboration based on PAI calls published by CONICyT (2009, 2010, 2011, 2012b)

In 2011, the program was open to R&D projects (at least 2 years long) from all industries, with no priority industries. In this round, participant researchers must have doctorate degrees and the age limit was removed. Similarly, the gross annual salary range granted to researchers was extended to between USD 36,000 and 54,000. Within the evaluation criteria, a bonus is offered to projects that include postgraduate students, to encourage the participation of such students.

The 2012 call for applications includes a new feature: "Consultancy for the installation of new R&D capacities," which is directed to MSMEs lacking the expertise or the qualified personnel to prepare a R&D project. The program offers a consultancy service for up to 4 months to prepare an R&D project, which is expected to be ready for submission at the end of that period. It consists of an additional support to encourage the presentation of R&D projects, providing a previous orientation for its suitable design and preparation.

9.4 Comparative Analysis of Brazilian RHAE and Chilean PAI Programs

9.4.1 Demand for Program Resources

Observing the paths of both "RHAE: researchers in firms" and "PAI: Placement of researchers in the productive sector", it can be noted that both programs have experienced successive amendments to improve their design and fully meet their objectives. In addition, Brazil has made a significant and increasing investment in the RHAE program, which saw its resources triple between 2007 and 2010 (see Table 9.1). Chile, in turn, has shown a constant investment in the PAI program, and the contributions of the firms have followed government investments (see Table 9.2).

Considering the Brazilian case in detail, data from the 2011 Technological Innovation Survey (IBGE 2013) shows that of the 15,696 surveyed innovative firms that received public resources for innovation purposes, 74.9 % sought funds for equipment and machinery acquisition. Regarding those firms that benefited from RHAE resources in the two first funding rounds (a total of 1315), they represent only 8.4 % of all applications for public resources to foster innovation activities. Thus, the demand of firms for RHAE resources remains low.[6] It is hoped that the steady increase of RHAE resources will to reduce the excessive importance assigned to physical capital, at the expense of human capital (i.e., postgraduate human resources dedicated to innovation activities).

Nevertheless, it should be taken in account that the number of firms that compete for RHAE funding has significantly increased in the recent years, especially between 2008 and 2009, when the number of applicants increased by 46.9 %. However, the project approval rate (see Table 9.1) has not increased during the five published calls for applications, standing at just 18 %. It could be assumed from these data that applicant firms are not designing and preparing their projects according to the program requirements, and therefore, the mechanisms of the calls for applications should be improved.

In the case of Chile, despite the modifications and gradual adjustments taken to relax the requirements and facilitate the submission of projects, PAI did not succeeded to attract a significant demand between 2009 and 2012 (see Table 9.2). The data from the eighth Innovation Survey (Chilean Government 2014) shows that there are approximately 38,700 innovative firms in Chile.[7] From this population, 21 %, or 8,100 engage in R&D activities (in-house or externally), and 6 % (or 2300) rely on a formal R&D department. Firms of this kind are eligible to submit R&D

[6] It is important to note that this picture refers to the latest data available from the Technological Innovation Survey 2009–2011, and may have changed in recent years. However, this is valid to stress the problem in the context of the demand for public resources for innovation.

[7] This measurement is based on a representative sample of 144,141 firms at a national level. According to the survey, 26.9 % of the sample firms have undertaken innovation activities.

Table 9.1 RHAE program numbers: projects applied, projects approved, approval rate, participant firms, scholarships granted, amount invested by the government and by the firms, 2007–2012

Year (Call)	Number of projects submitted	Number of projects approved	Approval rate (%)	Number of participant firms	Number of scholarships granted by education level	Amount invested by government in USD millions	Amount invested by firms in USD millions[a]
2007	710	131	18.4	124	*385* (93 doctorate holders, 112 master's, 96 graduates, and 84 graduate students)	10.5	2.1
2008	727	173	23.7	166	*690* (130 doctorate holders, 182 master's, 155 graduates, and 223 graduate students)	13.6	2.7
2009	1068	188	17.6	177	*621* (112 doctorate holders, 166 master's, 180 graduates, 152 graduate students, and 11 experts)	15.7	3.1
2010	1125	211	18.7	209	*694* (117 doctorate holders, 198 master's, 189 graduates, 113 graduate students, 23 experts, and 54 technicians)	20.9	4.1
2012	1558	265	17.0	247	*932* (125 doctorate holders, 255 master's, 280 graduates, 178 graduate students, 28 experts, and 66 technicians)	31.4	6.2
Totals	*5188*	*968*	*18.6*	*923*	*3322* (577 doctorate holders, 913 master's, 900 graduates, 750 graduate students, 62 experts, and 120 technicians)	*92.1*	*18.2*

Source: Authors' elaboration based on data from CNPq (2013b)
Note: USD exchange rate from 2012: R$1.91
[a]The amount was estimated over the 20 % counterpart contributions of the projects

Table 9.2 PAI program numbers: project submissions, projects approved, approval rate, scholarships granted and government and firm invested amounts (2009–2012)

Year (call)	Number of projects submitted (equals number of firms)	Number of projects approved (equals number of participant firms)	Approval rate (%)	Number of scholarships granted by education level	Amount invested by government in USD millions 2012	Amount invested by firms in USD millions 2012
2009	15	09	60	*16* (7 doctorate holders, 2 master's, 7 graduated technicians)	0.75	1
2010	26	14	54	*28* (17 doctorate holders, 11 master's)	1.2	1.5
2011	14	09	60	*17* (doctorate holders)	1	0.82
2012	35	19	54.2	*35* (doctorate holders)	1.4	1.4
Totals	*90*	*51*	56.6	*96*	*4.35*	*4.72*

Source: Authors' elaboration based on data from CONICyT (2012a)

projects in this mode of PAI program, and could attract the approximately 500 doctorate graduates that Chile produces each year. However, this is not occurring.

As also noted in the Brazilian experience, innovative firms in Chile confer greater importance to the acquisition of equipment and machinery than to the attraction of qualified human resources for innovation activities. The aforementioned Innovation Survey stresses that the main innovation activities undertaken by the firms is the acquisition of equipment, machinery, and software (53 % of the innovative firms in 2012). For these firms, their strategy is the absorption of embodied knowledge via physical capital, as the knowledge embodied in personnel plays a less significant role.

9.4.2 Distribution of Participant Firms by Industry

Regarding the industries of the participant firms, since the inception of both the Brazilian and Chilean programs, they have intended to match the various priority industries with those from national industrial and S&T policies. Furthermore, and in an effort to increase the demand of firms for their public resources, both programs extended the number of applicable industries, and in the Chilean case, the 2011 round removed all industry constraints. Interestingly, the paths and the results obtained in this investigation are very similar in both cases.

In the Brazilian case, in the beginning, the RHAE program defined the strategic industries based on PITCE and PDP (see Frame 9.2). Therefore, for the 2007–2010 period, the industry distribution of participant firms mainly concentrated in ICT, biotechnology, and pharmaceuticals industries, as shown in Table 9.3. This trend does not change with the increase of RHAE resources, even after the extension of priority industries in the 2008 round.

Regarding the PAI program, the two first calls for applications (2009–2010) defined a set of priority industries (see Frame 9.2). In the third round in 2011, these priority industries were withdrawn, given the low demand for the program resources in general. Despite the opening up of the program to all the industries, demand decreased in 2011: only 14 firms presented R&D projects. In the 2012 round this trend changed, and the number of firms that submitted projects increased to 35, and 19 were approved. The number of doctorate holders awarded with a scholarship doubled, reaching a total of 35. Despite this increase, as previously mentioned, the number of firms that applied for PAI funding is still low.

Again, as observed with the RHAE program, the approved projects are still concentrated in the previously defined priority industries, and even the withdrawal of these priority industries in 2011 did not work to increase the number of participant firms from other industries. As shown in Table 9.3, approved projects, up to 2012, are concentrated in aquaculture (20 %), mining (18 %), ICT (14 %), and health (12 %) industries.

9.4.3 Regional Distribution of Participant Firms

The design of both the Brazilian and Chilean programs shows an effort to decentralize the distribution of resources, reflected in special mechanisms to ensure a balanced regional distribution. Regarding RHAE, it was always intended that at least

Table 9.3 Distribution of participant firms by industry: Brazilian RHAE and Chilean PAI programs in selected years

RHAE program 2007–2010	PAI program 2009–2012
ICT (29 %)	Aquaculture (20 %)
Biotechnology (21 %)	Mining (18 %)
Pharmaceuticals (10 %)	ICT (14 %)
Capital goods (8 %)	Health (12 %)
Nanotechnology (8 %)	Food (6 %)
Oil and gas (5 %)	Energy (6 %)
Other industries (19 %)[a]	Biotechnology (4 %)
	Engineering (4 %)
	Other industries[b] (16 %)

Sources: Authors' elaboration based on data from CNPq (2008a, 2013c), CGEE (2011), and CONICyT (2012a)

[a]The following industries were considered: biofuel, aeronautics and aerospace, agroindustry, nuclear energy, food, services, personal hygiene, and plastics

[b]The firms of the following industries presented only one project: environment, management, education, biofuel, structural engineering, agriculture, viniculture, and fuels

Table 9.4 Regional distribution of participant firms in RHAE and PAI programs in selected periods

RHAE program 2007–2010	PAI program 2009–2012
Southeast region (57 %)	Santiago metropolitan region (64 %)
Southern region (25 %)	South region (24 %)
Northeast region (13 %)	North region (6 %)
North and center-west regions (5 %)	Central region (6 %)

Source: Authors' elaboration based on data from CNPq (2008a, 2013c), CGEE (2011), and CONICyT (2012a)

30 % of the resources of each funding round should be assigned to firms from north, northeast and center-west regions, establishing partnerships with state S&T and innovation fostering agencies. Thus, 11 state calls for applications were published between 2007 and 2010, focusing on industries that presented leadership in these regions, such as agriculture in the northeast and center-west regions. However, given the low demand obtained by these calls for applications, the initiative was discontinued in 2009.

Table 9.4 shows that the RHAE program has a strong concentration of participant firms in the southeast and southern regions of Brazil. Furthermore, it was not even possible to comply with the condition to allocate 30 % of the resources to firms in those regions, given the low demand aforementioned.

Regarding the PAI program, the decentralizing mechanism used was the publication of five regional calls for applications to place researchers in firms located in specific regions, according strategic industries, with resources from the "Innovation Fund for Competitive Edge." The results were disappointing: the first three rounds received no submissions and the two others received one project submission and it was approved (from the Magellan, southern, and central O'Higgins regions).

Therefore, despite the existence of greater innovative dynamics in regions located outside central Chile, as shown in data from the Chilean seventh Innovation Survey,[8] there is a clear trend that participant firms tend to be from the Santiago metropolitan area, the most developed region industrially and with greater scientific, technological, and institutional capacities.

9.5 Conclusions

The attempts of Chile and Brazil to boost the demand for highly qualified human resources in strategic industries via successive design and operational modifications have shown a degree of convergence in their respective programs. Although both countries show an increase in the demand for government resources for the hiring of

[8] The southern region of Araucanía presents the highest innovation rate in the country: 30 % of firms conducted some type of innovation project between 2009 and 2010. The Magellan region, in Chilean Patagonia, comes in second with an innovation rate of 26.6 %, followed by the northern Antofagasta region with 26.2 %.

researchers by firms, this growth is occurring at a slow and reactive pace, given the programs modifications. One of the main challenges of both RHAE and PAI is to foster business-sector interest in researchers.

Similarly, some adjustments are needed in both programs to reinforce its possible contributions to generate innovational capacities for firms at the local level and in regional specialization via knowledge-based activities. To date, the distribution of government resources for both programs is highly concentrated in regions where economic dynamics are more intense because of existing S&T structures. The present conditions in these regions already favor a higher incorporation of qualified human resources for innovative activities in firms. This effort clearly transcends the scope of these programs and must involve other policies that encourage the local training of human resources related to the strategic industries of regional economies.

One central issue is to determine whether the trend of a lack of researchers in firms—which relates to the weak emphasis placed on human resources to boost firms' innovative capacity—is actually reversing. For this result to be measured, it is necessary to verify how many researchers are being effectively incorporated into firms. The preliminary results obtained thus far cannot be used to draw any useful conclusions. An evaluation of the RHAE program conducted by CNPq (2013a) and regarding to the first three rounds of the program revealed that an average of 42 doctorate and 64 master's holders were hired per call for applications. Taking into account the average participation in the program for this period of 112 doctorate and 153 master's holders, the uptake rate is 37.5 and 41.8 %, respectively. For the PAI program, there are currently no evaluations available on this issue (2014 data are not yet available).

Another important aspect is whether these programs have contributed to increase the linkages between universities (or public research centers) and the productive sector of these countries; this has been an objective of both programs. Data from the Brazilian Technological Innovation Survey (IBGE 2011, 2013) indicates that the proportion of innovative firms that have cooperated with other institutions increased from 35.7 % to 38.7 % (an increase of 8.3 %) between 2006 and 2011. However, one can hardly tell whether this increase is due to the use of RHAE resources, given that the survey also indicates that, in this period, the proportion of innovative firms that received any kind of government support and might have requested scholarships[9] decreased from 32.4 % to 23.2 % (a 28 % decrease). For Chile, nevertheless, the seventh and eighth Innovation Surveys (Chilean Government 2012, 2014) suggest an increase of the proportion of innovative firms cooperating with universities and research centers: from 23.6 % to 38.4 % (a 47.4 % increase) for universities and from 10.2 % to 16.3 % (a 59.8 % increase) for research centers. In addition, the proportion of innovative firms that received any kind of public funds for innovation increased from 18.7 % to 28.8 % (a 54 % increase). Thus, it is possible that some part (even a small one) of this increase in collaboration between firms and universi-

[9] The data refer to "other forms of support," among which are RHAE program scholarships (IBGE 2013).

ties and research centers is due to PAI program resources. Thus, the contribution of these programs to stimulate linkages between academia and productive-sector firms varies between the countries.

There are two other aspects that deserve special attention and, although they were not considered in the objectives of this chapter, are important to improve the analysis of programs of such nature. The first is the need to identify what kind of innovation activities require postgraduate personnel, especially if they are really involved in research activities and participating in new-knowledge generation with business potential. In addition, these programs need to extend the information regarding their eventual contribution and involvement in the creation of formal R&D departments within the firms. The second aspect to be considered for further analysis is the efficacy of these incentives and the details of the employment contracts. In both studied cases, the programs do not require a formal commitment from firms to retain these researchers once the R&D projects are finished. This could be a barrier the placement of human resources into the business sector and discourage longer-term innovative activities. However, the linkages between universities (or public research centers) and firms in Latin America are typically short-term and oriented towards the production needs of firms. Firms though will assign greater importance to their relations with universities and research centers over matters concerning innovation activities (Dutrénit and Arza 2010). Thus, the demand for researchers in firms tends to be of a short-term nature, where project researchers do not become permanent staff members. If this is indeed the case, a policy that comprises the absorption of qualified researchers into firms must also arouse the interest of these firms in solving technological problems of an innovative character.

Finally, the experiences studied in this chapter also point to the need to establish new policy agendas for the training of human resources in higher education. This must be in line with the productive and technical-scientific reality, a step further than that taken in Latin American S&T policies. Only an integrated and coordinated approach with these policies can make program resources appear more attractive for firms to welcome qualified researchers into the productive sector.

References

Argentinian National Council for Scientific and Technical Research (CONICeT). (2013). Researchers within firms. http://web.conicet.gov.ar/web/conicet.ciencia.vinculacion/investigadores-en-empresas. Accessed 29 May 2013.

Brazilian Council for Improvement of Higher Education personnel (CAPES). (2013). GEOCAPES Statistical data. http://geocapes.capes.gov.br/geocapesds/#. Accessed 20 May 2013.

Brazilian Federal Government – Strategic and Management Studies Center (CGEE). (2011). RHAE Program Researchers into Firms. Projects Directory, calls #67/2008 and #62/2009. Brazil. http://www.cnpq.br/documents/10157/56ba93de-6cf3-4bf8-9e84-93f76b12f30f. Accessed 18 Apr 2013.

Brazilian Institute of Geography and Statistics (IBGE). (2011). Technological innovation survey 2008. Rio de Janeiro. http://www.pintec.ibge.gov.br/index.php?option=com_content&view=category&layout=blog&id=15&Itemid=19. Accessed 20 May 2013.

Brazilian Institute of Geography and Statistics (IBGE). (2013). Techonolgical innovation survey 2011. Rio de Janeiro. http://www.pintec.ibge.gov.br/index.php?option=com_content&view=category&layout=blog&id=27&Itemid=43. Accessed 12 Apr 2014.
Brazilian Ministry of Science, Technology and Innovation (MCTI). (2013). National science, technology and innovation indicators. http://www.mct.gov.br/index.php/content/view/740.html?execview. Accessed 20 May 2013.
Brazilian National Council of Scientific and Technological Development (CNPq). (2007). Public call # 32/2007 of RHAE Researchers into Firms program. http://www.cnpq.br/web/guest/chamadas-publicas?p_p_id=resultadosportlet_WAR_resultadoscnpqportlet_INSTANCE_0ZaM&filtro=encerradas&buscaModo=textual&tmp=1401395344754. Accessed 30 May 2013.
Brazilian National Council of Scientific and Technological Development (CNPq). (2008a). RHAE Researchers into Firms: Results and project catalogue. Brazil. http://www.cnpq.br/documents/10157/6629d145-d10c-40ca-bacc-faf7a51b0ee5. Accessed 30 Aug 2013.
Brazilian National Council of Scientific and Technological Development (CNPq). (2008b). Public call # 67/2008 of RHAE Researchers into Firms program. http://www.cnpq.br/web/guest/chamadas-publicas?p_p_id=resultadosportlet_WAR_resultadoscnpqportlet_INSTANCE_0ZaM&filtro=encerradas&buscaModo=textual&tmp=1401395344754. Accessed 30 May 2013.
Brazilian National Council of Scientific and Technological Development (CNPq). (2009). Public call #62/2009 of RHAE Researchers into Firms program. http://www.cnpq.br/web/guest/chamadas-publicas?p_p_id=resultadosportlet_WAR_resultadoscnpqportlet_INSTANCE_0ZaM&filtro=encerradas&buscaModo=textual&tmp=1401395344754. Accessed 30 May 2013.
Brazilian National Council of Scientific and Technological Development (CNPq). (2010). Public call #75/2010 of RHAE Researchers into Firms program. http://www.cnpq.br/web/guest/chamadas-publicas?p_p_id=resultadosportlet_WAR_resultadoscnpqportlet_INSTANCE_0ZaM&filtro=encerradas&buscaModo=textual&tmp=1401395344754. Accessed 30 May 2013.
Brazilian National Council of Scientific and Technological Development (CNPq). (2012). Public call #17/2012 of RHAE Researchers into Firms program. http://www.cnpq.br/web/guest/chamadas-publicas?p_p_id=resultadosportlet_WAR_resultadoscnpqportlet_INSTANCE_0ZaM&filtro=encerradas&buscaModo=textual&tmp=1401395344754. Accessed 30 May 2013.
Brazilian National Council of Scientific and Technological Development (CNPq). (2013a). *Impact assessment of RHAE Researchers into Firms program*. Brazil.
Brazilian National Council of Scientific and Technological Development (CNPq). (2013b). RHAE numbers. http://www.cnpq.br/web/guest/blog/-/blogs/numeros-do-rhae?_33_redirect=http%3A%2F%2Fwww.cnpq.br%2Fweb%2Fguest%2Fblog%3Fp_p_id%3D33%26p_p_lifecycle%3D0%26p_p_state%3Dnormal%26p_p_mode%3Dview%26p_p_col_id%3Dcolumn-3%26p_p_col_pos%3D1%26p_p_col_count%3D2. Accessed 23 Aug 2013.
Brazilian National Council of Scientific and Technological Development (CNPq). (2013c). RHAE Researchers into firms: Results and project catalogue. Brazil. https://docs.google.com/file/d/0B2_zuiazbQCKQ0ZMczM5dldtNWs/edit. Accessed 30 Aug 2013.
Chilean Government. (2012). 7th Innovation Survey (2009–2010). Ministry of Economy. http://www.economia.gob.cl/wp-content/uploads/2012/08/Presentaci%C3%B3n-Resultados-7ma-Encuesta-de-Innovacion.pdf. Accessed 12 Sept 2013.
Chilean Government. (2014). 8th Innovation Survey (2011–2012). Ministry of Economy. http://www.economia.gob.cl/wp-content/uploads/2014/02/Presentacion-Resultados-8va-Encuesta-Innovacion.pdf. Accessed 12 Apr 2014.
Chilean National Committee for Scientific and Technological Research (CONICyT). (2009). Attraction and placement of advanced human capital (PAI): Researchers into productive sector program. 2009 Contest. http://www.conicyt.cl/pai/2009/05/18/concurso-nacional-de-insercion-de-capital-humano-avanzado-en-el-sector-productivo-chileno-extiende-plazo-al-250809-a-las-1600-hrs/. Accessed 30 Jun 2013.
Chilean National Committee for Scientific and Technological Research (CONICyT). (2010). Attraction and placement of advanced human capital (PAI): Researchers into productive sector program. 2010 Contest. http://www.conicyt.cl/pai/2010/04/12/concurso-nacional-de-insercion-

de-capital-humano-avanzado-en-el-sector-productivo-chileno-convocatoria-2010/. Accessed 30 Jun 2013.

Chilean National Committee for Scientific and Technological Research (CONICyT). (2011). Attraction and placement of advanced human capital (PAI): Researchers into productive sector program. 2011 Contest. http://www.conicyt.cl/pai/2011/05/09/concurso-de-insercion-de-capital-humano-avanzado-en-el-sector-productivo-chileno-segunda-convocatoria-2011/. Accessed 30 Jun 2013.

Chilean National Committee for Scientific and Technological Research (CONICyT). (2012a). *Statistical summary 2008–2012*. Santiago, Chile.

Chilean National Committee for Scientific and Technological Research (CONICyT). (2012b). Attraction and placement of advanced human capital (PAI): Researchers into productive sector program. 2012 Contest. http://www.conicyt.cl/pai/2012/02/29/concurso-de-insercion-de-capital-humano-avanzado-en-el-sector-productivo-chileno-convocatoria-2012/. Accessed 30 Jun 2013.

Chilean National Committee for Scientific and Technological Research (CONICyT). (2013). Scholars statistics. http://www.conicyt.cl/becas-conicyt/estadisticas/informacion-general/. Accessed 20 Aug 2013.

Chilean National Educational Council. (2012). Indexes: statistics and databases. http://www.cned.cl/public/Secciones/SeccionIndicesEstadisticas/indices_estadisticas_posgrado.aspx. Accessed 20 June 2013.

Colombian Administrative Department for Science, technology and innovation (Colciencias). (2011). Application call to encourage Colombian and foreign doctorate holders placement into Colombian firms, 2011. http://www.colciencias.gov.co/convocatoria/convocatoria-para-estimular-la-inserci-n-de-doctores-colombianos-y-del-extranjero-las-e. Accessed 6 July 2013.

Dutrénit, G., & Arza, V. (2010). Channels and benefits of interactions between public research organizations and industry: Comparing four Latin American countries. *Science and Public Policy, 37*(7), 541–553.

Iberoamerican and interamerican Science and Technology Indicator Network (RICyT). (2013). Indicators: Higher education graduates. http://www.ricyt.org/index.php?option=com_content&view=article&id=149&Itemid=3. Accessed 20 May 2013.

Luchillo, L. (2009). Impacts of Mexican CONACyT scholarship grants program: An analysis on the careers of former scholarship owners (1997–2006). *Iberoamerican Journal of Science, Technology and Society, 5*(13), 175–205.

Mexican National Council of Science and Technology (CONACyT). (2013). Scientific and technological indicators. http://www.siicyt.gob.mx/siicyt/cms/paginas/IndCientifTec.jsp. Accessed 20 Aug 2013.

Organization for Economic Cooperation Development (OECD). (2012). Main science and technology indicators 2011/2012. http://www.oecd.org/sti/msti.htm. Accessed 20 May 2013.

Schoenecker, T., Daellenbach, U., & McCarthy, A. M. (1995). Factors affecting a firm's commitment to innovation. *Academy of Management Proceedings, 1*, 52–56.

Souitaris, V. (2002). Firm-specific Competencies determining technological innovation: A survey in Greece. *R&D Management, 32*, 61–77.

Velho, L. (2007). The role of the formation of researchers within innovation system. *Science and Culture, 59*(4), 23–28.

Viotti, E. B. (2010). Doctorate holders 2010: Demographic studies of Brazilian technical and scientific staff. In: Brazilian Government – Strategic and Management Studies Center (CGEE), *Doctorate holders 2010: Demographic studies of Brazilian technical and scientific staff* (pp. 15–50).

Chapter 10
Itaipu Technology Park: A Territorial Development Tool

Juan Carlos Sotuyo and Maria Angélica Jung Marques

10.1 Introduction

The Itaipu Technology Park (*Parque Tecnológico Itaipu*; PTI) in Brazil is a prime example of the creation of a successful science and technology (S&T) park. It was created in 2003 by the world's largest generator of renewable clean energy, Itaipu Binacional. The company's mission is to generate quality electricity via socially and environmentally responsible practices, and to foster sustainable economic, tourism, and technological development in Brazil and Paraguay. The influence of the PTI on the region provides a multidimensional overview of local development, especially its contributions regarding capacity building and the formation of local critical mass. Its implementation model and methodology differ from conventional views on technology parks development; that is, it did not satisfy the typical prerequisites such as a consolidated university and high-quality research and development groups. The PTI goes beyond conventional high tech condominiums including typical local actions and characteristics.

This chapter concerns a retrospective view of the creation of the PTI, highlighting the context of its political creation, its peculiarities, and the role of its management organization, the Itaipu Technology Park Foundation (FPTI). The FPTI has been transformed from its basic structure as a "business condominium administrator" to a broader based structure, a "territorial articulator," creating a Quadruple Helix where government, industry, academia, and civil participants work together to co-create the future and drive structural changes in an integrated development process, far beyond the scope of what any one organization or person could do alone.

J.C. Sotuyo (✉) • M.A.J. Marques
Itaipu Technology Park Foundation, Foz do Iguaçu, Paraná, Brazil
e-mail: sotuyo@pti.org.br; angelica@pti.org.br

H. Horta et al. (eds.), *Trends and Challenges in Science and Higher Education*, Knowledge Studies in Higher Education 3,
DOI 10.1007/978-3-319-20964-7_10

This chapter also emphasizes that state companies could follow the Itaipu Binacional example by enlarging their initiatives regarding "social responsibility and sustainability" (such programs are typically tenuous and simplified as commonly occurs in supplying the exclusive expectations of the market), and play a strategic role in the development of the territories where it has been established. The goal is to contribute to and directly benefit from the tangible and intangible results of the PTI.

10.2 Technology Parks and the Promotion of Territorial Development

10.2.1 Context

The levels of inequality in Latin American countries—the lack of access to quality education, integrated health care services, and basic sanitary services—constitute one of the region's greatest development problems. Structural inequality is also observed in companies' productivity, differences in competitiveness, institutional capacities, and environmental sustainability. Territorial inequalities require a multidimensional and integrated focus, as well as the implementation of public policies to reduce existing gaps (Ilpes/Cepal 2012).

Regional imbalances provoke serious social exclusion problems, and these were aggravated by the failure of neoliberal models (based on financial speculation) prevalent in most Latin American countries in the 1990s. The privatization of strategic sectors, negative impacts on technical and university education, and poor investment in S&T have made the situation of industries even more critical because of their loss of competitiveness internationally. Furthermore, the reliance on the export of primary agricultural, livestock, and natural resource commodities has resulted in negative impacts on Brazil's development.

The globalization phenomenon is based on financial and commercial networks, via the integration of value chains. These have deepened as time goes on because the effects of S&T (Ferrer 2010) have required Latin America to make profound changes to its productive systems to incorporate knowhow and to strengthen the territorial systems of innovation adapted to the characteristic processes of the territories.

Thus, universities must study development alternatives to bring about reflection and the assessment of new models for S&T parks that are different from conventional models. Furthermore, they could introduce alternatives to conventional polygons and industrial parks, which can undergo a transformation. Thus, the essential components can be integrated to transform the old into new, incorporating new dynamics to combine knowhow and innovations to established companies and, therefore, to the territories in which they are located (Sotuyo 2014).

During the first decade of this century in various countries, governments were elected with policies to succeed neo-liberalism. These policies promoted heterodox economic policies, and sought to increase employment levels, social inclusion processes, and reduce poverty despite the 2008 crisis. Brazil is perhaps the most concrete example of such quests.

10.2.2 Territorial Development

The concept proposed by Santos (2007) considers that a territory must be understood as the junction of land together with its identity. Thus, each inhabitant belongs to the space where he/she lives, a place where there are social, political, cultural, and environmental institutions.

Alburquerque (2013) distinguishes between generated regional policies as "top-to-bottom," to address a country's regional imbalances, and generated development policies from "bottom-to-top," within a concerted process of the participants in a particular territory. The existence of complementariness between the two focal points must be emphasized so that the national policies of a region or territory help to bring about favorable factors for the proposals and initiatives created from the territory.

Alburquerque (2011) also considers territorial development as a new way of thinking about economic development. He proposes a characterization that opposes conventional analyses of economic theory, stressing that each territory has its own peculiarities. This is a critique of the macroeconomic view that reduces development to quantitative economic growth in the weakening of qualitative aspects, social transformation, and environmental development. He contrasts the conventional view by considering the productive and enterprising reality as based on the logic of large enterprising and financial groups, stating it is the only way to explain this reality. He also proposes a heterogeneity view regarding the present economy and the role of local economies, studying their context, opportunities, and limitations exposed to the globalization process.

In this context, Latin American S&T parks could be constituted as agents in this territorial developmental process, articulating, fomenting, and executing structuring and innovative actions via strategic alliances. It is necessary to prepare park models for this purpose, to play an active role in the articulation of development processes, and not merely as isolated production units in the territory.

To enhance territorial development, a number of problems should be addressed to ensure the inclusion of all actors. Specifically, the following issues: the company–university relationship; the strengthening of the role of the federal government in the development process of education, science, technology, and the definition of public policies in the innovation systems; and incrementing investments in innovation from private initiatives, generating legal frameworks to establish alternatives to address the permanent complaint for decreased burdens, labor flexibility for tax breaks, free zones, among others. These circumstances do not contribute to local development but instead to the importation of technology and the impoverishment of Brazil's territories.

Conditions must be created so that small and medium-sized companies can access S&T platforms at universities and research centers. Therefore, reforms should be promoted to ensure that research results on products, processes, and services that can be used to generate income and, via university involvement, supply social demands.

Another issue is the lack of medium- and long term projects, a result largely based on the perspective of the public administration associated with changes in government instead of state policies. Finally, there is the low level of participation and linkage to development actors, who should establish government agreements among the social components in the territory. Therefore, in general, there is a low level of cooperation, collaboration, and coordination in convergent actions on problems resolution. These are the challenges that FPTI has been facing since its creation.

10.3 PTI Creation and History

10.3.1 The Role of Itaipu Binacional

In 2003, when Luis Inácio Lula da Silva was elected president of Brazil, he held a meeting soon after with the leaders of various governmental companies at *Furnas Centrais Elétricas* (a Brazilian federal power company) in Rio de Janeiro. He looked out of the building facing the slums and asked his guests to look beyond their offices, as just a few meters away there were many people living in poverty. He requested the company heads to think about what they could do in their companies to solve the community problems surrounding them. It was this guidance that impelled the Brazilian General Director of Itaipu Binacional, Jorge Miguel Samek, to take on a set of concrete action plans to encourage social inclusion in Foz do Iguaçu, Paraná State, Brazil.

Up until 2003, the mission statement of Itaipu Binacional was: "Hydroelectric use of the hydro resources of Paraná River in cooperation with both countries", from and including Salto Grande de Sete Quedas or Salto Guaíra to Foz do Iguaçu. Following the directive of the new government, the Itaipu mission became: "Social and environmental responsibility, dependable generation and supply of electric energy, and furthering sustainable, economic, touristic and technology development, in Paraguay and Brazil."

However, Itaipu Binacional did not envisage at that time that this statement would soon result in the appointment of specialized staff to attend to the necessities stemming from its new mission, and to the creation of a technology park. Assessing the development of the Institute of Applied Technology and Innovation (ITAI) with The Western University of Paraná State (UNIOESTE), Itaipu Binacional articulated the creation of an appropriate environment to develop the programs and projects included in its mission. As a result, in May 2003, a Letter of Intentions for the creation of Itaipu Technology Park (Itaipu Binacional 2003) was signed, defining the concept of the park as follows: "An innovative area joining projects and programs to promote social inclusion, job and income generation, and knowledge distribution, for all socioeconomic levels, as well as the development and transference of technologies, favoring experience exchanges and integration between people for a better understanding and changing realities." The park's goals were to "Promote the integration of Latin American countries, through the mobilization of government bodies, representative entities of the organized social society, academic entities, research institutions and entities of promotion for the sustainable economic, social and cultural development, based on education, science and technology, generation of jobs, work and income."

In August 2003, President Luís Inácio Lula da Silva approved the creation of the park, formally named *Parque Tecnológico Itaipu*. In October of the same year, various activities started at the temporary PTI headquarters on the campus of UNIOESTE, in Foz do Iguaçu, with operational support from ITAI.

In March 2004, to validate the concept and define the goals of the new park, a participative planning process was initiated: the Technology Regional Development Plan. Five meetings with more than 500 participants were held; more than 100 Brazilian, Argentinean, and Paraguayan entities participated, defining the various areas of the park's operations.

The PTI concept included economic, tourist, and technology development, resulting from the new mission of Itaipu Binacional, merging programs, and projects turning to education, S&T, as well as actions to generate jobs and income, contributing to the social inclusion process. As an agent of integration and innovation, and a development inductor, the park started operating in partnership with support and promotion entities PTI also worked with Brazilian, Paraguayan, and Argentinean universities, with a focus on Região Tri Nacional do Iguaçu (tri-national Iguassu region).

10.3.2 PTI Evolution

In 2004, the PTI began participative planning to renovate a housing area (approximately 50,000 m^2 in an area of 360,000 m^2) that had initially housed the workers involved in the construction of the Itaipu Hydroelectric Power Plant. The houses were gradually converted into classrooms, laboratories, spaces for incubated companies, and S&T research and development centers. October 2004 marked the opening of the first newly renovated 3000 m^2.

Projects in the fields of education, culture, and tourism were initiated first. These included the Tri-National Citizen Network (promoting literacy to approximately 5000 young people and adults), the Art and Cultural Popularization Program (aimed at children aged 6–14, providing classes on theater, poetry, street dancing, classical ballet, keyboards, guitar, plastic arts, and singing lessons), and the Eirete Eiru-I[1]

[1] Eirete Eiru-I is a Guarani word that means a small honey bee.

Project (for public school students in the 1st to 4th grades in Argentina, Brazil, and Paraguay, to encourage students to improve their self-worth, integration, respect for national diversities, and tourism awareness).

In 2005, new educational projects were launched, such as college preparatory courses for low-income students, partnering with UNIOESTE and ITAI, and in October 2005, another important step was achieved in the PTI construction: the Engineering Center and UNIOESTE Exact Science Department moved into their new spaces in the park. Equipped with modern classrooms, teaching and research laboratories, four college degree courses were implemented in the PTI: computer science, electrical engineering, mechanical engineering, and mathematics.

In December 2005, Itaipu Binacional created the FPTI, a private, non-profit institution, to manage and operate the PTI. The foundation was responsible for providing economic and financial support for laboratory infrastructure, seeking sponsorship for research, and attracting partners from other regions, with the aim to complement local capacity via scientific development (Sotuyo and D'alkaine 2010).

In 2006, the Santos Dumont Company Incubator was created to provide support to generate, develop, and implement innovative product and service companies in the region, especially those involved in generating employment and income. Since the creation of the incubator, it has performed various activities in enterprising development in over 220 companies in the western region of Paraná State, including consultancy and advisory assistance, participation in trade shows and events, training courses on diverse subjects for enterprises and company employees, negotiations with investors, support for prospecting resources, and the technology development of innovative products and services. In 2014, it began working on business leveraging, which implemented a differentiated perspective on time versus scalability, as well as enlarging the market in stimulating new businesses.

The Science Station was also created in 2006—a space focused on the popularization, dissemination, and development of scientific education among teachers and elementary, middle school, and high school students in the region. This was followed by the Young People's Trail project to train and assist low-income young people in the tourism job market sector.

The FPTI, even back in 2006, began to stimulate research for students, professors, and researchers in the region, granting scholarships to encourage scientific production. In that year, other important spaces were also established: the Paulo Freire Library, which housed Itaipu Binacional, UNIOESTE, and FPTI bibliographic collections, the Cesar Lattes Auditorium, and the Florestan Fernandes Event Space.

Former President Lula visited the park on August 28th 2006 and he declared PTI to be an exceptional space to implement the Latin American Integrated Federal University (UNILA), officially launched in September 2010.

The Open University of Brazil (UAB) was established in 2007, teaching undergraduate and post-graduate courses in the hybrid (e-learning/classroom) modality, sponsored by partnerships with existing public universities in Brazil.

Foz do Iguaçu is famous for its tourist attractions (e.g., the Iguassu Waterfalls and the Itaipu Power plant), and in 2007, the FPTI became responsible for operating and managing the Itaipu Tourist Complex. Various improvements (regarding quality and innovations) were made to the attractions and the tourist services provided by Itaipu Binacional. As well as contributing to the strengthening of tourism, these changes also expanded the opportunities for employment and income in the region. Thus, funds became available to invest in the Technology Fund, assigned to financing educational, research, and development projects.

In 2007, a number of new research and development centers were established in the PTI: the Hydro-Informatics International Center; the Intelligence and Technology Innovation Office, working with different fields of Basic Industrial Technology; the Center of Advanced Studies on Dam Safety; and the Automation and Electric System Laboratory (the latter two in partnership with the Itaipu Corporate University). These centers have consolidated the park as an important environment for scientific production and technology innovation, providing an infrastructure of research centers and laboratories that are now noteworthy on such themes as geoprocessing, open technologies, dam safety, automation, and biogas.

In 2009, yet another space was inaugurated: the Casimiro Montenegro Filho Astronomical Center, including a planetarium and observation tower to disseminate information on astronomy science among students, teachers, and professors. In 2013, the center participated in the discovery of the first system of rings around an asteroid, published in 2014 in *Nature* (Braga-Ribas et al. 2014).

The PTI has quickly evolved and encouraged the attraction of new projects and partnerships. In 2011, for example, the Water Project: Knowledge for Management was established, which resulted from a partnership between the Brazilian National Water Agency, FPTI, and Itaipu Binacional. That same year, the Battery Development Project began to enhance the technology surrounding the production of nickel/sodium batteries, including the creation of a laboratory for mobility and storage applications for large amounts of power. The Hydrogen Research Laboratory was also established, coupled to a pilot project plant for generating hydrogen. The plant was launched in December 2014.

In the field of education, the Municipal Technology Center was created in 2011 to promote the digital inclusion of teachers and students from public schools. In 2012, the first doctorate courses (in law and numerical methods) were established in Foz do Iguaçu.

In 2013, the FPTI was one of the founding partners of the International Center on Renewable Energy–Biogas (CIBiogás-ER). CIBiogás-ER encourages the production, diffusion, and use of biogas, generated via the biodigestion of animal dejecta and agriculture residues. Biogas is an important energy asset that produces electric, thermal, and automotive energies, in addition to bio-fertilizers and enhancing rural sanitation. The Latin American Center of Open Technologies was also opened in 2013, with a focus on the development of innovative solutions using exclusively open-source software and hardware, launching the Innovation Program in Information Technology and Communications.

The PTI also provides support for research focused on sustainable urban mobility. One example of such support was the creation in 2014 of the Intelligent Electric Mobility Center via a partnership between Itaipu Binacional and the Portuguese

Excellence Center of the Mobility Industry. The partnership seeks to engage in a number dynamic activities ranging from research and development to technology experiments. Business models will also be defined for the industrialization of new electric mobility solutions to benefit people.

ITAI, in partnership with the FPTI, installed a system to register disturbances at the Itaipu dam. This system was entirely developed in the park by interconnecting generating units for a real time assessment to provide maintenance support.

The project *Vira Vida* (Life Turning) was also launched with support from the Industrial Social Service, assisting socially vulnerable young people. Through the promotion of training courses and socio-educational support services, the program seeks to guarantee and develop people's self-worth, as well as prepare and refer them to the job market.

The FPTI manages the Western Development Program, launched in 2014 to accelerate the economy and encourage competitiveness in the region. Its cutting-edge difference is in its participative governance, with cooperation among public and private participants to implement a developmental strategy to integrate 54 municipalities in the western region of Parana State. The current program participants are as follows: Itaipu Binacional; PTI; Sebrae/PR; Western Paraná Municipalities Association; Paraná State Commercial and Industrial Association Coordination; Paraná State Industrial Federation; and the Paraná State Cooperative Organization.

Regarding culture, the PTI opened its doors to cultural activities in 2014, promoting free-of-charge cultural attractions for the community, including theater, ballet, and musical productions. The aim was to attract more live shows to Foz do Iguaçu, strengthen cultural ties, and to offer further leisure options for residents and tourists. This initiative supports the PTI territorial development plan, which includes cultural incentives as a consideration.

10.4 PTI Building Capacities

10.4.1 PTI and the Territorial Innovation System

The FPTI disseminates information on the park model by hosting delegations from various Latin American countries, holding events and meetings, and encouraging and providing support for the implementation of parks. The foundation also helps to gather territorial participants to contribute in planning a multidimensional developmental process, considering the following dimensions: economic; human and social; environmental; institutional, political and cultural; and technology. For this to occur, the administrative entity of the park (as well as administrating business condominiums, universities, and the park's many diverse initiatives) encourages the articulation of the entities residing in the park, establishing communication processes and discussions with territorial agents.

One FPTI activity is to engage with qualified teams to conduct research, development, and innovation projects. Thus, the foundation operates based on an organizational structure (in an inverted pyramidal format), and functional and support sectors provide support for the operation and performance of projects. Within these sectors there are teams skilled in technical-scientific fields to obtain results, which are then applied to specialized products and services and in generating knowledge.

Operation is put into effect via projects that provide data input for an information and knowledge management system, which is propagated throughout the territory. Thus, the FPTI also promotes connections and interfaces within the territory and from the park, seeking to identify problems and opportunities and contribute to developing solutions, as well as stimulating and promoting science, technology, and innovation (ST&I) at PTI and in the territory. PTI is, and has been since its conception, a mechanism to promote innovation and territorial development, placed into the territorial innovation system as a driving element in the system, bringing together Quadruple Helix participants. Thus, it is important to understand the park and how it fits into the system, and how it seeks to fulfill its mission.

Brazil's innovation systems are institutional arrangements that have become extremely complex due to the country's large geography. Furthermore, Brazil's regional and inter-territorial specificities are an important factor in the innovative performance of companies. For this reason, it is necessary to consider regional innovative systems in the development process of a country such as Brazil.

The concept of a regional innovation system is relatively new. It first appeared at the beginning of the 1990s (Cooke 1992); it was jointly disseminated with the idea of innovation systems. Edquist (1997), Freeman (1987), Lundvall (1992), and Nelson (1993) define innovation as a systematic process based on the interdependence of different types of "units": business or other organizations that society is connected to in a structure where "multilateral" exchanges of information are intense. By way of these exchanges, new knowledge is created, and this tends to generate innovation. Thus, this concept expresses that innovation results from a territory's interactive and systematic process. For this reason, many derivatives have emerged from this approach, considering the territory as the starting point for analysis (national, regional, and local) or specific technologies and sectors (Doloreux and Bitard 2005).

According to Doloreux and Saeed (2005), the concept of regional innovation systems has evolved into a broadly analytical framework used to generate an empirical basis to formulate innovation policies. This means that on a territorial level, it is necessary to strengthen inter-institutional relations in the coordinated actions of mutual interest for the development and generation of innovation. This point is also emphasized by Albagli and Maciel (2007). They state that the characteristics and development of local innovative systems not only depend on explicit policies (governmental and others), but, more specifically, on the forms of historically molded interaction among participants, as well as implicit coordination mechanisms.

Thus, PTI performs its role (exchanging knowledge) in the context of a territorial innovation system by stimulating and providing support to the interactive and integration processes among diverse Quadruple Helix participants. This operation occurs in the named science, technology, and innovation platforms.

10.4.2 PTI and its Relationship with Universities

In Brazil, especially in developing territories, higher education institutions play a fundamental role in the development of territorial innovation systems. The importance of national regional educational policies has been emphasized since 2003, when higher education institutions started expanding into these regions, even before they were served by the federal education system. According to the Commission Report Constituted by Ordinance # 126/2012 2012 on the analysis of the expansion of federal universities in Brazil, Article 205 of the 1988 Federal Constitution states, higher education is the right of everyone and the obligation of the federal government and the family. This precept has been the sustaining basis for defining public educational policies in Brazil. The document recognizes the role of the university as an instrument for social transformation, sustainable development, and inclusion in the country, and internationally, mobilizing claims to expand public and free higher education. The elitization of access to higher education has been criticized and considered to cause social exclusion. To overcome this discriminatory situation by the provision of greater opportunities to access higher education, the 2001 National Education Plan was drafted (2001–2010), demanding a considerable increase in investments in education, as well as setting targets to expand the number of students enrolled in all levels of higher education (Commission Report Constituted by Ordinance # 126/2012 2012).

The policy to expand federal higher education institutions has benefited the territory in which the PTI is located, notably beginning in 2010 with the establishment of UNILA. However, it is important to stress the benefits of the PTI; the park has not only benefited and made possible the creation and operation of UNILA in the park facilities, but it has also welcomed the UNIOESTE Exact Science Center (established 2006). Since the 1990s, the relationship between Itaipu Binacional and UNIOESTE has leveraged the creation of college courses, including computer science (1995), electrical engineering (1998), and mechanical engineering (2002), promoting an environment to prepare qualified personnel and academic research in these fields.

Another important public policy from the Education Ministry benefitting the territory and hosted by the park, was the building of a teaching center for the Open University of Brazil (UAB). The UAB is a system integrated by public universities higher education courses—targeting populations that find it difficult to access to university education—via e-learning. Although the general public is served, elementary school teachers have priority, followed by school administrators and employees in the elementary state education system (UAB 2014). The UAB Darcy Ribeiro University Center for Classroom Support has been operating since 2007 at the PTI. It was created in partnership with Foz do Iguaçu City Hall, FPTI, and the Ministry of Education. As of 2014, more than 1500 students have graduated.

PTI provides shared spaces for classrooms and laboratories, as well as environments for fellowship and active participation in cooperative projects shared by FPTI, universities, and Itaipu Binacional. This makes it possible for undergraduate and postgraduate students, originating from different educational institutions, to both learn and experience an appropriate educational preparatory environment, providing them access to the job market and the possibility to create their own businesses.

10.4.3 PTI Relationship Strategies

The PTI is best represented by a concept defined by Sotuyo (2014), "action in the intersection." That is, it is recognized that the administrative organization of the park points to a common interest in the territory, including those entities established in the park. Based on this, concrete actions are identified to strengthen cooperation and collaboration among entities on themes of intersecting points of interests. This is achieved by respecting the unique and exclusive corresponding interests of each entity.

It is necessary to consider that each entity has its own specific mission, objectives, established targets, and specific time horizons. For example, government time horizons are associated with the electoral cycle, which generally disrupt public actions that should in fact be state policies. Private companies must serve the interests of partners and investors, as the generation of profits becomes increasingly difficult. Educational institutions must abide by schedules related to annual graduation cycles. Scientific research entities have timeframes related to the complexity of the problems being researched, which can take decades, and concrete results are not always achieved. Finally, civil society entities must adjust their deadlines to the problems of the community they represent, generally on a short-term basis.

The importance lies in establishing a set of strategies capable of guiding alliances for the "action in the intersection," jointly agreeing to meet the short-, medium-, and long-term objectives and targets.

The FPTI is based on this concept, as it operates within two main relationship strategies: the "extra park" (outside of the park) and "intra park" (inside the park). Regarding the former, it acts as an articulator and cooperator and collaborative fomenter seeking awareness of the social and economic reality, productive structures, value chains, available social capital, the quantity and quality of the professional and technical training entities, universities, research centers, predominant cultural aspects, social organization, representative entities, territorial infrastructure, and financing mechanisms.

The demands of the territorial participants' are supplied, whenever possible, via continued training, promoting debate forums, applying problem analysis methodologies, solution modeling, training personnel in drafting projects, and teamwork. This is done so that the interaction results in the proposition and execution of concrete projects aligned to the particular institutional missions of each participant by "intersection actions" and are convergent to the objectives of territorial development.

The latter impels interaction among residing entities, organizing physical spaces, and constructing relationship mechanisms among them. Furthermore, the intra park strategy jointly defines the rules and conditions for sharing resources (laboratories, classrooms, spaces for events, libraries, restaurants), and other necessary services for a complete operation of activities in each entity. This strategy is also responsible for supporting the coordination of the cooperative networks and the development of S&T platforms in the park, and the generation and articulation of the park innovation system connected to the territorial innovation system.

As a complementary strategy of the model, the FPTI is also a research, development, and innovation entity, working in a coordinated mode, cooperating and collaborating with the Quadruple Helix. It incorporates various themes associated with development, and acts as an interface for university research groups and S&T institutes, supporting the growth of productive chains and social demands. It also performs in the preparation and improvement of public and private management, and in the enhancement of educational systems, supporting inclusive public policies.

10.4.4 Science, Technology, and Innovation Platforms

The ST&I platforms are logical organizational structures that have arisen based on the necessity to systematize interaction among participants or institutions, and a certain "formalization" of these interactions/intersections in joint actuation via cooperative projects. The platforms enable the establishment of a system to stimulate technology development, and promote innovation in an environment of shared usage of infrastructure and resources via streamlined mechanisms and cooperative functions, as an induction vector of strategic themes in "intersection actions."

The main objective of the ST&I platforms is to promote and empower the results from the ST&I produced from connections established in the intra and extra park settings. Thus, a dynamic system is created to serve the needs of the complexity of these relations while enabling the measurement and improvement of ST&I results.

The motives of the PTI include development within all of its multi-dimensions based on the demands and needs of the territory; the platforms must also comprehend the connections between society, the park, and their dynamics, as well as the FPTI and its environment. For this purpose, ST&I platforms operate as communication, development, and negotiation environments among participants and they must provide the necessary conditions for the development of themes of common interest. This should result in the formulation and implementation of cooperative projects in ST&I.

The platforms comprise sets of infrastructure (laboratories, classrooms, and equipment), qualified personnel (originating from diverse partnering institutions), and specialized services, and result in cooperative projects in ST&I, jointly with the enterprising ecosystem. The operation of the platforms is based on a systematic approach, where the relations among the diverse elements occur in multi-directional and fuzzy modes. The platforms seek to promote the development of innovative services, encouraging the generation of new businesses or enterprising. They also enable the permanent development of creative and innovative human resources, permeating and producing results for both the enterprising and social ecosystems, always following the strategic themes to benefit the interests of the park and/or territory.

The enterprising ecosystem is the environment of companies, whether they are companies based on technology, social, traditional, spin-offs, or startups, constituted by these interactions with the park and the territory and its demands for innovation (e.g., products, services, management, financing, scientific, Technology, socioeconomic, and legal/regulatory information).

The objectives of the ST&I platforms are to energize and enlarge results (products, services, and processes) in innovation for the enterprising ecosystem. Furthermore, the platforms aim to concentrate and optimize efforts and resources, creating favorable conditions to innovate, especially in micro and small businesses, as well as guiding knowledge management on a given theme. In addition, these efforts enable the operation (and accelerates the development of related competencies) of the strategic themes of the FPTI, creating and strengthening the bases supporting FPTI sustainability, connections and interactions among PTI, the territorial innovation system, and the national innovation system.

In parallel to the consolidation in the operation of the ST&I platforms, PTI has sought to establish a streamlined and efficient relationship between these platforms and the enterprising ecosystem, and for this purpose, it has developed an intelligence system for innovation.

The role of this intelligence system for innovation is to integrate the layers of the enterprising ecosystem and ST&I Platforms using data and information produced by different participants, generating sets of "intelligent" information, in the time, quality, and quantity required by each participant or set of participants to facilitate and/or promote the innovation process.

This system is intended to serve both the layers and benefit a feedback process that generates knowledge and intelligence for the entire territory, strengthening the motivation in a cooperative network. The information can be from prospection and S&T information, socioeconomic and marketing data, for the development of new businesses, products, or services.

Based on the evolution of the ST&I platforms and enterprising ecosystem, the intelligence system for innovation will also evolve and aggregate knowledge to the territory on a larger scale, becoming an "intelligent and creative territory," learning and developing its own social and productive capital. Thus, it will promote a favorable environment for investment and business activities on an educative basis, as well as equitable and solid R&D, a set of physical infrastructures to empower the developmental processes. Thus, it will create a fair and inclusive social and cultural framework (Marques and Sotuyo 2014).

10.4.5 Operation and Management Model of the PTI

The FPTI, the park administrator, operates via an "inverted pyramid" arrangement (Sotuyo 2014, p. 113) that modifies the old pyramidal hierarchical structure by inverting the relationships. The FPTI system is largely based on a lack of hierarchies, as hierarchy means "order and subordination" before a constituted power or authority. The foundation has replaced the concept of "authority" with "responsibility." This approach to management impels new modes of relationships among people, within an organizational environment guided by participation, respect, and democratic discussion based on argumentation, in cooperation and collaboration.

Under this management concept, it is necessary for each person in the FPTI to have a clear-cut concept of the mission of the organization and the role he/she must perform. Project managers and the operational sectors of FPTI do not command but lead, acting as facilitators for the teams, removing the "stones from the road," and solving technical issues. Moreover, they resolve conflicts and collaborate in the performance of activities, making the objectives of projects converge with the objectives of the organizations.

It is a great challenge to institutionalize this operating mode in the FPTI, to ensure that the managers and teams work in forthright cooperative, collaborative, and transparent environments; therefore, it is necessary to remove any fears, insecurities, and remaining legacies from previous experiences within hierarchical systems.

The FPTI is involved in the following areas: press, fundraising, corporate development, legal issues, security, purchasing and contracts, accounting and assets, agreements, project management, financial and budgeting, infrastructure and construction projects, human resources, general maintenance services and logistics, information and communication technologies, as well as more than 80 projects (as of December 2014). These tasks are all performed via a matrix organization that defines the collaboration of each operational area to provide the requirements of each project. This makes it possible to match the necessary skills to ensure that project requirements are satisfied, thereby avoiding the repetition of structural support and the rationalizing and specializing of non-core activities. This approach also enables project managers to primarily dedicate themselves to technical and management matters, avoiding, wherever possible, involvement in bureaucratic activities that can be performed by operational sectors.

One of the main operating problems of this matrix structure is the culture related to departmentalization, which tends to encourage people to only work within their operational sector. To resolve this problem, our operation permanently works on behavioral skills, increasing flexibility, interpersonal relations, and encourages cooperation, collaboration, and serving the overall mission of the organization.

10.5 Results and Conclusions

The results obtained by the PTI demonstrate that it is possible to impel projects from governmental companies to modify the conditions of a territory, aligning with articulation processes, proposals, and the implementation of public policies. For the purpose of analysis, these results can be divided into two groups: the tangible and intangible.

Tangible PTI results take into consideration the benefits stemming from Itaipu Binacional, as the park became the operational framework with which to address its expanded mission, executing projects to serve the ever-present needs of the technology renovation of the Hydroelectric Power Plant. The local support supplying these demands has resulted in the avoidance of costs, quality solutions, and fundamentally, the permanence and continuity of the dam knowledge management, and that of the park's S&T platforms. It has provided the training of human resources originating from the park, as year after year, there are public recruitment programs to increase to the power plant's workforce. The park has also resulted in the restoration of a degraded area; over 50,000 m^2 has been renovated, providing high-quality facilities.

The transfer of the Engineering Center and Exact Sciences from UNIOESTE to PTI has resulted in the growth of the park's laboratory facilities and bibliographic collection, and contact with an integrated environment of academic activities to develop solutions for companies and the community.

The establishment of the Latin-American Integrated Federal University represents investments in the construction of headquarters and annual maintenance. The establishment of the UAB has made it possible to grant social mobility to thousands of people to achieve access to higher education. Furthermore, the companies generated in the incubator facilities and business condominium have generated jobs, taxes, and have provided a new economic alternative based on knowledge. The park has also enabled hundreds of students to achieve access to improved infrastructure conditions in laboratories and has provided an environment for relations among companies and universities, granting scholarships, scientific initiation, trainee jobs, and post-graduate courses.

The FPTI project has also produced intangible results; these include the impact on the population who have direct access to the projects, as well as the thousands of children and teenagers who have visited the Science Station and the Astronomic Center. Furthermore, they include the thousands of teachers who have accessed the Municipal Educational Technology Center, using computerized tools and participating in continued educational programs, and the hundreds of young people who have accessed professional trainings, using differentiated methodologies to access the job market and university courses changing their lives.

These positive results are also due to the characteristics of PTI networking, incorporating national and international organizations, creating an integration of relationships with concrete projects and providing continuity through the setting up of offices and long-range programs, thus consolidating the credibility of its proposals.

The PTI, since its launch 11 years ago, has brought about significant contributions to the territorial development process, not only via the results presented here, but also in advances in the articulation of diverse fields of development: economic, social, institutional, cultural, technology, and political.

The PTI experience reveals various noteworthy components. Initially, it represented the political decision of a government to place a government company at the service of the territory. As a result, it has produced both tangible and intangible benefits for the respective company, in this case Itaipu Binacional. This initiative can encourage other companies to adopt similar stances and contribute to development processes in territories where they are located and perform. The institutionalization (via the decision of the Administrative Council of Itaipu Binacional) incorporated in the financial model is similar to that of Fraunhofer Institutes, which guarantee financing based on the possibility of developing enterprises within short periods of time.

The change in the paradigm in sharing resources among park inhabitants enables optimum usage and makes it possible to engage in collaborative work among residents. The innovative management model of the FPTI Foundation is also of note. The foundation not only acted as the administrator of a condominium, but also as an articulator and interface for PTI entities, adopting the role as a facilitator for entity relations.

Finally, the above results encourage us to consider PTI as a successful alternative park model, as an example to be observed and studied. It is a tangible experiment, a large-scale living laboratory, provoking real development with social inclusion and integration. Thus, it must be continually monitored so that it can be offered as a successful experiment, applicable to regions that require pre-tested actions on development implementation.

References

Albagli, S., & Maciel, M. L. (2007). *Information and knowledge of local innovation systems: a comparative perspective*. In: VIII ENANCIB—national meeting on information science research.

Alburquerque, F. (2011). Foundations in local development. Little is important, even if it is not beautiful. *Editorial Académica Española*.

Albuquerque, F. (2013). *Regional politics and territorial development in Latin America*.http://www.conectadel.org/biblioteca. Accessed 24 Nov 2014.

Association of University Research Parks. (1986). http://www.aurp.net. Accessed 8 July 2013.

Braga-Ribas, F., et al. (2014). A ring system detected around the Centaur (10199) Chariklo. *Nature, 508*, 72–75.

Commission Report Constituted by Ordinance # 126/2012. (2012). *Analysis on the expansion of federal universities 2003a, 2012*. http://portal.mec.gov.br/index.php?option=com_docman&task=doc_details&gid=12386&Itemid=. Accessed 23 Sept 2014.

Cooke, P. (1992). Regional innovation systems: Competitive regulation in the new Europe. *Geoforum, 23*, 365–382.

Doloreux, D., & Bitard, P. (2005). Les systèmes régionaux d'innovation: Discussion critique. *Géographie, Économie, Société, 7*, 21–36.

Doloreux, D., & Saeed, P. (2005). Regional innovation systems: Current discourse and unresolved issues. *Technology in Society, 27*, 133–153.

Edquist, C. (1997). *Systems of innovation: Technologies, institutions, and organizations*. London: Pinter.

Ferrer, A. (2010). Raul Prebisch and the dilemma of development in the global world. *Revista CEPAL, 101*, 8–14.

Freeman, C. (1987). Generic technologies, changes of techno-economic paradigm, and technology forecasting. In C. Freeman (Ed.), *Technology, policy, and economic performance: Lessons from Japan* (pp. 55–90). London/New York: Pinter Publishers.

ILPES/CEPAL. (2012). *Panorama of territorial development in Latin America and the Caribbean.* http://www.cepal.org/ilpes/publicaciones/xml/6/41686/Panorama_DT_nov_2010_final.pdf. Accessed 7 Jul 2014.

International Association of Science Parks. (2002). http://www.iasp.ws/publico/index.jsp?enl=2. Accessed 7 July 2014.

Itaipu Binacional. (2003). *Letter of intensions for the preparation of the "Parque Tecnológico Itaipu" (Itaipu Technology Park).* http://www.pti.org.br/system/files/file/carta_intencoes_PTI.pdf. Accessed 17 Nov 2014.

Lundvall, B. Å. (1992). *National systems of innovation: Towards a theory of innovation and interactive learning*. London: Pinter.

Marques, M. A. J. & Sotuyo, J. C. (2014). *ST&I platforms to generate innovation and promote the territorial development – The case of the Itaipu Technology Park.* IV International Congress of Technology Management and Innovation – COGESTEC, Colombia.

Melo, L. M., & Rapini, M. S. (2012). *Financing innovation in Brazil: Empirical evidence and implicit S&T policy.* Belo Horizonte: UFMG/CEDEPLAR.

Nelson, R. R. (1993). *National innovation systems.* Oxford: Oxford University Press.

Organization for Economic Co-operation and Development. (1997). *National innovation systems.* Paris: OECD.

Santos, M. (2007). Money and the territory. In M. Santos & B. K. Becker (Eds.), *Territory, territories: Tests on territorial ordering* (p. 14). Rio de Janeiro: Lamparina.

Soete, L., Verspagen, B., & Weel, B. (2010). *Systems of innovation.* CPB Discussion Paper No. 138, CPB Netherlands Bureau for Economic Policy Analysis.

Sotuyo, J. C. (2014). *Scientific and technology park model for contributing to territorial development.* Doctoral thesis, Universidad Nacional de Misiones [Misiones National University], Posadas, Argentina. 8 Aug 2014.

Sotuyo, J. C., & D'alkaine, C. V. (2010). *Itaipu Technology Park, an innovative model of STP to the sustainable development and regional integration of the triple frontier Argentina – Brasil – Paraguay.* In Proceeding XXVII IASP world conference on science and technology parks. Daejeon, Korea. 23–26 May 2010.

United Kingdom Science Park Association. (1984). http://www.ukspa.org.uk. Accessed 8 July 2013.

Part III
Networking and Industry-Science Relationships

Chapter 11
On the Dynamics of the Industry–Science Relationship: Brazil 1980–2012

Eva Stal and Asa Fujino

11.1 Introduction

The relationship between universities and companies, involving their advantages and difficulties, was a topic of great relevance between 1970 and 1990, and resulted in the introduction of the US Bayh–Dole Act in 1980. Subsequent studies have sought to highlight the strengths and weaknesses of the law via discussions concerning academic independence and universities' contributions to national economic development (Feller 1990; Etzkowitz 1989, 2008; Geisler and Rubenstein 1989; Armstrong 1992/1993; Mazzoleni and Nelson 2007; Thursby and Thursby 2011).

Although there is still some resistance concerning cooperation, especially in developing countries where most research universities are public (federal or state supported), there are several reasons to seek it. Taking into consideration their different values and goals, universities and companies can benefit from this interaction. For companies, this involves access to qualified human resources, the possibility of staying up-to-date on progress in their sectors, solving specific problems, and access to facilities and laboratories, in addition to contributing to their image and prestige (Lee 1996; Teece 1986; Rosenberg and Nelson 1994).

Three important articles (published within a few years) present the core activities involved in cooperation between universities and companies. Given its broad spectrum, and based on the classifications by Geisler and Rubenstein (1989) and Bonaccorsi and Piccaluga (1994), we highlight the main forms of collaboration:

E. Stal (✉)
Policy and Technology Management Group,
Universidade de São Paulo, São Paulo, Brazil
e-mail: eva.stal@terra.com.br

A. Fujino
Information Sciences, Universidade de São Paulo, São Paulo, Brazil

H. Horta et al. (eds.), *Trends and Challenges in Science and Higher Education*, Knowledge Studies in Higher Education 3,
DOI 10.1007/978-3-319-20964-7_11

(1) *Informal Personal Relations (university is not involved)*
Individual consulting (paid or free)
Informal workshops (meetings to exchange information)
Academic “spin-offs”
Publication of research results

(2) *Formal Personal Relations (agreements between a university and company)*
Scholarships and support to graduate programs
Students’ internships and in-company courses
Sabbaticals for faculty
Personnel exchange

(3) *Participation of an Intermediation Institution*
Liaison offices
Industrial associations
Applied research institutes
General assistance offices
Institutional consulting (university foundations)

(4) *Formal Agreements with a Determined Objective*
Contract research (proprietary)
Contract services (prototype development, tests, etc.)
Training of company employees
On-the-job training for students
Joint research projects

(5) *Formal agreements without a Determined Objective*
“Umbrella” agreements
Industrial sponsorship of R&D in university departments
Donations and research grants

(6) *Creation of Special Structures*
Association contracts
University–industry research consortia (or cooperative research centers)
Company incubators
Scientific parks
Mergers

Relations with universities do not entirely replace companies’ investment in basic research. However, the costs of maintaining cutting-edge research are increasingly high, and research is becoming multidisciplinary. Thus, maintaining strong relationships with universities configures an important business technological strategy (Bonaccorsi and Piccaluga 1994).

Regarding barriers to cooperation between universities and firms, they stem from a historical division between the missions of each institution. In the past, these differences were quite clear—universities should contribute to the creation of knowledge and industry should use this knowledge. However, this distinction has led to a misperception that universities are only oriented to long-term research, while companies are concerned only with product development (Geisler and Rubenstein 1989). Furthermore, this supposed exclusivity of operation has resulted in differences in style and behavior that hinder closer collaboration.

In the United States, the relationship between academic research and industrial research has been shaped by the decentralized structure and financing of higher education. Many public universities are state supported and governments influence both curriculum and research, linking them to the needs of the states and their local industry (Mowery and Rosenberg 1993). As for private universities, which depend on tuition revenues and donations, they were also inclined to respond to the changing needs of the society by providing professional education in different fields and contributing to the economic development of the country (Mazzoleni and Nelson 2007).

Pavitt (2005) observed that the possibilities of interactions between firms and universities are extensive. At one end there is basic research conducted at a university; this leads to a discovery, and its practical importance is recognized by a company that can work with the scientist to develop it further. This occurs most frequently in the chemical, biotechnology, and pharmaceutical industries.

At the other end there is the supply of trained researchers familiar with the latest techniques and integrated in international networks. This is of great importance for companies, and many consider it the greatest benefit provided by universities. Between these two extremes there are several other possibilities for cooperation, such as direct industry funding for universities, the use of academic consultants, and the exchange of researchers.

Perkmann and Walsh (2007) summarized the main forms of cooperation, as shown in Table 11.1 below. It is important to note that academic entrepreneurship has appeared as a relevant form of collaboration since the 1990s, with the rise of business incubators located at universities.

Considering the degree of the partners' involvement in the various types of interactions, Perkmann and Walsh (2007) show that research partnerships or the hiring of academic research involves a strong relationship while technology licensing, via

Table 11.1 University-industry links

Research Partnerships	Inter-organizational arrangements for pursuing collaborative R&D
Research services	Activities commissioned by industrial clients including contract research and consulting
Academic entrepreneurship	Development and commercial exploitation of technologies pursued by academic inventors via a company they (partly) own
Human resource transfer	Multi-context learning mechanisms such as training of industry employees, postgraduate training in industry, graduate trainees, and secondments to industry
Informal interaction	Formation of social relationships and networks at conferences, etc.
Commercialization of property rights	Transfer of university-generated intellectual property (such as patents) to firms, e.g., via licensing
Scientific publications	Use of codified scientific knowledge within industry

Source: Perkmann and Walsh (2007)

the commercialization of intellectual property, involves a lower level of engagement between parties.

Historians have traced collaborations between European companies and university researchers back to the 1800s (Mazzoleni and Nelson 2007). The natural sciences became a generally accepted area of teaching and research at universities after the reform of German universities (beginning in 1810), based on the vision of Alexander von Humboldt. Despite the fact that Humboldt's vision was that "knowledge ought to be pursued for its own sake, without regard for its practical relevance to social and economic problems" (Mazzoleni 2005), some German universities started to collaborate with industry, engaging in research activities to investigate problems brought into focus as a result of scientists' observations of related practical problems.

According to Nelson (1996), universities are recognized as repositories of public scientific and technological knowledge, created and reproduced through education and enhanced through research. In the United States, universities have provided technical personnel and many ideas about innovations in products and processes to industry. Contradicting the notion that university and academic scientists remain separate from the productive sector (except to provide qualified personnel and published papers), academic science and industrial research have been very close in many fields.

Cooperation between universities and companies is a natural partnership in developed countries, where firms seek external sources of knowledge, despite having vast resources and R&D laboratories. The increasingly rapid pace of the creation of new goods and services demands sources of creativity that go beyond the boundaries of firms, and cooperation with customers, suppliers, research institutes, and even competitors is already common in many industries. In developed countries, collaboration with universities and research institutes is directed towards the pursuit of radical product innovations, which may open new markets or segments. Universities are the preferred partners in new technological fields or when the speed of technological change is high and business results are uncertain (Belderbos et al. 2006). Cooperation is most needed in emerging countries, where universities are the main source of knowledge that can lead to innovation, and the agents of economic and social changes. However, it is in these countries that cultural differences and prejudices can inhibit collaboration (ECLAC 2010).

Etzkowitz (1989) identified a new institutional role for universities. From their original mission of education for the training of qualified personnel, universities incorporated systematic research activities by the end of the nineteenth century. Research activities were once performed, in general, by individual inventors in domestic workshops, or existed only in some universities (e.g., the American Land Grant, having a strong interaction with the agricultural sector; cf. in Germany, where some universities collaborated with the chemical industry). Etzkowitz called attention to a new academic revolution, translated into "the capitalization of knowledge." This concept evolved into the Triple Helix, which explains the intersection of relatively independent institutional spheres, creating hybrid organizations like technology transfer offices (TTOs) in universities and research institutes, and new funding agencies such as venture capital firms and networks of angel investors (Etzkowitz 2008).

In the last decade, open innovation has been considered the preferred way for company growth, via collaboration among firms, universities, research institutes, suppliers, and customers. The famous quote by Henry Chesbrough (2003), "not all brilliant scientists work in your company," features well this new era of innovation. Before him, several authors (e.g., von Hippel 1988) had referred to the important contribution of different sources for business innovation.

Recently, international literature has discussed new concepts such as "academic science," "entrepreneurial university," "capitalization of knowledge," "academic entrepreneurship," and the differences in the underlying values of this transformation (Lam 2011; Siegel et al. 2007; Etzkowitz 2008). Critics mention "commercial science," "academic capitalism," and the "privatization of knowledge" (Oliveira and Velho, 2009; Chauí 1995). According to Chauí (1995), privatization "should not be understood only as equivalent to private sources of research funding, but mainly private determination of criteria, objectives, procedures, results, time and use of research." In the 1990s, this theme has achieved great importance in discussions on Brazilian science and technology (S&T), which sought to increase the competitiveness of Brazilian companies in the context of economic liberalization. The *USP Journal* published the *University–Company Dossier* (Mar/Apr/May 1995); the Brazilian Institute of Scientific and Technological Information published the book *University-Business Interaction* (in two volumes: 1998, 1999) and the *Journal of Management* (RAUSP) published a special issue in its October/December 1999 edition.

The first Brazilian governmental efforts to bring universities and companies together in innovation projects began in 1980, which resulted in the creation of technological innovation offices (NITs), a program introduced in 1981 by the National Council for Scientific and Technological Development (CNPq), with the support of FINEP—Innovation and Research. However, as can occur with public initiatives, resources became scarce over time, and many organizations failed to utilize them and most were later closed (Medeiros et al. 1987). Other programs and laws followed to stimulate cooperation: Act 8661/93 (tax incentives for innovation), Act 8248/91 (tax incentives for information technology companies), Partnership for Technological Innovation (a program of the São Paulo Foundation for Research Support, 1995), the Green–Yellow Fund (2000), various sectorial funds with calls for partnership projects, and the Innovation Act (2004). Article 16 of the Innovation Act requires that public S&T institutions establish NITs to manage their innovation policy, and lists their minimum competencies. Over three decades there has been progress, especially with the institutionalization of these offices and mechanisms for technology transfer. However, various ideas and attitudes contrary to cooperation still exist, which have hindered the process of innovation in Brazil.

This chapter analyzes the evolution of university–industry relations for innovation in Brazil, from a survey of articles (between 1980 and 2012) published in major Brazilian journals or presented at relevant national and regional conferences. The year 1980 marks the beginning of discussions regarding the creation of the first NITs in Brazilian universities. We seek to determine whether these papers portray the changing perceptions of the different stakeholders—universities, companies, and financing agencies—on the subject, from reports of individual experiences and case studies to reflections on more effective models and processes of collaboration.

We found two papers that evaluated Brazilian academic publications on this topic. The first is Zanluchi and Gonçalo (2007), who conducted a survey of the most important articles of the previous 10 years. However, they did not explain the criteria of relevance adopted and used an imperfect sample: they ignored major national events where such papers are presented, such as the National Meetings of Production Engineering, the Symposia on Management of Technological Innovation, and ALTEC (Asociación Latino-Iberoamericana de Gestión Tecnológica) conferences, which comprise the majority of Brazilian authors. However, the definition of categories for the classification of articles is relevant. The second study is that by Closs and Ferreira (2012); they examined articles published in national journals on the transfer of technology from universities to companies, covering 2005–2009.

This chapter is organized into five parts, starting with this introduction. The second part presents a brief review of the literature and key issues presently under discussion. To follow, we explain the methodological procedures of the research, the results and discussion, and conclusions.

11.2 Background and Theoretical Framework

11.2.1 University–Industry Relations in the International Literature

The landmark of relations between universities and companies for innovation is represented by the Bayh–Dole Act (Public Law 96-517), approved in the United States in December 1980 amid concerns about the loss of US competitiveness. The introduction of the act is considered a watershed on the issue of licensing technologies developed at universities. At the time, the US Congress was discussing the ways to promote private sector growth by making use of research conducted with public money, considered as an underutilized economic resource. Until then, academic research was seen as a good whose results should be distributed to all interested parties, in exchange for the autonomy to choose its direction—the only recognized property rights for inventors were the satisfaction and recognition stemming from their findings (Etzkowitz 1989).

The adoption of the law was not a smooth process, with several congressmen raising the possibility of foreign companies obtaining financial benefits from the results of years of government-funded research. One example mentioned at the time was the partnership between the German company Hoechst and the Massachusetts General Hospital, Harvard University, regarding the commercialization of biomedical research (Matkin 1990).

Before the act, government retained ownership of all patents resulting from public funding and the right to license them, without exception, to any interested company. To some extent, this hindered innovation because companies did not want to develop products over which they did not have exclusive marketing rights (Schacht 2006). Thus, only 5 % of government-owned patents were used by the private sector, although much of this portfolio had market potential. The law was created to overcome this issue, allowing universities, small businesses, and nonprofit organizations, as users of federal funds for research, to retain ownership of the results, licensing them, and stimulating innovation in companies (Thursby and Thursby 2011).

Since then, there has been a dramatic rise in the number of disclosed inventions, requested patents, and their exclusive licensing, via the payment of royalties. Other laws were created soon after, forming a coherent set of instruments to induce innovation (Stevenson-Wydler Technology Innovation Act, 1980; National Cooperative Research Act, 1984; Federal Technology Transfer Act, 1986; National Competitiveness Technology Transfer Act, 1989); however, the Bayh–Dole Act is still considered the most important.

Over time, criticisms arose concerning conflicts of interest and the emphasis on applied (rather than basic) research, particularly regarding research in the pharmaceutical and biotechnology industries (Schacht 2006). Some claimed that the Bayh–Dole Act "inserted the motivation for profit directly into the heart of the academic life," "diverted teachers from basic research motivated by curiosity," and "favored the execution of research projects with more immediate potential market" (Thursby and Thursby 2011). Such arguments have not been proven. A survey of 3,400 teachers from six US universities between 1983 and 1999 suggests that the share of basic research has not changed, while licensing has increased 10-fold. Although this is the best way to maximize social returns on public investment in R&D, it does not reduce the effectiveness of open channels of knowledge dissemination, such as publications, conferences, and consulting, even in the pharmaceutical industry where patents are extremely important (Thursby and Thursby 2003).

The development of the US economy in the second half of the 1990s was extraordinary, especially in fields like biotechnology and information and communication technologies, and relates to the significant increase in patenting and licensing (Mowery et al. 2001). Some universities, like California and Stanford, were already very active in patenting and licensing before the Bayh–Dole Act, while Columbia was directly influenced by its introduction (Mowery and Sampat 2005).

Foray and Lissoni (2010) summarized the two dilemmas related to this cooperation. The first concerns individual scientists and the potential trade-off between basic research activities and those activities required to develop and commercialize academic inventions. The second occurs at the system level, and deals with the tension between the industry's need to rely upon clear intellectual property rights, and the cumulativeness of university knowledge, which requires that the results of academic research should be freely accessible.

Lee (1996), in a national survey of nearly 1,000 academic researchers, noticed a stronger disposition to collaborate with companies compared with the 1980s. The majority supported the idea of active participation in local and regional economic development, promoting the commercialization of academic research and encouraging teachers to offer consulting to private companies. However, they were against the involvement of universities in partnerships with companies (e.g., assisting in the creation of start-ups or holding shares in firms) in exchange for the payment of patents' licensing fees or royalties. In a later study, two concerns emerged: the need for business financing, given the reduction of federal R&D budgets, and the impact of a closer cooperation with companies, which could interfere with academic freedom. Lee (1998) suggested that this tension could be resolved by granting tax incentives to companies, thus retaining basic research while also contributing to applied research.

Kenney and Patton (2009) criticized the model implemented by the Bayh–Dole Act. In that model they see information asymmetries, ineffective incentives, and contradictory motivations for the university, inventors, potential licensees, and technology transfer offices. They suggest two alternatives for commercializing research results. The first is gives the property to the inventor, who will choose the best way to license the invention—via the university office or another organization that may assist in the process. In exchange, the researcher would grant the university an interest in any returns arising from commercialization. The second alternative is to make all inventions available in the public domain or to license them without exclusivity. In a 2011 article, Kenney and Patton (2011) examined the creation of spin-offs from six universities—five US and one Canadian. The University of Waterloo was the only one to adopt a regime where the inventor holds the intellectual property rights, and it was significantly more successful in creating technology-based firms.

Several European countries were inspired by the Bayh–Dole Act to promote technology transfer from universities to industry. However, as Mowery and Sampat (2005) show, the structural differences between countries' higher education systems mean that simply transposing the law to another institutional environment does not guarantee its success. In the 1990s, European universities began to seek new sources of research funding in view of decreasing government budgets, which were replaced by competitive funds oriented to applied research and business financing (Geuna and Nesta 2006).

Over time, patenting by universities has grown, but this is heterogeneous across countries and sectors. Furthermore, licensing has not been profitable for most universities, although some have attracted significant additional revenue. The optimistic projections for income resulting from patent licensing did not consider the expense of the process and the maintenance of patents, not that concerning the operation of TTOs.

The end of the 1990s saw most European countries change the ownership of intellectual property rights. These went from *inventor ownership* (or *professor's privilege*) to *property of the institution*, as was the case in research institutes (Geuna and Rossi 2011). Denmark was the first country to make these changes in 2000, followed by Germany, Austria, Norway, and Finland between 2001 and 2007. France, Greece, Spain, Switzerland, and the United Kingdom, where intellectual property was already the property of the institution, were encouraged to reinforce these rights and increase technology transfer. Italy did the opposite, adopting professor's privilege in which the university would receive between 30 % and 50 % of licensing revenue. Unlike other countries, the 2001 Italian legislation considered inventors to be more capable of taking advantage of their inventions, as universities did not have the expertise or the culture to do so. Sweden retained the system based on professor's privilege, in which all revenue goes to the researcher, who bears the costs of patenting and licensing. In these countries, patenting and transfer of knowledge are increasingly recognized as legitimate and relevant academic activities.

One of the major problems identified by the directors of TTOs in American universities is to convince researchers to communicate their inventions—less than half of all inventions with market potential are revealed. This situation can occur for a number of reasons including scientists do not realize the value of their inventions or do not want to divert time from their research, as their involvement with the licensee is often necessary to ensure commercial success. In contrast, many inventions have questionable potential (Jensen et al. 2003). Greenbaum and Scott (2010) defended the establishment of regional NITs rather than them being located at every university. These are often inefficient, underfunded, and have few experts. Regional offices could operate with scale economies, with staff trained in technology transfer, assisting several universities. Geuna and Rossi (2011) suggested a modern version of the professor's privilege, which would give researchers the intellectual property rights; the licensing would occur through regional offices without university intervention. This would speed up the process of technology transfer, as well as encouraging entrepreneurship.

After the large number of articles praising the Bayh–Dole Act, and others showing that the initial concerns about the shift from basic research to applied research were not proven, some more recent articles have proposed improvements in the law, especially regarding the performance of TTOs. Furthermore, the literature on the creation of spin-offs or spinouts (companies established to explore the results of academic research) has expanded as a mechanism for the commercialization of intellectual property, as well as the traditional licensing mechanism (Siegel et al. 2007). This form of technology transfer requires a review of the structures and practices of universities because it takes longer, is more uncertain, involves the search for venture capital, and the development of marketing capabilities. In the case of rupture technologies (genetics and stem cells), the moral and ethical aspects require a more active role from the university.

Several papers have examined changes to the profile of researchers. Lam (2010) investigated the opinions of 734 scientists from five major UK universities. According to more or less favorable views on university–industry relations, forms of interaction, motivating factors, the perceived legitimacy of the commercialization of results, and work strategies and identities regarding their role, Lam suggests a typology of scientists: *traditional*, *traditional hybrid*, *entrepreneur hybrid*, and *entrepreneur*. In a subsequent paper, she uses the concepts of *gold* (financial reward), *ribbon* (reward for the reputation or career) and *puzzle* (intrinsic satisfaction) to analyze the internal and external aspects of scientists' motivation to engage in commercial activities. Lam concludes that financial rewards are secondary in relation to intrinsic satisfaction and earnings in reputation/career, and suggests that policies to encourage cooperation should be based on these motivations (Lam 2011).

Rothaermel et al. (2007) made a comprehensive review of 173 articles published between 1980 and 2005 in international journals under the broad topic "entrepreneurial activity in universities," including patenting, licensing, creating new companies, technology transfer through incubators and technology parks, and contributions to the regional economic development. They present a detailed assessment of the state of the art in this field and, ultimately, a conceptual framework with four main streams of research that resulted from the analysis of the papers of the last decade: (i) entrepreneurial university, (ii) productivity of technology transfer offices, (iii) creation of new companies, and (iv) innovation environment, including innovation networks. Despite covering a 25-year range, the vast majority of articles have been published in recent years, with a large increase of papers on academic entrepreneurship from the late 1990s. This exponential growth is also the result of the publication of journals' special issues—between 2000 and 2005, 127 articles in the sample were published.

Research from the United States and Europe starts from the natural reality of cooperation with companies and its relevance to business innovation. Much of the current literature, as shown above, discusses how to improve cooperation; how to better structure TTOs to be more efficient; how to examine the profiles, the motivations and justifications of researchers who cooperate (or not) with companies; and how to reflect upon the trend towards greater academic entrepreneurship as compared with the traditional form of technology transfer via patent licensing.

There are many quantitative studies that arise precisely from the practice of cooperation, and many use large data sets. Several articles discuss whether this relationship will have harmful effects on basic research and the production and dissemination of scientific knowledge; while there is a range of distinct results, most are favorable to cooperation (Larsen 2011). However, regarding biotechnology, given its strategic importance for the production of new drugs, the discussion is deeper, involving conflicts of interest (Glenna et al. 2011).

A very different situation exists in Brazil, where academic representatives still discuss whether cooperation should be encouraged, whether it is good or harmful for the university, and whether the developed knowledge should be patented to allow its transfer or the creation of new ventures, stimulating entrepreneurship.

11.2.2 The Evolution of Science & Technology in Brazil: Support for Innovation and University–Industry Relations

In the early 1950s, with the creation of the Coordination for Improvement of Higher Education Personnel (CAPES) and CNPq, the Brazilian government institutionalized the support for S&T activities, inspired by the linear model of innovation. These two agencies were designed to meet the demands of researchers, mainly considering the academic merit of research proposals without giving priority to themes or areas (De Negri and Cavalcante 2013). Companies were seen as external agents of the S&T system, as users or consumers of knowledge generated in universities and research institutes.

The National Bank for Economic and Social Development (BNDES) soon followed, established in 1952. It developed mechanisms for long-term financing such as credit supply and shareholding in national companies. In 1967, the Financing Agency for Studies and Projects (currently named FINEP Innovation and Research) started to support feasibility studies made by national engineering and consulting firms to stimulate investments in productive capacity. In 1976, it began to clearly support the technological qualification of domestic firms via a specific program (Bastos 2012).

FINEP expanded the role played by the Scientific-Technological Development Fund (established in 1964 by BNDES) to finance the introduction of graduate programs in Brazilian universities. In 1971, it became the Executive Secretariat of the National Fund for Scientific and Technological Development. During the 1970s, it financed the establishment of new research groups, the creation of thematic programs, the expansion of S&T infrastructure, and the institutional consolidation of research and graduate studies in Brazil. Its performance became quite comprehensive, including all stages of S&T development, including basic research, applied research, and the improvement and development of products, services and processes. FINEP also supports incubating technology-based companies and establishing technology parks, and operates its programs through refundable (loans) and non-refundable (grants) financial support, and shareholding investments in companies.

The Ministry of Science and Technology (now the Ministry of Science, Technology and Innovation) was created in 1985, strengthening the roles of CNPq and FINEP. In the 1990s, Brazil's S&T policy was geared towards the absorption, adaptation, and diffusion of imported technology—in a direct way via licenses and other agreements, or as technology embedded in machinery, equipment and systems—to raise the levels of productivity and competitiveness. There was a consensus that industrial policies should be horizontal, meeting the demands of all sectors without electing specific priorities.

Innovation became an important goal of the S&T policy and justified the creation of sectorial funds in the late 1990s, whose resources came from specific contributions from a number of industry sectors (e.g., electricity, telecommunications, and oil exploration). These funds were intended to be a source of stable funding for R&D in 14 strategic sectors, in addition to two special funds: for promoting university–industry interactions and for improving research infrastructure in universities and research institutes. Their role grew over time and came to represent one of the main instruments of innovation policy in Brazil (Araújo 2012).

The Industrial, Technological, and Foreign Trade Policy (PITCE), launched in 2003, opened the third period in the history of innovation incentives in Brazil, bringing two major advances: the Innovation Act in 2004, and the "Good Act" (*Lei do Bem*) in 2005. The first stimulated university–industry cooperation based on the Bayh–Dole Act, and provided the institutional basis for strategic alliances between research institutes and companies. The Innovation Act also established rules for sharing infrastructure and the economic benefits resulting from innovations, with rewards to researchers. Furthermore, the act facilitated technology transfer and the mobility of scientists between academia and the business sector. It first introduced in Brazil direct subsidies (grants) for company R&D, and created the possibility of government procurement oriented by technological criteria. The Good Act provides (see Chapter III of the act) a set of tax incentives for technological innovation, improving the 1993 legislation.

PITCE was replaced in 2008 by the Productive Development Policy, which extended its scope by including more sectors. However, the general direction was not changed, and innovation remained one of the basic pillars for economic growth. Currently, there are two programs in effect: the Bigger Brazil Plan (*Plano Brasil Maior*, 2011–2014) and the National Strategy for Science, Technology and Innovation (2012–2015). The first comprises a set of initiatives to support and protect the industry, with a broader extent than previous plans, and lists 10 goals for 2014 that refer to investments in R&D, labor qualification in industry, and a more efficient use of energy. It also combines instruments to support competitiveness, including the expansion of funding from BNDES, the reduction of the value-added tax, tax reductions for selected sectors with countermeasures against "de-industrialization," an increased use of safeguards, and increases in import tariffs.

The Brazilian innovation policy is managed by three ministries: the Ministry of Science, Technology and Innovation, through its agencies CNPq and FINEP; the Ministry of Development, Industry and Foreign Trade, through BNDES and the Brazilian Agency for Industrial Development; and the Ministry of Education, to which CAPES is subordinated.

From the early 1950s to the 1980s, Brazilian economic policy favored industrialization based on an import substitution strategy. The country grew at high rates with the protection of tariff and non-tariff barriers, which kept away international competition without the widespread practice of product and process innovation. Established multinationals innovated in their home countries, making it difficult to increase local technological capacity via learning. Hence, with some exceptions, Brazilian companies do not have a history of conducting R&D, which results in a low innovation capacity (Arruda 1994).

In contrast, the S&T policy aimed at technological autonomy by sending lecturers to get master's and doctorate degrees abroad to enable graduate courses to be established in Brazil. This dissociation between the two policies resulted (years later) in a group of highly trained professionals working at universities and public research institutes without concern for the needs of companies; these needs were met by technology imports.

The 1980s were years of low economic growth that did not stimulate innovation. Fiscal and macroeconomic unbalances, with low growth in income and domestic demand, along with less sophisticated consumption patterns, did not generate a demand for innovative products (De Negri 2012). This influenced innovation performance in the following decades. In 2011, the Brazilian private sector invested 0.59 % of GDP in R&D, far from the numbers of South Korea (2.68 %) and China (1.22 %).

Since the economic liberalization in the early 1990s, domestic companies have been forced to adopt modern actions to face competitors in Brazil (i.e., the arrival of new foreign companies). Innovation was necessary and technological capability became the main objective of Brazils' industrial policy, for which cooperation with universities was essential (Mello et al. 2011).

In 1993, a comprehensive study commissioned by the Ministry of Science and Technology and the World Bank concluded that a new policy for S&T should perform apparently contradictory aims: "to encourage freedom, initiative and creativity of researchers and, at the same time, establish a strong link between their activities and the needs of the economy, of the educational system and of the society as a whole" (Schwartzman 1993, p. ii). However, the second aim has not evolved unlike the first, and Brazil continues to favor support for academic research, reflecting the adoption of the linear model of innovation that can no longer adequately meet the innovation needs of the modern world.

It is acknowledged that innovation occurs in companies that can offer new products and services to the market. In all countries that have used knowledge to generate innovations, most scientists work in the R&D laboratories of enterprises. However, in Brazil, the latest official data (for 2010) show that of 234,797 researchers, only 42,000 work in companies (17.6 %), with the others being distributed in higher education and in governmental or private research institutions (Cruz 2010).

It is still said that the country lacks an innovative entrepreneurial culture and that this is the main cause of Brazils' low innovation rates and the reduced concern of companies for available support instruments (despite their significant growth in number, variety, and volume of resources). Some studies indicate reasons for little innovation (e.g., unstable economic environment, high taxes, poor infrastructure, low qualification of manpower, legal uncertainty, excessive bureaucracy, in addition to the high risk inherent to such activities), thus emphasizing the need for state support. In developed countries, government also supports innovation, and companies do not face most of the problems outlined above (Pacheco and Almeida 2013; De Negri 2012).

The reorientation of Brazil's S&T policy in the early 2000s enhanced the relevance of technological innovation, meeting the interests of industry to increase competitiveness (Bastos 2012). This change could have redirected the support to academic science and its extensive research infrastructure. However, the strong lobby of scientists, via their associations, ensured the continuity of significant support.

The private sector has also moved forward, demanding greater government support for innovation. Initiatives such as the Competitive Brazil Movement and more recently the Entrepreneurial Movement for Innovation, along with industry associations such as the National Confederation of Industry, the Federation of Industries of the State of São Paulo, and the National Association for Research and Development of Innovative Companies, have emphasized the need for stronger support for innovation.

Although there is currently a broad set of public instruments available to businesses, they reach only a small proportion of companies that claim to be innovative, and there is still a considerable ignorance regarding the legislation that benefits innovative companies. As of 2011, only 767 companies had taken advantage of the tax incentives provided in the Good Act (between the introduction of the act in 2006 and 2011, there were a total of 1,475 registered companies in Brazil). According to the Technological Innovation Survey conducted in 2008 by the

Brazilian Institute of Geography and Statistics (PINTEC/IBGE), approximately 30,000 companies reported R&D investments, while the government estimated that more than 4,000 would be able to claim the tax benefit. The law encourages firms to hire researchers, master's and PhD holders, whose wages may be deducted as R&D expenses.

The latest program to support business innovation was launched by the government in 2013, the *Inova Empresa* (Innovate Company) Plan, which supports firms of all sizes in the agriculture, industry, and service sectors, via finance at subsidized interest rates, economic grants, the promotion of projects in partnership with research institutions, and shareholding in technology-based companies. Seven strategic axes were defined: agriculture and agribusiness, energy, oil and gas, health, defense, information and communication technology, and socio-environmental sustainability. The creation of the Brazilian Company for Industrial Research and Innovation is one of the results of the program. It aims to promote cooperation between companies and public/private research institutions that carry out R&D activities.

With these initiatives the government expects to improve Brazil's performance in science and innovation. In science, Brazil is responsible for 2.4 % of all articles published in leading international journals, ranked 11th worldwide. The fact that the country concentrates its research in biological, health, and agricultural sciences has resulted in an extremely successful bio-energy program. However, the scientific fields most cited in world patents are engineering, physics, inorganic chemistry, and materials sciences (De Negri 2012). Furthermore, Brazil graduates a low number of engineers—23 per 10,000 inhabitants (30,000 per year), while Israel and Japan graduate 140 and 75 per 10,000, respectively. India and China together graduate 850,000 engineers per year (Feldmann 2011).

The same university that must be congratulated for the scientific production of its professors is partly responsible for poor business innovation in Brazil, given its difficulty in establishing relations with companies to provide them with technology. The Brazilian academic culture stems from the European model, in which the scientist works in the internal environment, while American universities encourage scientists to be an inventor and entrepreneur (Feldmann 2011).

Universities have taken on several different functions in recent years. Their mission, beyond education and the training of human resources, also includes research and extension, and they must ensure the advancement of science. However, the transfer of research results to companies is critical to guarantee a flow of information within the national system of innovation. In the case of Brazilian universities, there is an increased awareness of the need to transfer the results of publicly funded research to society; however, this has not resulted in a cooperative attitude with the private sector (Fujino et al. 1999). Significant barriers to collaboration relate to the organizational culture of public universities, and are sustained by conflicting ideological values with business partnerships and standards that maintain the bureaucratic hierarchy (Troyjo 2013).

11.3 Research Methods

This research employed a qualitative and exploratory approach, using bibliographic research and bibliometric study. After an extensive literature search of articles published in national journals and papers presented at relevant local and national conferences between 1980 and 2012, we conducted a content analysis (Bardin 2000) for the categorization of works by subthemes within the scope of university–industry cooperation for innovation.

We surveyed various proceedings from the Symposium on Technology Innovation Management (SGIT), Seminars of the Latin-Iberomerican Association of Technology Management, the National Meetings of Production Engineering, the Annual Meetings of the National Association for Research and Graduate Studies in Administration—Division of Science, Technology and Innovation Management, and the National Meetings of Economics, promoted by the National Association of Graduate Centers in Economics (EnANPEC). The research was limited by the availability of the proceedings of such conferences on the internet or at the library of the School of Economics, Administration and Accounting of the University of São Paulo, either printed version or on CDs.

The following journals were also examined: *Journal of Business Administration*, *Journal of Contemporary Management*, *Journal of Management*, *Journal of Management and Innovation*, *Brazilian Journal of Innovation*, *Management & Production*, and *Strategic Partnerships*. A special issue of the *USP Journal* (1995), which published the *University-Company Dossier*, was used in the discussion, but its articles were not included in the sample.

We also included EnANPEC meetings because in recent years they have received several studies on this topic, especially related to the results of the thematic project "Interactions of universities and research institutions with industrial companies in Brazil," funded by the São Paulo Foundation for Research Support from 2008 to 2012, and whose team includes many economists. Papers resulting from this project were found in the *Journal of Economics*, *Journal of Political Economy*, and *Contemporary Management*. Relevant information was used in the discussions, although they were not part of the sample.

We did not include any theses and dissertations in the bibliometric study because they had not undergone anonymous review (peer review). To be included, relevant results stemming these works must have been later published in academic journals or presented at conferences. Furthermore, papers related to specific arrangements for the promotion of innovation (e.g., incubators, technology parks, local clusters, national, regional and local innovation systems, or the creation of spin-offs) were not considered. In these cases, we only included studies that specifically focused on the relations between local universities and businesses located in incubators or technology parks.

To better understand the results, we searched the literature to limit the discussion to certain topics within the broad matter of university–industry relations, and turned to authors who are a reference on this subject, even if they are not part of the sample.

From a historical point of view, we situated the context in which NITs were created, in the early 1980s, via a government initiative in collaboration with public universities and research institutes. Regarding the sample, Tables 11.2 and 11.3 present the analyses of the papers, according to the categories below, which were adapted from a classification suggested by Zanluchi and Gonçalo (2007):

A: theoretical papers on university–industry cooperation in general, featuring the positive aspects, barriers, and challenges of the relationship; the Triple Helix; management tools and strategies; studies on intellectual property, licensing and marketing policies in universities;
B1: papers that address the mechanisms of interface that induce and facilitate cooperation in universities (offices, information services, service networks, technology transfer structures);
B2: papers that specifically deal with relations between universities and firms that are located in incubators, technology parks, and business networks;
B3: papers that focus on the aspects of technology transfer from universities and research institutes to companies. These include case studies or field research in companies, to diagnose general or specific aspects of the transfer;
C: papers that highlight the perspective of companies on cooperation with universities or research institutes; and
D: papers that highlight the perspective of government/society. They include studies on public policies or the evaluation of programs managed by agencies that foster innovation.

The articles were classified according to the content analysis technique (Bardin 2000), specifically thematic analysis. We collected and analyzed 249 papers, of which 201 were presented at conferences and 48 published in journals (Tables 11.2 and 11.3). Papers presented at events that were later published were counted in the first edition. The events with the largest number of papers were ALTEC Seminars and SGIT Symposia, representing 61.7 % of the papers presented at events. Considering the conferences, which show research results faster than journals, most papers on the subject were presented between 1999 and 2005. There were 30 relevant papers presented at the SGIT Symposia in 2000 (11) and 2002 (19), and we found 28 papers presented at ALTEC seminars (8 in 1999, 5 in 2001, 5 in 2003, and 10 in 2005).

It should be mentioned that the bibliometric method has its weaknesses and provides a partial view of the cooperation between universities and companies in Brazil. Other analytical evidence brought by alternative or complementary approaches could contribute to a more complete analysis of this topic.

11.4 Results and Discussion: University–Industry Relations in the Brazilian Literature

The period between 2003 and 2005 coincides with extensive discussions on the subject and the launch of the Brazilian Innovation Act in 2004 (regulated in 2005), which changed university–industry relations, especially regarding the requirement

Table 11.2 Distribution of papers in the sample, by event or journal, and thematic categorization

SGIT 1980–2012	ALTEC 1985–2011	EnANPAD 1980–2012	ENEGEP 1996–2011	EnANPEC 2001–2011	RAE 1980–2012	RAUSP 1980–2012	RAI 2005–2012	RBI 2002–2012	G&P 1994–2012	P.E. 2000–2012	RAC 1997–2012	Total
78	*46*	*35*	*33*	*9*	*4*	*23*	*5*	*5*	*3*	*6*	*2*	*249*
34 A	*16 A*	*14 A*	*12 A*	*6 A*	*2 A*	*13 A*	*2 A*	*5 A*	*2 A*	*5 A*	–	*111 A*
7 B1	8 B1	5 B1	5 B1	–	–	3 B1	–	–	–	–	–	28 B1
4 B2	1 B2	1 B2	–	–	–	–	1 B2	–	–	–	–	7 B2
28 B3	*17 B3*	*9 B3*	*10 B3*	*2 B3*	*2 B3*	*5 B3*	*1 B3*	–	*1 B3*	–	*1 B3*	*76 B3*
3 C	1 C	5 C	4 C	–	–	2 C	1 C	–	–	1 C	1 C	18 C
2 D	3 D	1 D	2 D	1 D	–	–	–	–	–	–		9 D

Note: SGIT: Symposium on Technology Innovation Management; ALTEC: Seminars of the Latin-Iberomerican Association of Technology Management; EnANPAD: Annual Meetings of the National Association for Research and Graduate Studies in Administration—Division of Science, Technology and Innovation Management; ENEGEP: National Meetings of Production Engineering; EnANPEC: National Meetings of Economics; RAE: *Journal of Business Administration*; RAUSP: *Journal of Management*; RAI: *Journal of Management and Innovation*; RBI: *Brazilian Journal of Innovation*; G&P: *Management & Production*; P.E.: *Strategic Partnerships*; RAC: *Journal of Contemporary Management*

A: Theoretical papers on university–industry cooperation in general, featuring positive aspects, barriers and challenges in the relationship; the Triple Helix; management tools and strategies; studies on intellectual property, licensing, and marketing policies in universities

B1: Papers that address the mechanisms of interface that induce and facilitate cooperation in universities (offices, information services, service networks, technology transfer structures)

B2: Papers that specifically deal with relations between universities and firms located in incubators, technology parks, and business networks

B3: *Papers that focus on the aspects of technology transfer from universities and research institutes to companies. These include case studies or field research in companies, to diagnose general or specific aspects of the transfer*

C: Papers that highlight the perspective of companies on cooperation with universities or research institutes; and

D: Papers that highlight the perspective of government/society. They include studies on public policies or evaluation of programs managed by agencies that foster innovation

Table 11.3 Distribution of papers in the sample, by period of publication

Period	SGIT 1980–2012	ALTEC 1985–2011	EnANPAD 1980–2012	ENEGEP 1996–2011	EnANPEC 2001–2011	RAE 1980–2012	RAUSP 1980–2012	RAI 2005–2012	RBI 2002–2012	G&P 1994–2012	P. E. 2000–2012	RAC 1997–2012
1980–84	1	–	1	–	–	–	1	–	–	–	–	–
1985–89	1	3	–	–	–	2	4	–	–	–	–	–
1990–94	6	1	3	–	–	1	7	–	–	–	–	–
1995–99	4	10	3	3	–	–	9	–	–	–	–	–
2000–04	38	10	6	12	–	1	1	–	1	1	5	–
2005–12	28	22	22	*18*	9	–	1	5	*4*	2	*1*	2
	78	*46*	35	33	9	*4*	23	5	5	3	6	2

to create NITs at federal universities and research institutes. It is important to mention a number of studies that reflect researchers' concerns on the subject and describe the prospects and challenges of cooperation. Stal and Fujino (2005) presented experiences of collaboration in a survey conducted with business members of the National Association of R&D in Innovative Companies, concerning technological partnerships with universities, the management of intellectual property, and their expectations regarding the potential of the Innovation Act to improve technology transfer mechanisms. The results indicate the need for improvements in the law and a better definition of university policies for cooperation with companies related to the management and sharing of intellectual property.

Matias-Pereira and Kruglianskas (2005) discussed policies for innovation management in Brazil, by analyzing the consistency of the law and the subsequent regulation decree, which they compared with the successful implementation of industrial and technological policies in other countries. Their conclusion is that despite some shortcomings, the act represented an important milestone by building a model of technological development for Brazil. Garnica et al. (2007) assessed the management of intellectual property at the University of São Paulo in view of the challenges imposed by the law. They identified the difficulties in regulating this matter and the need for its update, thus strengthening the corporate view reported by Stal and Fujino (2005).

Hence, we decided to analyze in detail the contents of category A papers (theoretical papers) presented and published as of 2005, after the Innovation Act was introduced, to observe new approaches to the topic. Furthermore, those in category B3 (technology transfer) were studied to verify, in empirical studies, whether there had been changes in the university–industry relationships given the mandatory establishment of NITs. We assumed that articles in these two categories could reflect new concerns in terms of research or more favorable assessments and prospects than papers published before the act. We also expected to observe in empirical studies changes in the infrastructure of universities to cooperate with companies, resulting from improvements in policies regarding intellectual property management, and thereby affecting the university's organizational culture by highlighting concrete results of technology transfer.

We found 114 articles in the period, presented at conferences and published in journals, which represent 45.8 % of the papers collected in the research (Tables 11.2 and 11.3). In category A, we found literature reviews; studies that analyze the academic environment to assess the potential operationalization of the Innovation Act; challenges for the management of intellectual property in universities; suggestions of best practices for innovation based on international experiences; proposals of management models; evaluation of implemented policies; analysis of management strategies of innovation networks and partnerships; mapping of capability building in universities; operation of the Triple Helix; proposals for setting up a national evaluation system; knowledge management for cooperation; and the motivations of scientists involved in cooperation with companies. In category B3, we found descriptive papers, such as studies in companies to identify the potential involvement of universities or research institutes in the development of innovation; identification of external sources used by firms or opportunities for technology transfer from universities; studies to identify barriers for transfer; relationship difficulties due to cultural differences; stages of interaction; evaluation of implemented projects; and the forms and mechanisms for technology transfer. We found only a few articles that confirm the assumption described above.

Carvalho et al. (2006) proposed a methodology to identify relevant research groups at universities and research institutes and their expertise in certain technologies and industries; they called this method a "strategic search." The project developed, tested, and implemented procedures to map and classify the competencies of several groups. While Renault requested the project to identify existing capabilities in the automotive industry, the methodology can be applied in other sectors. Garnica and Torkomian (2009) studied the institutional policies and challenges for technology transfer at the five public universities located in the state of São Paulo (*Universidade de São Paulo, Universidade Estadual Paulista Júlio de Mesquita Filho, Universidade Estadual de Campinas,* and *Universidade Federal de São Paulo e Universidade Federal de São Carlos*) to identify the difficulties and successes in technology transfer processes. The study involved interviews with university and partner company professionals about contracts, thus allowing a comparative analysis. In all five universities, there was an increase in patenting and licensing, but these procedures are very recent.

Póvoa (2006) discussed whether a university should patent its inventions (following the logic of scientific production) based on criticisms found in the literature and the results of a survey conducted by the Directory of Research Groups of the National Council on Scientific and Technological Development (CNPq 2004 census) on technology transfer from universities and public research institutes. He concluded that patenting must be combined with knowledge dissemination, and suggests that licenses should not be exclusive, being an invention available to any interested party; only in the case where there are no interested parties should the university offer an exclusive license, but researchers can still use the protected knowledge in new studies. Noveli and Segatto (2009) presented a model applied to university–industry cooperation in technology parks. Field research was conducted with companies located at the Scientific and Technological Park of the Pontifical University of Rio Grande do Sul and university researchers, and involved three cases of cooperation. Results showed that other elements must be added to the proposed taxonomy, such as motivators, barriers and facilitators, and informal connections that occur within the park.

Dias et al. (2011) identified the organizational skills that differentiate universities' NITs. Using qualitative research, they defined the criteria for selecting cases for interviews. In the quantitative phase, a survey was conducted with those responsible for the offices, chosen among the members of the Minas Gerais Network of Intellectual Property and of the National Forum of Innovation and Technology Transfer Managers. The following differential competencies were identified: intellectual property, national patenting, consulting provided by researchers, and available information on the areas of excellence in research at the institutions, based on the number of registered patents, publications and distinguished research groups.

The methodological proposal for quantitative measurement of these skills and their constituent factors is noteworthy.

Toledo et al. (2011) presented the results of a training project for NITs, to prepare or update their internal policies of intellectual property management, as well as the establishment of bodies and procedures for its transfer, negotiation and licensing. The InovaNIT project, funded by the Ministry of Science, Technology and Innovation, promotes the exchange of employees, contributes to the professionalization of technology transfer in Brazil, and stimulates the creation of new NITs and the improvement of existing ones, via free training for university and research institute personnel. Between August 2007 and December 2010, 833 professionals were trained, 279 institutions visited, and over 20 offices created, showing that government initiatives aligned to the needs of organizations may have a positive impact on the national innovation system.

Public universities still have a defensive attitude towards enterprises, often considering cooperation as the transfer of public resources to private activities (Oliveira and Velho 2009). Many academic researchers see their social role as limited to forming qualified human resources to meet the needs of the state (Rosa and Hemais 2005). In an interview with *VEJA* magazine (27 March 2013), the president of the National Council of Scientific and Technological Development, Glaucius Oliva, stated (Weinberg 2013):

> Fortunately, there are more and more researchers bent over concrete problems, devoted to applied science. Yet there are still university centers that are lost in ethereal themes, some of them with attitudes biased by their own beliefs and still clung to old ideological flags... It's a minority, but yet there are people in the academia who do not see with sympathy closer ties with the private sector. They repeat the same old catchphrase: 'we will end by putting public resources at the service of the capital'. These centers of resistance, sustained on their ideological speech, have historically contributed to keep companies away from the academic world, and the Brazilian innovation, therefore, far from the top.

A similar speech was made by the minister of Science, Technology and Innovation, Marco Antonio Raupp, at the annual meeting of the Brazilian Society for the Advancement of Science, in July 2013:

> We need to encourage the relationship between universities and companies in Brazil in order to develop the national scientific production. There are Brazilian businessmen who seek universities in the United States to make partnerships because here in Brazil, everything is very complicated... We need to look for these other customers, the entrepreneurs.

Presently, Brazil has a complex production system and a comprehensive research system, measured by the number of master's, PhDs, and international papers, and these two systems remain separate. Companies have a limited capacity to absorb technology and develop innovations, and universities have not yet accepted the new role required by the Innovation Act (Mello et al. 2011). However, cooperation with companies has not turned scientists from basic research (Suzigan et al. 2011). Most groups that interact with firms improved their performance, generating more theses and dissertations, research projects, and articles.

The project "Interactions of Universities and Research Institutes with Companies in Brazil," conducted between 2008 and 2012, presents a favorable view of university–industry relations (although it is still the practice of just a few groups). It was based on questionnaires sent to representatives of teams registered in the Directory of Research Groups of the National Council of Scientific and Technological Development that declared to maintain partnerships (1,005 researchers responded). The study identified 1,687 companies, of which 326 R&D professionals (approximately 20 %) answered the questionnaires. The results show that the level of interaction is still low, and the graduation of qualified professionals is considered the main contribution of universities. There were also distinct interests among agents, and a prevalence of short-term actions. Furthermore, universities were found to seek academic results while companies aim to develop new products and processes (Puffal et al. 2012).

The technological demands of Brazilian companies are different from those in developed countries. They do not involve radical innovations, and partnerships for adaptation, improvement and adjustments to local conditions of processes and products predominate in the medium-low and low technological-intensity industries (pulp and paper, metal products, steel, food, and textiles). There are few partnerships with high-tech enterprises, resulting in an industrial system based on process innovations (Gouveia 2013).

This confirms the observations of Belderbos et al. (2006), and perhaps is one of the reasons for the low interest of Brazilian academics to cooperate with firms, whose research demands are not considered to be very challenging. Brazilian companies invest little in innovation, both internally and in collaboration. The national survey titled Technological Innovation Research (De Negri and Cavalcanti, 2013), conducted every 3 years by the Brazilian Institute of Geography and Statistics (PINTEC/IBGE), shows in its 2011 edition a decrease in the percentage of companies that perform innovation activities (innovation rate): the rate was 38.11 % in 2008 and 35.56 % in 2011. Innovative industrial companies with internal R&D totaled just 4.1 % in 2008 and 5 % in 2011. Brazilian firms tend to engage in innovation through the purchase of machinery and equipment for new or improved products and processes rather than by internal performance or the acquisition of external R&D.

Eight years after their rebirth, NITs are still trying to achieve legitimacy and acceptance, while promoting a culture of innovation in universities to mediate relationships with companies, stimulate entrepreneurial activities, and manage developed technologies (Castro and Souza 2012). The Innovation Act has encouraged patenting, but the next step, the licensing of patents to firms, is still restricted. This activity requires new capabilities from NITs in marketing, advertising, and partners search, which are more difficult to achieve than patenting.

This chapter has highlighted a number of problems regarding the university–industry relationship in Brazil, problems that have been overcome in the United States and Europe. The international literature on this subject is extensive, and many models and structures are proposed to improve it. There are still articles that present the pros and cons from the point of view of academia, and many papers try to justify them via surveys and quantitative analysis. Some articles discuss the most appropriate model for intellectual property rights and technology transfer—the individual property (professor's privilege) model or the institutional model; others debate whether TTOs should be located in every university or merely at a regional level, increasing their scope and effectiveness.

However, the debate concerning the value of cooperation no longer exists in the international literature, despite its prevalence in Brazilian articles. The international literature shows that there are academics who cooperate, those who do not, and no university fully encompasses either position. They highlight the characteristics of researchers more inclined to cooperation: scientific entrepreneurs (e.g., entrepreneurial scientists, entrepreneurial academics) who create spin-off companies out of academic research results, and academic or expert scientists, who only envisage collaboration with other scientists.

11.5 Conclusions

This study aimed to evaluate the evolution of university–industry cooperation in Brazil, via a survey of papers published since 1980 in major national and regional journals and presented at relevant events. We examined 249 articles; of these, 114 published since 2005 were assessed in greater detail. However, it was not possible to identify significant changes in the general behavior of universities and companies, be it in culture, negotiations, or licensing procedures that could demonstrate an improvement or a trend towards new practices and cooperation management models. The vast majority of studies did not utilize individual cases to offer general propositions that would substantially allow for a change in the character of these relations. It is worth mentioning that as far as academic papers are concerned, lecturers'/researchers' views about this relationship prevail, be they positive or negative. Some case studies have company professionals as co-authors or show the results of interviews made with them. The government's view is present in studies on public policies, in programs to encourage innovation, or in excerpts from speeches and published interviews.

University–industry relations do not yet constitute a regular and widely accepted process in public universities, despite the existence of several programs to support this partnership. Cooperation is not institutionalized, and it is carried out by some teachers who believe in its potential to leverage innovation. In some way, the Innovation Act of 2004 has interrupted the endless discussions about the university's role in the country's development, requiring a concrete action on their part—the creation of NITs.

The more relevant topics found in international literature (Rothaermel et al. 2007) are also studied in Brazil. However, the main difference is that university–industry cooperation is a natural and widespread phenomenon in developed countries, and the articles (many of them of a quantitative nature) test the most commonly used forms of technology transfer, examine the profiles of scientists, and discuss the results of cooperation. These papers also debate whether there really was a reduction of basic research in favor of applied research, assess the performance of the technology transfer offices, and suggest new types of organizations to better operate the transfer of academic research results.

However, the majority of Brazilian articles are composed of the individual experiences of universities and companies, descriptions of the obstacles to cooperation, the assessment of incentive policies that do not work, and the performance of NITs. Few studies suggest more appropriate structures to increase cooperation, and some papers even discuss whether university should cooperate with companies, and whether the results of academic research should be patented and licensed, topics that are no longer of interest in developed countries.

Although new programs and fiscal mechanisms have been created by federal and state financing agencies to stimulate innovation in companies and cooperation with universities and research institutes, the response was not the intended one. Institutional voids can be partly blamed for this lower demand. These voids can be a problem internal to firms, such as scant attention paid to available instruments and programs, a shortage of skilled personnel to seek funding opportunities for innovation (often the financial area is responsible for this task), and difficulties in presenting projects that meet the patterns of the agencies. These problems can also be external, and concern government agencies or the general business environment in Brazil. Thus, they may include conflicting norms, exaggerated demand of documents and certificates, poor communication, red tape, lack of transparency, legal insecurity, flexible regulatory framework, and economic conditions. Universities lack clear rules on intellectual property and the licensing of results in the case of successful partnerships. Furthermore, the practices of public financing agencies are in opposition to the Bayh–Dole Act, requiring financial participation in the licensing of patents that result from research that they support.

References

Araújo, B. C. (2012). *Políticas de apoio à inovação no Brasil: uma análise de sua evolução recente [Policies for innovation support in Brazil: An analysis of its recent evolution]*. Brasília: IPEA, *Texto para Discussão* [*Discussion Text*], 1759.

Armstrong, J. A. (1993). University research: New goals, new practices. *Issues in Science and Technology, 9*(2), 50–53.

Arruda, M. F. M. A. (1994). Indústria e o desenvolvimento tecnológico Nacional [Industry and national technological development]. In *Ciência e tecnologia: Alicerces do desenvolvimento [Science and technology: Foundation of development]* (pp. 23–44). São Paulo: CNPq/ COBRAM.

Bardin, L. (2000). *Análise de conteúdo* [*Content analysis*]. Lisboa: Edições 70 Ltda.

Bastos, V. D. (2012). 2000–2010: Uma década de apoio federal à inovação no Brasil [2000–2010: A decade of federal support for innovation in Brazil]. *Revista do BNDES [BNDES journal], 37*, 127–176.

Belderbos, R., Carree, M., & Lokshin, B. (2006). Complementarity in R&D cooperation strategies. *Review of Industrial Organization, 28*(4), 401–426.

Bonaccorsi, A., & Piccaluga, A. (1994). A theoretical framework for the evaluation of university-industry relationships. *R&D Management, 24*(3), 229–247.

Carvalho, R. Q., Consoni, F. L., Quintão, R., & Vieira, G. (2006). Mapping out technological capabilities in research institutions as a tool for prospecting R&D outsourcing opportunities: A methodology developed for the R&D centre of a major car assembler. Gramado: XXIV Simpósio de Gestão da Inovação Tecnológica [XXIV Symposium on Technological Innovation Management], *Annals...*

Castro, B. S., & Souza, G. C. (2012). O papel dos Núcleos de Inovação Tecnológica (NITs) nas universidades brasileiras [The role of technological innovation offices in Brazilian universities]. *Liinc em Revista, 8*(1), 125–140. Rio de Janeiro.

Chauí, M. (1995). Em torno da universidade de resultados e serviços [About the university of results and services]. In Dossiê universidade – empresa [Dossier university–industry]. *Revista USP, 25*, 54–61.

Chesbrough, H. W. (2003). *Open innovation: The new imperative for creating and profiting from technology*. Cambridge: Harvard Business School Publishing.

Closs, L. Q., & Ferreira, G. C. (2012). A transferência de tecnologia universidade–empresa no contexto brasileiro: uma revisão de estudos científicos publicados entre os anos 2005 e 2009 [University – industry technology transfer in the Brazilian context: A revision of scientific studies published between 2005 and 2009]. *Gestão & Produção [Management & Production], 19*(2), 419–432.

Cruz, C. H. B. (2010). Ciência, Tecnologia e Inovação no Brasil: desafios para o período 2011 a 2015 [Science, technology and innovation in Brazil: Challenges for the period 2011–2015]. *Interesse Nacional* [*National Interest*], *3*(10). http://interessenacional.uol.com.br/index.php/edicao/edicao-numero-10/. Accessed 23 Feb 2013.

De Negri, F. (2012). Elementos para a análise da baixa inovatividade brasileira e o papel das políticas públicas [Elements for the analysis of Brazilian low innovation capacity and the role of public policies]. *Revista USP [USP Journal], 93*, 81–100.

De Negri, F., & Cavalcante, L. R. (2013). Sistemas de Inovação e Infraestrutura de Pesquisa: considerações sobre o caso brasileiro [Innovation systems and research infrastructure: Considerations in the Brazilian case]. *Radar Tecnologia, Produção e Comércio Exterior* [*Radar Technology, Production and Foreign Trade*], *24*, 7–18.

De Negri, F., & Cavalcanti, L. R. (2013). Análise dos dados da PINTEC 2011 [Analysis of PINTEC 2011 data]. *Brasília: IPEA. Nota Técnica [Technical Note], 15.*

Dias, J. C., Balbinot, Z., & Souza, R. B. (2011). As competências organizacionais diferenciadoras dos núcleos de inovação tecnológica Brasileiros [The distinguishing organizational competences of the Technological Innovation Offices]. Rio de Janeiro, XXXV Encontro Nacional da ANPAD [XXXV ANPAD National Meeting], *Annals....*

Economic Commission for Latin America and the Caribbean. (2010). *Espacios iberoamericanos: vínculos entre universidades y empresas para el desarrollo tecnológico* [*Iberoamerican spaces: Links between universities and companies for technological development*]. Santiago: CEPAL/AECID.

Etzkowitz, H. (1989). Entrepreneurial science in the academy: A case of the transformation of norms. *Social Problems, 36*(1), 14–27.

Etzkowitz, H. (2008). *The Triple Helix—university-industry-government: Innovation in action*. New York: Routledge.

Feldmann, P. (2011). Bons em ciência, ruins em tecnologia [Good in science, bad in technology]. http://www.fecomercio.com.br/ImprensaArtigosInterna/Artigo/2685. Accessed 21 Feb 2013.

Feller, I. (1990). Universities as engines of R&D-based economic growth: They think they can. *Research Policy, 19*(4), 335–348.

Foray, D., & Lissoni, F. (2010). University research and public–private interaction. In B. H. Hall & N. Rosenberg (Eds.), *Handbook of the economics of innovation* (Vol. 1, pp. 275–314). Chapter 6.

Fujino, A., Stal, E., & Plonski, G. A. (1999). A proteção do conhecimento na universidade [Knowledge protection at the university]. *Revista de Administração [Journal of Management], 34*(40), 46–55.

Garnica, L. A., & Torkomian, A. L. V. (2009). Gestão de tecnologia em universidades: uma análise do patenteamento e dos fatores de dificuldade e de apoio à transferência de tecnologia no

Estado de São Paulo [Management of technology at universities: Analysis of patenting and of the factors of difficulties and support to technology transfer in the State of São Paulo]. *Gestão & Produção [Management & Production], 16*(4), 624–638.

Garnica, L. A., Prado, F. O., Entorno, D. D., & Massambani, O. (2007). A gestão da propriedade intelectual na Universidade de São Paulo [The management of intellectual property at the University of São Paulo]. Buenos Aires: XII Seminário ALTEC [XII ALTEC Seminar], *Annals...*

Geisler, E., & Rubenstein, A. H. (1989). University-industry relations: A review of major issues. In A. N. Link & G. Tassey (Eds.), *Cooperative research and development: The industry-university-government relationship* (pp. 43–62). Boston: Kluwer Academic Publishers.

Geuna, A., & Nesta, L. (2006). University patenting and its effects on academic research: The emerging European evidence. *Research Policy, 35*(6), 790–807.

Geuna, A., & Rossi, F. (2011). Changes to university IPR regulations in Europe and the impact on academic patenting. *Research Policy, 40*(8), 1068–1076.

Glenna, L. L., Welsh, R., Ervin, D., Lacy, W. B., & Biscotti, D. (2011). Commercial science, scientists' values, and university biotechnology research agendas. *Research Policy, 40*, 957–968.

Gouveia, F. (2013, March 27). Parcerias entre pesquisadores e empresas geram conhecimento [Partnerships between researchers and firms generate knowledge] *Agência FAPESP*, Especiais [Specials].

Greenbaum, D., & Scott, C. (2010). Hochschullehrerprivileg—A modern incarnation of the professor's privilege to promote university to industry technology transfer. *Science, Technology and Society, 15*(1), 55–76.

Jensen, R. A., Thursby, J. G., & Thursby, M. C. (2003). Disclosure and licensing of university inventions: 'The best we can do with the s**t we get to work with'. *International Journal of Industrial Organization, 21*(9), 1271–1300.

Kenney, M., & Patton, D. (2009). Reconsidering the Bayh-Dole Act and the current university invention ownership model. *Research Policy, 38*(9), 1407–1422.

Kenney, M., & Patton, D. (2011). Does inventor ownership encourage university research-derived entrepreneurship? A six-university comparison. *Research Policy, 40*(8), 1100–1112.

Lam, A. (2010). From 'ivory tower traditionalists' to 'entrepreneurial scientists'? Academic scientists in fuzzy university–industry boundaries. *Social Studies of Science, 40*(2), 307–340.

Lam, A. (2011). What motivates academic scientists to engage in research commercialization: 'gold', 'ribbon' or 'puzzle'? *Research Policy, 40*(10), 1354–1368.

Larsen, M. T. (2011). The implications of academic enterprise for public science: An overview of the empirical evidence. *Research Policy, 40*(1), 6–19.

Lee, Y. S. (1996). Technology transfer and the research university: A search for the boundaries of university-industry collaboration. *Research Policy, 25*(6), 843–863.

Lee, Y. S. (1998). University-industry collaboration on technology transfer: Views from the ivory tower. *Policy Studies Journal, 26*(1), 69–84.

Matias-Pereira, J., & Kruglianskas, I. (2005). Gestão de inovação: a lei de inovação tecnológica como ferramenta de apoio às políticas industrial e tecnológica do Brasil [Innovation management: the Technological Innovation Act as a tool to support Brazilian industrial and technological policies]. *RAE Eletrônica [Electronic RAE], 4*(2), 1–21.

Matkin, G. W. (1990). *Technology transfer and the university*. New York: Macmillan.

Mazzoleni, R. (2005). The development of universities and public research institutions: A historical overview of its role in technological and economic catch-up. In G. D. Santangelo (Ed.), *Technological change and economic catch-up. The role of science and multinationals*. Great Britain: Edward Elgar Publishing Limited.

Mazzoleni, R., & Nelson, R. R. (2007). Public research institutions and economic catch-up. *Research Policy, 36*(10), 1512–1528.

Medeiros, J. A., Stal, E., & Souza Neto, J. A. (1987). A difícil relação pesquisa-produção: experiência brasileira dos núcleos de inovação tecnológica (1981–1987) [The difficult relationship of research and production: The Brazilian experience of the Technological Innovation Offices]. II Seminario ALTEC, Ciudad de México, September, *Annals...*

Mello, J. M. C., Maculan, A. M., & Renault, T. B. (2011). Brazilian Universities and their contribution to innovation and development. In B. Göransson & C. Brundenius (Eds.), *Universities in transition—The changing role and challenges for academic institutions* (pp. 53–76). New York: Springer/IDRC. Chapter 4.

Mowery, D. C., & Rosenberg, N. (1993). The U.S. national innovation system. In R. R. Nelson (Ed.), *National innovation systems—A comparative analysis* (pp. 29–75). New York, 10016–4314: Oxford University Press. chapter 1.

Mowery, D. C., & Sampat, B. N. (2005). The Bayh-Dole Act of 1980 and university–industry technology transfer: A model for other OECD governments? *Journal of Technology Transfer, 30*(1/2), 115–127.

Mowery, D. C., Nelson, R. R., Sampat, B. N., & Ziedonis, A. A. (2001). The growth of patenting and licensing by U.S. universities: An assessment of the effects of the Bayh-Dole act of 1980. *Research Policy, 30*(1), 99–119.

Nelson, R. R. (1996). *The sources of economic growth*. Cambridge: Harvard University Press.

Noveli, M., & Segatto, A. P. (2009). Proposição de um modelo conceitual para o estudo do processo de cooperação universidade-empresa para a inovação tecnológica [Proposition of a conceptual model to study the university – industry cooperation process for technological innovation]. São Paulo: XXXIII Encontro Nacional da ANPAD, setembro, *Anais....*

Oliveira, R. M., & Velho, L. (2009). Benefícios e riscos da proteção e comercialização da pesquisa acadêmica: uma discussão necessária [Benefits and risks of protecting and commercializing academic research: A necessary discussion]. *Ensaio: avaliação e políticas públicas em Educação [Essay: assessment and public policies in Education], 17*(62), 25–54.

Pacheco, C. A., & Almeida, J. G. (2013). *A política de inovação*. Instituto de Economia Unicamp, Texto para Discussão, nº 219. www.eco.unicamp.br/docprod/downarq.php?id=3268&tp=a

Pavitt, K. (2005). Innovation processes. In J. Fagerberg, D. C. Mowery, & R. R. Nelson (Eds.), *The Oxford handbook of innovation*. New York: Oxford University Press. Chap. 4.

Perkmann, M., & Walsh, K. (2007). University-industry relationships and open innovation: towards a research agenda. *International Journal of Management Reviews, 9*(4), 259–280.

Póvoa, L. M. C. (2006). A universidade deve patentear suas invenções? [Should the university patent its inventions?]. *Revista Brasileira de Inovação [Brazilian Journal of Innovation], 9*(2), 231–256.

Puffal, D. P., Ruffoni, J., & Schaeffer, P. R. (2012). Características da interação universidade – empresa no Brasil: motivações e resultados sob a ótica dos envolvidos [Features of university – industry interaction in Brazil: Motivations and results by those involved]. Porto Alegre, *Gestão Contemporânea* [*Contemporary Management*], Edição Especial [Special Edition], n. 1 (pp. 71–94).

Rosa, E. O. R., & Hemais, C. A. (2005).A dinâmica do relacionamento universidade-empresa na visão de seus atores: um estudo de casos [The dynamics of university–industry relationship from the viewpoint of the players: Case studies]. Brasília: XXIX Encontro Nacional da ANPAD, *Anais...*

Rosenberg, N., & Nelson, R. R. (1994). American universities and technical advance in industry. *Research Policy, 23*(3), 323–348.

Rothaermel, F. T., Agung, S. D., & Jiang, L. (2007). University entrepreneurship: A taxonomy of the literature. *Industrial and Corporate Change, 16*(4), 691–791.

Schacht, W. H. (2006). The Bayh-Dole Act: Selected issues in patent policy and the commercialization of technology. *Congressional Research Service Reports*. Paper 23. http://digitalcommons.unl.edu/crsdocs/23

Schwartzman, S. (Ed.) (1993). *Ciência e tecnologia no Brasil: uma nova política para um mundo global* [*Science and technology in Brazil: A new policy for a global world*] São Paulo: EAESP/FGV.

Siegel, D. S., Wright, M., & Lockett, A. (2007). The rise of entrepreneurial activity at universities: Organizational and societal implications. *Industrial and Corporate Change, 16*(4), 489–504.

Stal, E., & Fujino, A. (2005). Aprimorando as relações universidade-empresa-governo no Brasil: A lei de inovação e a gestão da propriedade intelectual [Improving university – industry –government relations in Brazil: The Innovation Act and the management of intellectual property]. Salvador: XI Seminário ALTEC, *Anais*...

Suzigan, W., Albuquerque, E. M., & Cario, S. A. F. (Eds.). (2011). *Em busca da inovação: interação universidade-empresa no Brasil [In search of innovation: university–industry interaction in Brazil]*. Belo Horizonte: Autêntica Editora.

Teece, D. (1986). Profiting from technological innovation: Implications for integration, collaboration, licensing and public policy. *Research Policy, 15*(6), 285–305.

Thursby, J. G., & Thursby, M. C. (2003). University licensing and the Bayh-Dole Act. *Science, 301*, 1052.

Thursby, J., & Thursby, M. (2011). Has the Bayh-Dole Act compromised basic research? *Research Policy, 40*(8), 1077–1083.

Toledo, P. T. M., Hourcade, V., Magalhães, A., Lotufo, R. A., Salles Filho, S. L. M., & Bonacelli, M. B. (2011). Difusão de boas práticas de proteção e transferência de tecnologias no Brasil: a contribuição do Projeto InovaNIT [Diffusion of best practices for protection and transfer of technologies in Brazil: The contribution of Project InovaNIT]. XIV Seminar ALTEC, Lima, Octubre, *Anales*...

Troyjo, M. (2013). Por que inovamos tão pouco? [Why do we innovate so little?] *Folha de São Paulo*, 18 October. http://www1.folha.uol.com.br/colunas/marcostroyjo/2013/10/1358347-por-que-inovamos-tao-pouco.shtml

Von Hippel, E. (1988). *The sources of innovation*. New York: Oxford University Press.

Weinberg, M. (2013). A maioria quer ser inovadora (entrevista com Glaucius Oliva) [The majority want to be innovative (interview with Glaucius Oliva)]. São Paulo, *VEJA, 46*(13), 15–19.

Zanluchi, J. B., & Gonçalo, C. R. (2007). Relação universidade-empresa: diferentes perspectivas de estudos no Brasil [University–industry relation: different perspectives from studies in Brazil]. XXXI Encontro Nacional da ANPAD, Rio de Janeiro, September, *Annals*...

Chapter 12
On the Evolution of Research Networks: The Case of Micro Technologies in Mexico

Eduardo Robles-Belmont

12.1 Introduction

With the emergence of new science and technologies, it is important to understand their characteristics and to evaluate their development; these are two important issues in social studies of science and technology, feasibility studies, and technology management. In the academic literature and institutional reports, various indicators have been proposed that enable us to evaluate available scientific and technological capabilities (OECD 2007; Moed et al. 2004). In general, this involves quantitative indicators expressed in inputs and outputs.

However, there is some disagreement regarding the proposed indicators, and this is evident, for example, in cases where there is a gap in the information required to create these indicators, preventing the collection of valid and comparable quantitative data (Albornoz and Fernandez-Polcuch 1996). In the field of scientometrics, there have been many initiatives and advances in the analysis of bibliometric data that enables us to characterize and evaluate scientific activity from the point of view of publications. In the last decade, the application of social network analysis to co-authors, citations, and knowledge areas (where scientific publications are ranked) has become more common. These academic works provide us with landscapes or maps of the science under investigation, enabling us to identify and graphically illustrate scientific communities (Morillo and Aparicio 2011) and the multi- or inter-disciplinary character of the science (Rafols et al. 2010; Waltman and van Eck 2012). These studies are mainly quantitative and the results obtained enable us to characterize the relationships between the different actors involved in the production

E. Robles-Belmont (✉)
Networks Laboratory, Institute for Research in Applied Mathematics and Systems, National Autonomous University of Mexico, Mexico City, 04510, Mexico
e-mail: roblesbelmont@yahoo.fr

H. Horta et al. (eds.), *Trends and Challenges in Science and Higher Education*, Knowledge Studies in Higher Education 3,
DOI 10.1007/978-3-319-20964-7_12

of new knowledge. However, the representations obtained in this way tell us little about the dynamics of the relationships revealed. To adequately assess these dynamics it is necessary to undertake qualitative studies.

A number of studies on science and technology across various disciplines now apply the concept of "network." There is a wide academic literature surrounding this concept, in which two distinct approaches can be distinguished. The first focuses on social network analysis (SNA) and the second on the actor–network theory (ANT) (Grossetti 2007). SNA is generally applied to scientific and technological publications, where it has the appeal of enabling us to identify the scientific actors and their relationships, as well as record some measures of these to determine the position of the actors within the scientific structure. Furthermore, different ways of visualizing collaboration networks have been proposed, utilizing different methods (Kamada and Kawai 1989; Van Eck and Waltman 2010). In contrast, the interest in ANT lies in its capability to identify heterogeneous actors (beyond the scientists and other actors involved in the production of new knowledge), characterize and assess the relationships between them, and describe the dynamics that describe scientific and technological networks in the production, use, and dissemination of new knowledge. In fact, according to the type of data and methodological focus, SNA provides the quantitative aspect and ANT reveals the qualitative. In studies on networks in the scientific realm, there seems to have been a tendency for both approaches to be in opposition. However, according to Grossetti (2007), it would be interesting to create a methodological and conceptual framework that enables us to apply both approaches to "network" to enrich and broaden the explanatory and interpretive capabilities of studies on science and technology. This involves the implementing studies based on mixed methods that obtain relational data to better understand the dynamics within networks (Kolleck 2013).

It is in this context of the evaluation of the production, use, and dissemination of new knowledge and the application of the idea of networks that this research is embedded. The results of our proposal to employ the idea of network from both approaches are presented here. The study is particularly interested in the case of the development of Electromechanical Micro Systems (MEMS) technology in Mexico. In effect, it attempts to characterize and evaluate the network of heterogeneous actors within the process of implementing and developing this emerging technology. It is worth mentioning here that our interest in studying MEMS development in Mexico stems from the fact that it deals with a technology in the early stages of development. Furthermore, by observing the dynamics that take form in its implementation and dissemination, we can identify key actions and elements. In examining the development of micro- and nanotechnologies in Mexico, our attention is also drawn to the rapid implementation of scientific infrastructure and the formation of collaboration networks for MEMS development. Our study covers 2001–2010, during which the implementation and dissemination of these technologies occurred, and are described in the following pages.

In this study, both ANT and SNA are applied, to graphically represent the networks of sociotechnical actors that have undertaken for the most part, the implementation of these new technologies in Mexico. Therefore, the concept of technical–economic networks within the scientific, technological, and market poles (Callon et al. 1991) is proposed to evaluate the scientific and technological networks in the different dynamics observed in this study. The results obtained enable us to see a broad panorama of the scientific and technological collaboration networks related to MEMS devices in Mexico; this in turn helps us to identify the relevant scientific actors and the relationships among them. Additionally, the results of a survey enable us to broaden the view of scientific collaboration networks, as they show the diverse sociotechnical actors in the development of this emerging technology. This empirical study also reveals some identifying factors in the dynamics of the creation of the networks we examine, and thus we can better understand the implementation of this technology in Mexico. Finally, the study's proposed methodological framework can be replicated in the examination and evaluation of other emerging technologies, identifying the key elements to better understand the dynamics of the processes observed and to formulate strategies for their development.

In the first part of this chapter, we offer a review SNA and ANT, including their scope, limitations, and shared complementary features. We then explain the methodology employed in this study. In the section on findings, we first provide a brief introduction to the emergence of microtechnologies in Mexico and then the results are presented in three parts, corresponding to the different dynamics in the development of these new technologies. Finally, we provide some conclusions.

12.2 Network Analysis in Studies on Science and Technology: Scope and Limitations

As mentioned above, the application of SNA in studies on science and technology has broadened in recent years. Mullins performed pioneering work using this kind of analysis in the field of social studies of science, studying the development of molecular biology (Mullins 1972). In that study, the author found that the construction of this specialty developed in four phases, which gave shape to the social structure of the molecular biology scientific community. Mullins identified the basis of SNA applied to the study of scientific communities, proposing different categories of relationships. Today, it is even more common for SNA to be used in studies across various disciplines in social sciences, where the data uncovered in the analysis are principally scientific publications, with the aim of forming an image of the social structure of a particular scientific community. These studies focus on co-authorship (Hou et al. 2008), citations (Wang et al. 2012; Meyer 2000), and the categorization of the journal disciplines in which articles are published (Rafols et al. 2010; Porter and Rafols 2009). Within the findings of these analyses, the nodes symbolize scientific actors (authors), institutions, and disciplines. Their relationships are based on the co-occurrence of these nodes in bibliometric references, arising from co-authorship. The findings are then translated as a reflection of the social structure of scientific communities, or as a profile of scientific knowledge in the case of disciplines.

The analysis of scientific articles using the SNA method enables us to obtain a picture of the social structure of scientific communities. This kind of study shows us the key scientific and institutional actors and their relationships; the results are interpreted as scientific collaboration networks. Furthermore, as mentioned in the introduction, in recent years this type of analysis has been applied to other data relating to scientific documents, which has made SNA even more attractive in studies on science and technology. Despite the differing applications of SNA, the findings obtained from such analysis show us what lies beneath social relationships. However, the use of SNA does not enable us to consider how these relationships arise and are transformed, nor does it show us the elements that influence and motivate actors to initiate, maintain, or end a relationship with other actors.

The alternative approach studied here, ANT, emerged from the research of Latour (1989) and Callon (1995). They proposed an examination of the "non-human" actors in the construction of scientific facts. Callon (1989) considered that work in sociology and anthropology had shown us "the heterogeneous diversity of the elements that are employed by scientists in the pursuit of their activities." These "elements" include researchers, technicians, and the administrative staff that form part of the laboratory's workforce, who are categorized by their abilities and competencies. Likewise, a laboratory's infrastructure is composed of scientific and technological instruments that scientists use to perform their experiments. The scientific documents that researchers consult, whether to communicate their own studies or to learn that of their colleagues, are a further element used in laboratories. Furthermore, Callon makes reference to the funding obtained by institutions as a necessary element for personnel recruitment, the acquisition of new instruments, and to obtain essential scientific documents for the development of scientific activities. On this point, a scientific laboratory (whether in natural or social sciences) is seen as a network because within that laboratory necessary, heterogeneous actors are employed and mobilized to produce a scientific outcome.

An ANT analysis enables us to broaden the identification of actors who intervene in these processes: heterogeneous actors. Additionally, from the description and ethnographic analysis, the studies that make use of ANT show us elements that identify the dynamics within the scientific and technological networks under study. It is for this reason that the proposal in this study is to make use of both approaches in a complementary fashion. Undertaking a study that incorporates both approaches is no easy task, as in academic environments *research* is presented with the label of the sociotechnical network study (the ANT approach) and *findings* are presented visually within a network (the SNA approach), and both without providing elements that enable us to comprehend the dynamics observed in the scientific networks.

The visual representation of networks is attractive because we can then optimize the explanatory capability of network studies. However, at this visual level, on occasion, the findings of some studies reveal networks with nonheterogeneous actors. The primary criticism of those studies is that they do not deal with sociotechnical networks, as we do not see heterogeneous actors reflected in the network graphics. Arellano (2010) attributes this kind of error to SNA computer programs, which according to the author are not capable of handling heterogeneous data that enable us to map out their relationships. Despite the fact that these programs have been designed to manipulate heterogeneous data, this does not rule out their use in performing the analysis. The errors we see in the use of these programs are caused in part by a poor implementation of ANT and the accompanying theoretical and methodological assumptions.

This is seen, for example, in the moment that relationships between the actors are formed. In fact, the difficulty of identifying and measuring the relationships is a topic for discussion within the framework of network studies (Grossetti 2009; Schmidt 2009). This is a methodological problem, and to overcome it, this study proposes the application of three methodological principles in the sociology of translation: (a) researcher's agnosticism; (b) generalized symmetry; and (c) free association (Callon 1995). When performing studies on sociotechnical networks, these concepts should be considered from the moment of ethnographic observation, such as in the construction of the graphic representing the networks. This enables us to establish a category of actors and relationships from the same dynamics and to conduct a symmetric analysis using either of the two approaches: ANT or SNA.

12.2.1 *Objectives and Heterogeneous Interests Inside the Networks*

Returning to the main focus of this chapter, the characterization and evaluation of scientific and technological networks via the course of their implementation as a new technology, we consider the social, technical, and economic actors identified in our empirical study. To do so, the concept of sociotechnical networks for this kind of study is proposed. This notion lies within the ANT approach, in which Callon (1992) notes that the different actors interacting within the networks "have, of course, heterogeneous and even contradictory goals, projects and interests." In the course of our empirical study on the development of MEMS technology in Mexico, those divergent interests are revealed. The question that has emerged here concerns how to categorize these heterogeneous networks to evaluate them. One study on technology management described a tool for the evaluation of public programs of technological development with a focus on technical–economical networks (Callon et al. 1991). The authors showed that in these networks, there are heterogeneous actors that represent diverse interests, cooperate, and organize themselves around three poles: a scientific pole (S), technical pole (T,) and market pole (M). The distinction of the three poles is useful to characterize the state of the development and implementation of a new technology, as it allows us to show the diversity of collaboration network configurations where actors with heterogeneous interests interact. Furthermore, it enables us to evaluate the state of each pole within the processes of technological change.

In this study, the definition of these three poles is revisited, with the aim to identify the actors and their relationships through different stages during the development and implementation of MEMS technology in Mexico. Furthermore, it is important to remember that this not only involves a visual representation of the networks at each pole via SNA, but likewise the use of ANT seeks elements to characterize the networks, revealing their dynamics.

12.3 Methodology

This study is based upon a series of qualitative and quantitative data. It employs mixed methods, in which initial data were obtained from a survey conducted at various research laboratories, via interviews with various actors, direct nonparticipant observation in some forums, and meetings in the field of micro- and nanotechnologies in Mexico between 2008 and 2010. Throughout the survey, the methodological principles of the sociology of translation (Callon 1995) were considered. The relationships were not established a priori but were defined and identified by the human actors during the interviews. Furthermore, various actors and their relationships were deduced at some observed academic events. The interviews were conducted according to a general interview guide for all actors from the first interview performed, with questions concerning four axes to characterize the dynamics of the development of micro- and nanotechnologies in Mexico. In a number of cases, two or more interviews were held, in which interview guides with new questions were created to deepen and verify data. Using the record of relationships, and cross-referenced with the relationships of the actors we interviewed, the relationships and key elements to understand their dynamics were verified and confirmed.

The second data type was obtained via a search of databases of scientific and technological publications (articles and patents). Other sources were also consulted, such as research and technological development projects, institutional reports, and websites. Finally, the data underwent SNA using the NetDraw computer program.

12.4 MEMS Technology and Its Emergence in Mexico

Before discussing the findings of this study, it is germane to outline MEMS technology and its emergence in Mexico to contextualize our study. MEMS are devices smaller than 1 mm in length and greater than 1 μm, combine electrical and mechanical components, and are manufactured with electronic circuit technologies (Gad-el-Hak 2001). This technology is also known as microtechnology or microsystems technology, and has its origins in the 1960s with the publication of the first scientific article in this field. A decade later, with the development of the first prototypes, this technology grew in importance. The first MEMS were commercialized in the 1980s, in the form of micro-accelerometers that were introduced to activate air bags in the automotive industry. The development of nanotechnology and advanced materials since the beginning of this century has driven MEMS development. This is due to the new properties that these materials acquire at the nanometric level; these have increased their technical capabilities and as a consequence have led to new applications and the miniaturization of MEMS devices.

Today, these devices are found in a wide range of equipment commonly used in daily life and have found a significant niche in the microelectronics industry. Similarly, the development of MEMS plays an important role in the health sector, with the integration of biocompatible material devices, which we know as BioMEMS. Furthermore, within the market we find many products that include this kind of technology with applications in health, as in the case of the "lab-on-a-chip," which is the miniaturization of clinical analysis laboratories onto electronic chips. These new devices have enabled bio-analysis and clinical testing without requiring the patient to visit the hospital, and the results are immediate. The development of such devices in the health sector has drawn the attention of government agencies and the pharmaceutical industry, both of which are searching for new devices to administer medications. The growth in this sector has been significant: in 2001, Britain's Royal Academy of Chemistry created a scientific journal titled *Lab on a Chip*.

This technology has also entered the telecommunications sector, where devices have been developed for mobile phones. The use of MEMS not only has civilian applications but also military uses, where there is significant interest in the development of this technology. In fact, one of the largest laboratories specializing in the development of MEMS with military applications is found in the United States: Sandia National Laboratories is located in New Mexico, engaging in scientific and technological developments in weaponry.

The emergence of this technology in Mexico occurred at the beginning of this century, with the Mexico–United States Foundation for Science (*Fundación México Estados Unidos para la Ciencia*; FUMEC) its primary proponent. Since then, cross-border, philanthropy-oriented, nongovernmental organizations have implemented various initiatives for the creation of scientific and technological infrastructure, the training of human resources, and the commercialization and spread of MEMS technology in Mexico. FUMEC's microsystem initiatives and its role in the implementation and development of this technology in Mexico have been examined in earlier studies (Robles-Belmont 2011; Robles-Belmont and Vinck 2012), which note that at the beginning of this century, there were no human resources nor any scientific infrastructure for the development of MEMS in Mexico. In 2008, when this research began, there were already approximately 14 research centers in which scientific and technological activities were underway for the prototyping and development of MEMS devices. The first article on this topic was published in Mexico in 2002; as of 2009, at least 29 articles on this topic had been published throughout the country.

12.4.1 Scientific and Technological System in Mexico

The Mexican scientific and technological system is made up of academic institutions and public and private research institutions. The public institutions are in the form of universities, technological institutes, and research centers. The creation, application, adaptation, and dissemination of scientific and technological knowledge developed in Mexico primarily occur in public institutions. In fact, scientific production is highly concentrated in just a few institutions located in the capital. The participation of private institutions in scientific and technological development in Mexico is less evident. However, in some fields of knowledge, the importance of private institutions is quite valuable.

In 1970, the National Council on Science and Technology (CONACYT) was founded in Mexico to stimulate and promote scientific and technological development. One year later, it began performing its functions, of which the central goal was the creation of a science and technology policy. Other public entities were also involved in the creation and evaluation of science and technology policies, such as the Science and Technology Advisory Forum, the President's Science Advisory Council, and the Mexican Academy of Sciences.

In the production of new knowledge in Mexico, the institutes and research centers belonging to the CONACYT System of Centers are worthy of note. There are 27 member institutions, of which ten are dedicated to the exact and natural sciences, eight to social sciences and the humanities, eight to technological and services development, and one that specializes in postgraduate studies funding. The several universities, and more recently, technological institutes and technological universities employ diverse groups of researchers who create projects and scientific topics and technologies across a wide variety of areas of scientific knowledge. Furthermore, the public institutes in the health sector and the food and farming sector have their own scientific and technological research systems.

The funding of scientific and technological research in Mexico comes from various sources, and for the most part they are public. CONACYT, via the issuing of various public funds, is the key source of funding in Mexico for basic and applied research. In recent years, it has created specific funds for the financing of projects of innovation, in which the participation of private sector companies is a requirement. Key funding also comes from other state ministries, which, in tandem with CONACYT, allocate funding to applied research in specific sectors. Similarly, there are shared funds operating among various local governments. In contrast, the top Mexican universities maintain their own funds to support research and the Ministry of Public Education (SEP) also maintains dedicated funding for research. Some research projects are the recipients of international funding for the development of science and technology.

The installation of scientific and technological infrastructure in research institutes is primarily funded by the research projects of researchers and funds received by public institutions. In fact, CONACYT has issued calls for proposals to fund the installation of research laboratories and in the majority of the cases those funds are shared with universities or research institutions (in 2006, two micro- and nanotechnology laboratories were established with these funds). Regarding the private sector, research laboratories for research scientific and technological development are few and far between.

The training of researchers occurs in all public and private universities throughout the country. CONACYT also offers a program of grants for doctoral training in practically all areas of knowledge. Universities have their own grant systems for this kind of academic training. Furthermore, various specialist training programs in different knowledge areas have been implemented in cooperation with businesses and the support of these same universities, and in some cases with CONACYT. The training of scientists abroad has been very important for Mexico, and grants for this kind of training generally come from CONACYT (on some occasions such funding is shared with government programs and/or institutions where Mexican researchers study abroad).

Individual incentives for research also exist in Mexico. One of those is offered via the National System of Researchers (SNI), created in 1984 to provide economic stimulus. Currently, this system supports 21,364 researchers. Furthermore, top Mexican universities have their own systems to encourage research, and SEP has yet another program of research stimulus.

12.4.2 Scientific Pole in the Development of MEMS Technology in Mexico

In characterizing the scientific pole, it is common to only consider scientific articles published within that field of study. In this study, we considered scientific articles and other data related to the path taken in the installation of MEMS technology in Mexico, where we identified actors beyond those coming from the academic world. Thus, we included politicians, public servants, and businesspeople.

Of the initiatives created by FUMEC for MEMS development in Mexico, two programs were oriented toward the creation of favorable conditions for MEMS development. This involved two key initiatives for the installation of scientific infrastructure: the networks of MEMS Design Centers (CD-MEMS) and MEMS Innovation Laboratories (LI-MEMS).

The CD-MEMS network is a network of MEMS design centers created by FUMEC in 2003 to create an infrastructure in the country for the design of MEMS devices. During the creation of this collaboration network, 18 academic research bodies were identified. Of these, 12 were Mexican and the remainder were based in the United States. The Mexican entities were those that responded to the call for proposals from FUMEC to form the CD-MEMS network; however, only ten were able to complete the installation of the design centers. To participate in the CD-MEMS network, the departments or research groups require the economic support of their research institutions, as the installation of the design equipment demanded the modification or construction of laboratory space. This required the involvement of various departments within the academic institutions.

In contrast, at the launch of the CD-MEMS network, the training of Mexican researchers called for the participation of researchers and specialists in MEMS design from various academic and research institutions from the United States. To mobilize researchers for training, it was necessary to make economic resources available. For Mexican researchers, their institutional mobility was financed by the same institutions where they worked. Participating US researchers were financed by FUMEC. The purchase of computing equipment was funded by Mexico's Ministry of the Economy (SE), with FUMEC responsible for negotiations to secure funding and its management.

The LI-MEMS network was created in 2004 by FUMEC for the continuing installation of infrastructure for MEMS development in Mexico. In this second network, the laboratories were oriented toward the manufacture and characterization of MEMS devices. A total of three laboratories were installed: one in the Faculty of Engineering at the Autonomous National University of Mexico (UNAM-FI), another in the facilities of the National Institute for Astrophysics, Optics and Electronics (INAOE), and the third in the Center for Research in Applied Science and Technology at the Autonomous University of Ciudad Juárez (UACJ-CICTA). The construction of these three laboratories was financially supported by the SE, the Government of the State of Puebla (Gouv-PUEBLA), CONACYT, and the Government of Ciudad Juárez (Gouv-Cd-Juarez).

These two collaboration networks (albeit their early efforts) are represented by a yellow triangle in Fig. 12.1. The ten CD-MEMS are located in the entities (red cir-

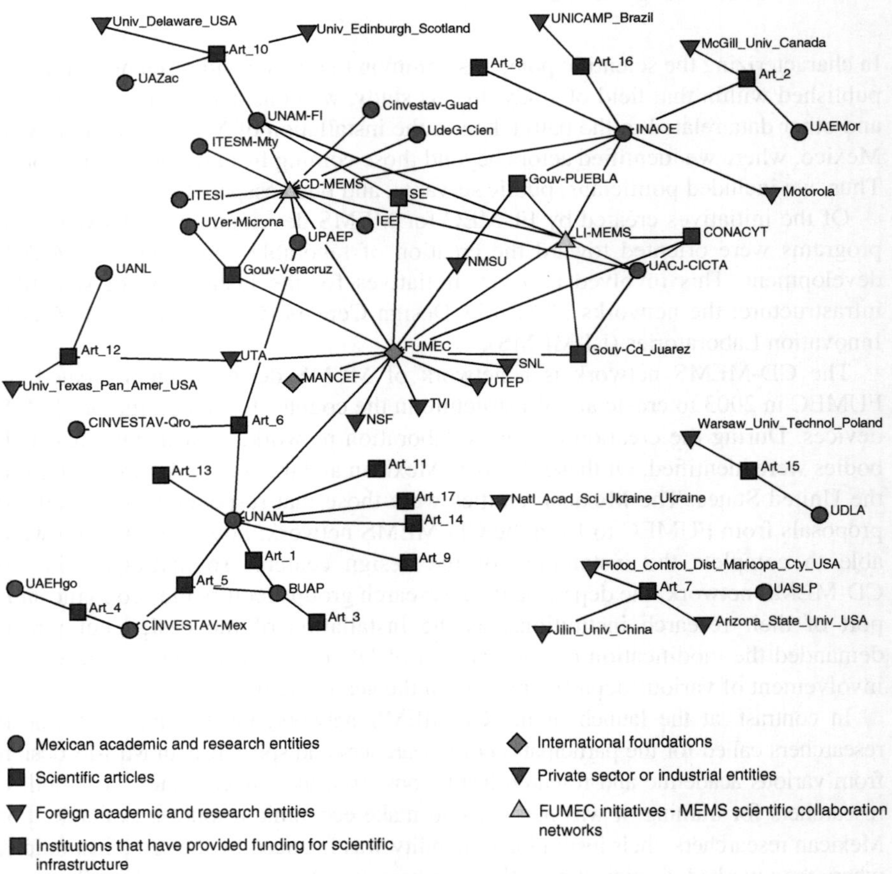

Fig. 12.1 Collaboration networks in the scientific pole in the development of MEMS technology in Mexico (Source: Author's own construction)

cles) related to the node that represents this network. The three entities where LI-MEMS were installed were also part of the CD-MEMS. These two networks are directly related in Fig. 12.1 with FUMEC (represented by the orange rhombus). The Mexican public entities that provided funding for the installation of this infrastructure are represented with a blue square. Researchers who participated in the training of Mexicans in these initiatives come from US academic and research institutions (represented by an inverted green triangle), which are directly related to FUMEC.

In the scientific pole, actors with economic interests are also present. For example, the international organization Micro, Nano and Emerging Technologies Commercialization and Education Foundation (MANCEF) is one such actor within the pole. FUMEC is an active member of this international organization and has used reports, market studies, and roadmaps, among other documents produced by MANCEF, to argue that the development of MEMS technologies represents a window of economic opportunity for Mexico. Furthermore, this international organization is considered an economic actor because one of its objectives is the commercialization of products based on micro- and nanotechnologies, among which are found in MEMS, and this is an economic goal.

A further economic actor is the Motorola Company, which donated an assembly line of integrated circuits to the INAOE under the framework of an initiative called "Latinchip," created in 1990. Motorola contributed to the creation of Mexico's technological infrastructure between 2000 and 2003, announced as a gesture of support for the technological development of Latin America (INAOE 2003; Gomez 2004). Although the assembly line of integrated circuits was already obsolete for the microelectronics industry at the time of its donation, for academic purposes, the donated equipment represented an excellent technological apparatus upon which to base activities in the manufacture of MEMS prototypes. The donation of this assembly line occurred after Motorola sold its production facility (in Zapopan, Jalisco) to the On Semiconductor company, and the factory then closed in 2002 (Jaén 2005). (Note: the donation of that equipment by Motorola also represented a tax break for the company.) Motorola also made donations to the São Paulo University and the Campinas State University in Brazil (ISTEC 1999).

Within the scientific pole, we also considered scientific publications (articles published by at least one Mexican institution). Figure 12.1 shows the publications for the period under study and the academic and research institutions involved. The scientific networks revealed here show the presence of institutions that have also pursued activities for the development of microtechnologies and that have not participated in FUMEC's MEMS program initiatives.

12.4.3 Technical Pole in the Development of MEMS Technology in Mexico

According to Callon et al. (1991), the technical pole is the space where there is room for "the conception and development of physical devices endowed with their own coherence ... and capable of performing specific services." FUMEC's microsystems program is oriented toward the development of MEMS devices. At the time of our survey, few MEMS device projects had been put into development. However, the initiatives that make up the FUMEC microsystems program are devoted to the formation of relationships between actors in the academic and productive sectors to further the development of MEMS applications.

The first of these initiatives was the creation of the MEMS Productive Articulation Center (CAP-MEMS). This initiative was created in 2004 and funded by the SE via three grants within the framework of the PYMES Fund between 2004 and 2006. We identified the following goals of CAP-MEMS: identify and support projects leading to new MEMS, identify windows of opportunity for MEMS applications in Mexico, and identify joint projects with private companies in this field.

These objectives were confirmed in a FUMEC report, stating that the objectives of CAP-MEMS included "facilitating ties between businesses, academics and decision-makers with the aim of forming collaborations that allow for the development of new products and business" (FUMEC 2006). The activities carried out under this initiative have been oriented toward fulfilling those objectives, and include forums that encompass both businesspeople and technologists in the field of microtechnology. Such forums were organized to serve the strategic sectors identified in FUMEC's feasibility studies and technology surveillance efforts. Furthermore, much planning has gone into these forums. For example, in 2004, a total of 11 forums were convened: three in the telecommunications sector, four in health (BioMEMS), two in energy, and two in MEMS packaging. Currently, CAP-MEMS is presented as, and considered to be, a spinoff of FUMEC. By examining CAP-MEMS presentations, reports on FUMEC activities, and information gathered in interviews and laboratory visits, it was possible to identify those actors who participated in the activities of this initiative to promote the development of MEMS and the collaborations between the academic and productive sectors. The resulting networks drawn from this analysis of the data are shown in Fig. 12.2.

The CAP-MEMS network initiative, shown in Fig. 12.2, is made up of four organized forums. The actors participating in these networks are heterogeneous: national research institutions (Institute for Electric Research, IIE; Monterrey Technological Institute and Higher Studies, ITESM-Mty) and one international institution (University of Texas at El Paso, UTEP), public bodies (Government of the State of Nuevo León, Gouv-NvoLeon; Government of the State of Jalisco, Gouv-Jalisco; National Institute of Public Health, INSP; Ministry of the Economy, SE; Federal Electricity Commission, CFE; Equipment and Materials Testing Laboratory, LAPEM), private bodies (National Chamber of the Electronics, Telecommunications and Information Technologies Industry, CANIETI; Delphi company), and a non-governmental organization with philanthropic origins (Mexican Health Foundation, FMS). In these networks we can see that FUMEC, via CAP-MEMS, has facilitate access to resources for the creation of a meeting space for diverse actors, at an auspicious time for the creation of new relationships around the conceptualization and development of MEMS devices.

Moreover, within the networks of the technical pole shown in Fig. 12.2, a technical–economic network is also shown, concerning the conception and development

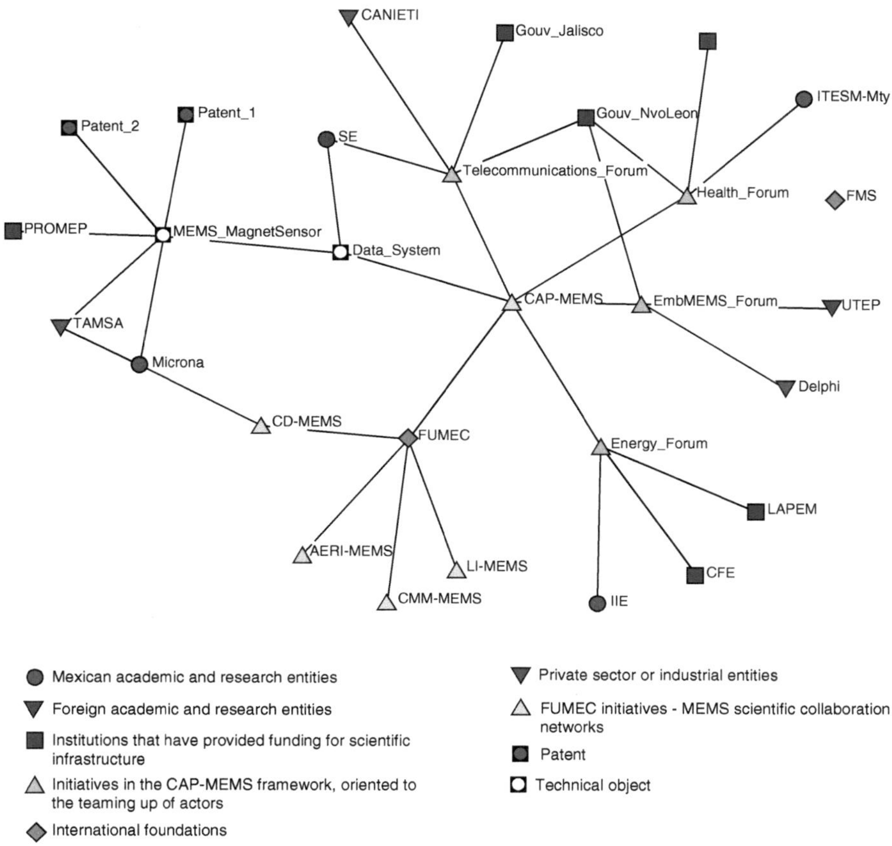

Fig. 12.2 Collaboration networks within the technical pole in the development of MEMS technology in Mexico (Source: Author's own construction)

of a MEMS device requested by a private business for a particular application. This was a project for the conception and manufacture of a magnetic microsensor and a system to monitor the magnetic field in the process of manufacturing steel tubes for the petroleum extraction industry. The device was developed at the Microna Laboratory, located in the Faculty of Engineering at the Veracruz University. After an initial meeting with the Tenaris Tamsa Company, the firm showed an interest in the capabilities of MEMS. A series of meetings between Tamsa engineers and Microna researchers were held, during which the technical requirements of the device were specified. The result has been the conception and development of a magnetic sensor (MEMS_MagnetSensor, see Fig. 12.2), based on two patents (represented in Fig. 12.2 as "Patent_1" and "Patent_2"). Later, Microna researchers developed a system for the management of data obtained in the monitoring of magnetic fields in the construction of seamless tubes for petroleum extraction, their translation and display on a monitor for plant operators. Funding to achieve the development of both technical objects (the sensor and the informatics program for data monitoring) was provided by four entities: Tamsa, SEP-Promep (Professor Improvement Program of the SEP), FUMEC and the SE. Financial support from FUMEC and the SE was provided within the framework of CAP-MEMS.

12.4.4 Market Pole in the Development of MEMS Technology in Mexico

Callon et al. (1991) define the market pole as corresponding "to the universe of the users. This is not the market of economic theory, that of the meeting of offer and demand, but rather the market of those practices that essentially describe the state of the demand." According to those authors, via the networks that exist in this pole, "information more or less legible, more or less explicit regarding the identity of the users and what they desire" is produced and circulated. In this case, the interest of this study was to situate the location of these actors in the relationships and interactions that exist between the producers of this kind of information and the potential users (or consumers) of MEMS technology.

In the technical pole, the discussion concerns the meeting space between producers of knowledge and the potential users of that knowledge, a space created in the framework of FUMEC's CAP-MEMS initiative. These spaces are concentrated in the various forums held, where scientific and technological actors expound upon their abilities and scientific and technological capabilities for the development of MEMS to potential users (actors in the productive sector), who likewise follow by expressing their needs.

Other initiatives of the FUMEC microsystems program have been oriented to create such meeting spaces between knowledge-producing actors and potential users. Such is the case of the Mexican Microsystems Consortium (MMC), created in 2007: its primary goal is the "promotion of business and research center competitiveness through the use of microsystems." The MMC is made up of a number of institutional actors oriented toward the production of knowledge in the MEMS field (21 centers and academic centers and research institutes) and various Mexican actors in the productive sector (5 private businesses). The MCC's activities go beyond microsystems technology; via feasibility studies and technological surveillance, FUMEC identified two emerging technologies very similar to MEMS: FPGA technology (Field Programmable Gate Array) and embedded systems. In Fig. 12.3, the technical–economic networks characterized in the market pole are shown, in which we see heterogeneous actors and their relationships. Thus, we can take account of other initiatives of the microsystems program that have also occurred within the market pole.

Another initiative suitable for analysis in this pole is the Strategic Alliance and Innovation Networks in MEMS (AERI-MEMS). The initiative was launched in 2008 and falls under CONACYT's AVANCE program, for the creation of alliances and collaboration networks between research institutions and companies to increase

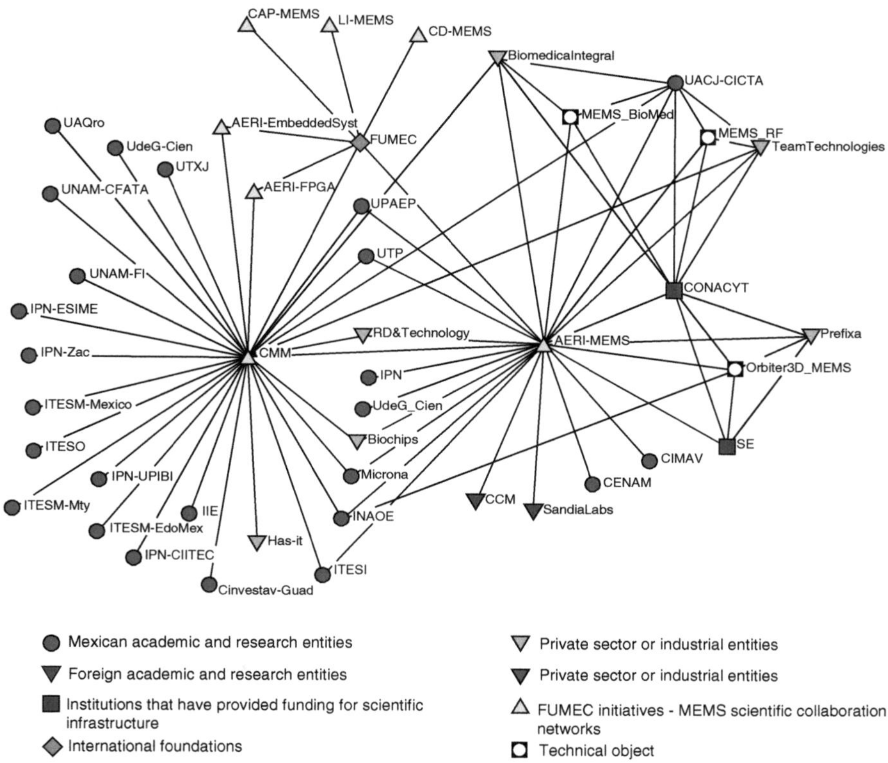

Fig. 12.3 Collaboration networks in the market pole in the development of MEMS technology in Mexico (Source: Author's own construction)

competitiveness in their sector. The proposal to create AERI-MEMS arose from the MCC and FUMEC, and includes the creation of two other AERI networks. In the AERI-MEMS network shown in Fig. 12.3, the alliance shows 10 academic and research institutions and five private companies. The initiative also includes collaboration with Sandia Laboratories (United States) and the Canadian Consortium for Microsystems (CCM). Within the AERI-MEMS framework, the development of three projects has been supported. One of these is a camera to digitize (scan) objects in 3D, a product development by Prefixa Vision Systems, a company created in 2006 by young technology graduates of the INAOE. Two years later, within the folds of that company, a new camera project was undertaken. This new camera is called the Orbiter 3D, for which Prefixa obtained the support and advice of TechBA-Silicon Valley to present their project to CONACYT's Technological Innovation Fund and the SE. The project began with this funding and succeeded in commercializing the camera in question. Later, in 2010, a third project for a camera intended to digitize MEMS devices was created and funded by CONACYT and the SE. It is also important to mention that since its creation, Prefixa has enjoyed the scientific support of INAOE researchers (who also participated in Prefixa's third project).

The third project identified is the development of an interrupter device based on RF-MEMS technology. One of the applications of this device is in the mobile phone industry. The MEMS device created here has been patented, and the funding for its development came partially from CONACYT. It was developed in collaboration with researchers from the UACJ-CICTA and a private company, Team Technologies, located in Ciudad Juárez, Chihuahua.

A further project in MEMS innovation, developed within the AERI-MEMS framework, is a biomedical system based on microtechnology to monitor respiratory fluids in babies. The project has its origins in the technology-based company Biomedical Integral, which specializes in the manufacture and commercialization of prenatal incubators. Funding for this project was provided by CONACYT, with technological development coming from UACJ-CICTA.

Figure 12.3 shows the technical–economic networks created and mobilized in the development of the MEMS devices mentioned above. In these networks, we see the technical objects (MEMS devices) and their relationships with the various actors that played a part in their development, both from supply and demand perspectives. It is worth mentioning that in these networks, government organizations have played a funding role and FUMEC has served as an intermediary between the institutions on one side and businesses and academic and research institutions on the other.

12.5 Conclusions

This chapter has presented the results of a study on scientific and technological collaboration networks in the development of MEMS technology in Mexico. The framework of analysis in this study was based on a mixed methodology employing the notion of network within two different approaches: ANT and SNA. The idea of technical–economical networks has been used with the goal of making a distinction between the different dynamics in the processes of the implementation and development of a new technology. The results have been expressed, in part, from a visual perspective (network graphics), which permitted the establishment of the location of heterogeneous actors in the different networks shown. Results were also based on a qualitative description of the technical–economical networks, which revealed certain elements to better understand the dynamics observed in the three poles analyzed. The methodological aspects have been central to this study, to succeed in obtaining these results. This has enabled us to characterize the authors and their relationships separately from the definitions of the three dynamics studied, expressed as scientific, technical, and market poles. It is worth mentioning that these results are not claimed to be exhaustive, noting that other research on the emergence of new technologies may enrich our understanding of these dynamics.

The methodology employed in this study as a tool for evaluating the implementation and development of new technologies can be applied in other cases. Performing an analysis in each pole is interesting in that it reveals the state of development of new technologies in its different levels and dimensions. The results constitute information that could serve in the strategic decisions made during the development of new science and technologies.

Furthermore, the characterization of the different collaborative network configurations across the three poles enables us to not only identify the diversity of the actors involved, but to also identify factors or elements that reveal the patterns of dynamics in the various stages of the installation and development of a new technology. As the results show, the networks identified here represent sociotechnical structures in different dynamics. However, it is beyond the scope of this research to undertake a dynamic study of these networks to evaluate their evolution over time, such as in the application of some structural measures of SNA, for example, different measures of centrality. This issue remains to be developed in a future investigation on the emergence of these new sciences and technologies.

References

Albornoz, M., & Fernández, P. E. (1996). Indicadores de ciencia y tecnología Iberoamericanos/ Interamericanos. *Revista Hispana para el Análisis de redes Sociales, 3*(7), 193–215.

Arellano, A. (2010). Mapeando las redes de investigación tecnocientífica de la Universidad Autónoma del Estado de México. In A. Arellano (Ed.), *Tramas de redes sociotécnicas: Conocimiento, técnica y sociedad en México* (pp. 221–247). México: UAEM y Porrua.

Callon, M. (1989). *La science et ses réseaux: Genèse et circulation des faits scientifiques*. Paris: La découverte.

Callon, M. (1992). Sociologie des sciences et Economie du changement technique: l'irrésistible montée des réseaux technico-économiques. In B. Latour (Ed.), *Ces réseaux que la raison ignore* (pp. 53–78). Paris: L'Harmattan.

Callon, M. (1995). Algunos elementos para una sociología de la traducción. La domesticación de las vieiras en la bahía de Saint Brieuc. In J. M. Iranzo, J. R. Blanco, T. Gonzales de la Fe, C. Torres, & A. Cotillo (Eds.), *Sociología de la ciencia y la tecnología* (pp. 259–282). Madrid: CSIC.

Callon, M., Laredo, P., & Rabeharisoa, V. (1991). Gestion des programmes publics et réseaux technico-économiques. In D. Vinck (Ed.), *Gestion de la recherche: Nouveaux problèmes, nouveaux outils* (pp. 280–307). Bruxelles: De Boeck.

FUMEC (2006). Reporte de actividades 2004–2005. http://www.fumec.org.mx. Accessed 10 Feb 2013.

Gad-el-Hak, M. (2001). The MEMS Handbook. CRC Press, p. 1368.

Gomez, M. C. (2004, January 26). Costará 18 mdd el laboratorio nacional de nanotecnológica. *La Jornada*. http://www.jornada.unam.mx/2004/01/26/037n1soc.php?origen=soc-jus.php&fly=2. Accessed 10 Feb 2013.

Grossetti, M. (2007). Reflexiones en torno a la noción de red. *REDES—Revista Hispana para el Análisis de Redes Sociales, 12*(25), 85–108.

Grossetti, M. (2009). ¿Qué es una relación social? Un conjunto de mediaciones diádicas. *REDES—Revista Hispana para el Análisis de Redes Sociales, 16*(2), 44–62.

Hou, H., Kretschmer, H., & Liu, Z. (2008). The structure of scientific collaboration networks in scietometrics. *Scientometrics, 75*(2), 189–202.

INAOE. (2003). Anuario 2003. CONACYT. http://www.conacyt.gob.mx/siicyt/index.php/centros-de-investigacion-conacyt/centros-de-investigacion-conacyt/coordinados-por-conacyt/ciencias-exactas-y-naturales/instituto-nacional-de-astrofisica-optica-electronica/anuarios-1998-2009-6/1118-anuario-inaoe-2003. Accessed 16 Feb 2013.

ISTEC. (1999). *Informe Trimestral ISTEC*. IX Asamblea General de ISTEC, 17–19 novembre, Ft. Lauderdale. www.istec.net/archive/quarterly-messages/qm1-esp.doc. Accessed 22 Sept 2010.

Jaén, J. B. (2005). El clúster de la electrónica de la región metropolitana de Guadalajara: Un análisis comparativo. *Carta Económica Regional, 17*(91), 39–50.

Kamada, T., & Kawai, S. (1989). An algorithm for drawing general undirected graphs. *Information Proceeding Letters, 31*(1), 7–15.

Kolleck, N. (2013). Social network analysis in innovation research: Using a mixed methods approach to analyze social innovations. *European Journal of Futures Research*. doi:10.1007/s40309-013-0025-2.

Meyer, M. (2000). Does science push technology? Patent citing scientific literature. *Research Policy, 29*(3), 409–434.

Moed, H. F., Glänzel, W., & Schmoch, U. (2004). *Handbook of quantitative science and technology research: The use of publication and patent statistics in studies of S&T systems*. Dordrecht: Kluwer Academic Publishers.

Morillo, F., & Aparicio, J. (2011). Características y disparidades entre sub-especialidades: Un estudio de caso con grupos de Biotecnología. *Revista Española de Documentación Científica, 34*(4), 563–580.

Mullins, N. (1972). The development of a scientific specialty: The Phage Group and the origins of molecular biology. *Minerva, 10*(1), 51–82.

OECD. (2007). *Main science and technology indicators* (Vol. 2007/1). Paris: OECD.

Porter, A. L., & Rafols, I. (2009). Is science becoming more interdisciplinary? Measuring and mapping six research fields over time. *Scientometrics, 81*(3), 719–745.

Rafols, I., Porter, A. L., & Leydesdorff, L. (2010). Science overlay maps: A new tool for research policy and library management. *Journal of the American Society for Information Science and Technology, 61*(9), 1871–1887.

Robles-Belmont, E. (2011). *Les Fondations, acteurs de l'émergence des nouvelles technologies dans les pays non hégémoniques : le cas des micro et nanotechnologies au Mexique*. Thèse de Doctorat, Université de Grenoble.

Robles-Belmont, E., & Vinck, D. (2012). Organismos filantrópicos y no gubernamentales en el desarrollo de nuevas ciencias y tecnologías: El caso de las micro y nanotecnologías en México. *Revista Tecnologia e Sociedade, 15*, 45–70.

Schmidt, S. (2009). La dificultad de medir. *REDES—Revista Hispana para el Análisis de Redes Sociales, 17*(7), 163–183.

Van Eck, N. J., & Waltman, L. (2010). Software survey: VOSviewer, a computer program for bibliometric mapping. *Scientometrics, 84*(2), 523–538.

Waltman, L., & van Eck, N. J. (2012). A new methodology for constructing a publication-level classification system of science. *Journal of the American Society for Information Science and Technology, 63*(12), 2378–2392.

Wang, F., Qiu, J., & Yu, H. (2012). Research on the cross-citation relationship of core authors in scientometrics. *Scientometrics, 91*(3), 1011–1033.

Chapter 13
Research Networks and the Relevance of Higher Education: Lessons Learned from Veterinary Sciences in Argentina, 2006–2010

Maria Fernanda Di Meglio, Luisa Mayoral, and José María Araya

13.1 Introduction

The formation of cooperation networks has been progressively incorporated into the processes of knowledge generation and the development of technologies and innovation, and is now considered an intrinsic component. Thus, research groups that support their own development via networking have a greater potential to strengthen research capabilities and to increase their institutional presence.

Within this framework, this chapter begins with a presentation on the general position of research and development (R&D) at the National University of Central Buenos Aires (UNCPBA) in Argentina. The following is a closer look at Veterinary Sciences in the context of Argentinean national universities and in comparison with other disciplines at UNCPBA. Subsequently, the characteristics of the analyzed case study are detailed, with a focus on academic and extra-academic networks in which the research groups in Veterinary Sciences at UNCPBA participate.

13.2 Establishment of Cooperation Networks and Their Impacts

Cooperation networks are increasingly common in studies related to the organization and generation of scientific knowledge (Gibbons et al. 1994), technological development, and innovation (Sábato and Botana 1968; Etzkowitz and Leydesdorff 1997).

M.F. Di Meglio (✉) • L. Mayoral • J.M. Araya
Centro de Estudios Interdisciplinarios en Problemáticas Internacionales y Locales (CEIPIL), Universidad Nacional del Centro de la Provincia de Buenos Aires (UNCPBA), Tandil, Argentina
e-mail: di_megliofernanda@hotmail.com; mayoral.luisa@gmail.com; araya@rec.unicen.edu.ar

H. Horta et al. (eds.), *Trends and Challenges in Science and Higher Education*, Knowledge Studies in Higher Education 3,
DOI 10.1007/978-3-319-20964-7_13

According to Gibbons et al. (1994), networks have the potential to contribute to the definition of problems, formation of research agendas, execution of research, and the transfer of results. By favoring flexible structures, networks can extend and diversify activities and experiment with new developments, boosting the profiles of universities and generating the supply of new material resources and competences (Gibbons et al. 1994).

Currently, networks are evolving from an instrument of cooperation to one that organizes research for the production of scientific knowledge and technology (Callon et al. 1992; Albornoz and Sebastián 1993; Albornoz and Estébanez 1998). Networks are also becoming a fundamental element in the creation and dissemination of reputations for institutions (Almuiñas Rivero and Galarza López 2010). As indicated by Muro (2009), research networks are not only restricted to connecting and reinforcing bonds between institutions and researchers with interests in the same problems. In the present context, networks also constitute public and institutional politics oriented to the strengthening and implementation of science, technology, and innovation into national systems (Muro 2009). Furthermore, Sebastián (2000) argues that the recognition of the role of national networks and its promotion is one purpose of scientific and technological politics to favor the complementarity of scientific capabilities and grant greater efficacy and efficiency in national innovation systems.

Thus, research groups that hold a competitive position are open to the outside world and maintain strong relationships with key players (García Hernández 2013), incorporating new resources, information, and knowledge, (Ahuja 2000) and establishing new ways of institutional articulation (Zarur 2008).

However, despite networks obtaining special relevance in recent times, it is not possible to find an integrated theoretical framework that allows the comparison of the various studies performed because they differ in several aspects, including their adopted approach, unit of analysis, variables used, methods employed, disciplines involved, temporal frames, and geographical contexts. Thus, it is apparent that most of these studies refer to diverse disciplines (and therefore comparisons are difficult), looking at the following perspectives in their analysis of networks: constitution (intrinsic properties), internal functioning dynamics (specifically referring to organization, collaboration processes, interactions between partners and group dynamics), and results (Sebastián 2000). Studies that analyze the establishment of networks as a broader process for the development and institutional projection of a specific group are limited.

Indeed, the results differ because of the different realities studied. Therefore, if Latin America is considered, Brazil is currently working on the generation of networks to improve undergraduate teaching (Pirróe Longo et al. 2000). However, in Mexico, higher education institutions are seen to be promoting the development of research (Muro 2009). Nevertheless, if the geographic field is moved to Europe, it is possible to observe how Finland (in what can be considered the other end of network evolution) is integrating its already existing networks into the academic and extra-academic fields (Pelkonen et al. 2008). Furthermore, several European countries are now debating how best to intervene (Láredo 2003).

Obviously, some scientific–technological systems are more complex and articulated than others, and this involves different stages of development in the conformation of networks. More specifically, in a study on the situation of R&D networks in Latin America, cooperation was found to be closely related to the degree of advance or delay of national systems of science and technology and their cooperation relationships (Royero 2005).

Finally, and despite the variability of the units of analysis, a greater involvement of the researcher or department is observed in centering the variables under study. Studies that analyze networks in terms of research groups are scarce because, according to García Hernández (2013), there is a lack of information at this level of analysis. At a national level, studies on cooperation networks are also very scarce and follow these general trends (Sebastián 2000, 2003).

13.2.1 The Network Concept and its Typology

Networks, understood as a method of association between universities, or between them and public and private sectors, are a fundamental tool for the improvement of the quality and relevance of training and research, as well as for institutional development (Sebastián 2000). Networks can be classified according to several criteria: for example, the area of knowledge in which the network operates, the type of activity it performs, the type of funding it receives, and the type of institutions involved. Accordingly, it is important to consider the distinction proposed by Casas (1991, 2001), which considers the type and variety of players participating in the configuration and objective of the network.

Thus, there are networks created with players from the scientific–technological field, such as institutions, research centers, and universities, whereas others are established with players from outside the academic system, including industries, commerce and service enterprises, usually from the private sector. Nevertheless, the possibility of articulating bonds with government cannot be excluded (Cassaigne 1997; Vacarezza 2000). In the building of networks within the scientific–technological field (which may well be called *academic networks*), research centers and universities are the key players in the training of qualified human resources and in the transfer of knowledge. Furthermore, and in agreement with their objective, those networks are predominantly based on generating joint projects for the training of human resources, continuing education and distance learning, academic exchange, the promotion of knowledge, information, methodology, ideas, and innovative approaches, the transfer of information for the academic recognition of qualifications, degrees, and diplomas, and thematic networks of cooperative research.

In contrast, *extra-academic networks* may be constituted with diverse players and their objectives may be oriented to solve joint problems and to collaborate in productive projects, counseling, and consultancies. As stated by Casas (2001), the processes of network formation with nonacademic players, the inter-institutional connection, and formal and informal collaboration are necessary in the current context of the integration of capabilities for the generation of economic and social development. It is true that the development of joint projects with nonacademic players enables a focus on the generation of knowledge, as well as the use of existing knowledge. This idea of shared interests is at the base of networking, through which it circulates and flows (Gibbons et al. 1994, Bianco and Sutz 2005).

However, there are other classification criteria. One refers to the geographic environment, making a distinction between national and international networks (Sebastián 2000; Zirene and Mejia 2011). Another criterion refers to the type of relationship, which classifies networks as formal and informal. The former are established via agreements or contracts, whereas the latter arise from tacit agreements and volunteers with no legal framework to support them (Sebastián 2000). These classifications are used in the case analysis presented below.

13.3 Methodology

This exploratory analysis was conducted via a case study based on secondary sources and from the perspective of research centers or cores (which are used as unit of analysis) to explore the presence and type of networks, as well as their objectives and main features. As an exploratory case, the intention here is to achieve a rapprochement between the theory and reality. This methodology has two stages: the first is based on secondary data and gives rise to the present study and the other (ongoing) consists of interviews with the researchers involved.

Case studies are typically used to address contemporary phenomena over which the researcher has no control. In general, regarding the stages of descriptive or explanatory research, there is some debate concerning the usefulness of the case and its scientific rigor (Stoeker 1991; Venkatraman and Grant 1986; Bowen and Wiersema 1999); however, in the case of exploratory research, there is consensus as to its usefulness (Yin 1989; Eisenhardt 1989; Chetty 1996).

The present study is a case study of four research cores at the Faculty of Veterinary Sciences at UNCPBA: the Consolidated Research Core in Veterinary Physiology and Pharmacology, the Biological Research Group, the Research Core in Veterinary Animal Production, and the Research Core in Animal Health and Preventive Medicine.

Secondary initial data were sourced from "Academic Memories,"[1] which are presented annually by research cores and groups to the Secretariat of Science, Art and Technology (SeCAT). This study covers the 5-year period of 2006–2010; within this period, 20 academic memories (4 for each research core) were surveyed.

[1] SeCAT regulations state that Cores of Scientific–Technological Activities (NACT) must submit their academic memories in which their research, teaching, and extension activities developed over a 1-year period are reported. Within these activities, bonds with academic and extra-academic cooperation are also reported.

With this material, a database was built, which was then supplemented with other documentary resources such as statistical yearbooks, reports from public institutions of higher education, and published national and international academic investigations. The analytical focus was centered on identifying existing collaborations and interactions between "research cores"[2] and various academic and extra-academic players (universities, research centers, government agencies, and private enterprises) at both international and national levels.

Thus, academic and extra-academic networks were identified and addressed in terms of participating players and the objective on which the relations of exchange and respective flows were based. Particularly, and as shown below, networks were analyzed by "research core" according to the type of player, geographical scope (national and international), and considering the purpose of the network, namely scientific collaboration, generation of joint research projects, and the training of human resources. It is important to note that because of the characteristics of the methodological design selected, this case analysis will only address institutional networks. Therefore, personal relationships between researchers and other players, as well as relations among institutions that were not reflected in agreements, endorsed projects, or formally supported collaborations, are excluded from the study.

Furthermore, it should also be noted that the results presented here represent the first stage of this research, consisting of the analysis of secondary data. As this book goes to print, interviews with researchers involved in the networks are also being conducted, to investigate the motivations for generating and participating in such networks, the networks' general and particular dynamics, and the impacts and precise results in teaching, research, and transfer to the socioeconomic environment.

13.4 A Brief Characterization of the Argentinean University System

Before addressing the case study, it is necessary to briefly describe the Argentinean university system and to mention to the contribution of the policies and programs promoted by the Ministry of Education through the Secretariat of University Politics and the Ministry of Science, Technology, and Innovation (MINCYT) to the constitution of the networks

In Argentina, the public university[3] is one of the most important institutions from the point of view of scientific–technological development. Indeed, most of the tasks and staff involved in research are concentrated in public universities (Emiliozzi

[2] The concept of NACT refers to the set of researchers and assistants who articulate, plan, and implement scientific and technological activities in a specific subject, sharing physical spaces, facilities, and technical and administrative services under guidelines (Resolution No. 2254/2003) stipulated in university regulations.

[3] The Argentinean university system has both national and private universities. National universities are considered public entities because they depend on state funds. Universities are considered

2011). The university structure is highly diversified and distributed throughout the country as a result of a process that began in the early 1970s with the creation of public universities in all provinces and in several regions; the decentralization process has been largely supported by Argentinian governments ever since.[4] Presently, the university system consists of 47 public universities, 50 private universities, seven state university institutes, 14 private university institutes, three provincial universities, one foreign university, and one international university.

13.4.1 Current Scientific–Technological Policy Guidelines

Argentinian scientific policy has more recently focused on rebuilding scientific–technological capabilities and achieving a greater willingness to establish networks, cooperative programs, and articulations among the various components of the system (Pérez Lindo 2005; Albornoz 2009).

In this context, the evolution of Argentina's scientific–technological policy seemed to follow a pattern that went from supporting the training of human resources to the underpinning of innovation, relying on the efforts of public entities in science and technology. It then went through a stage primarily aimed to reinforce the constitution of networks and articulations between the components of the scientific–technological system. Therefore, new instruments were designed to encourage research, train human resources, and strengthen bonds between research centers and production (Di Meglio and Avendaño 2012).

During this period, as noted below, there was a clear interest in scientific–technological research, such programs were promoted, and investments in R&D were increased. In this regard, there was a significant increase in government spending on R&D, with a noticeable expansion of human resources in science and technology. Within the scientific–technological structure stands the presence of a broad and highly qualified base of human resources. In terms of spending on R&D, the government promoted sustained growth from 0.41 % of the GDP in 2003 to 0.62 % in 2010 (RICYT 2010) (Fig. 13.1).

In relation to the training of human resources, the expansion and qualification of human resources is highlighted in this period. Support for graduate training and

autonomous and self-governing, as they elect their own authorities (with on government involvement) and set their own regulations and study programs.

[4]Marquis (1996) stated that considering the time in which they were created, national universities may be organized into three large groups: (i) nine "mature" universities, created in the first half of the twentieth century and with clear expectations at national and regional levels; (ii) 17 "young" universities, created in the 1970s and with a more restricted geographic scope, clearly provincial, and (iii) 10 "new" universities created in the 1990s, with a scope restricted to the municipality or local community. According to the author, the terms mature, young, and new are far from being clear-cut definitions; instead, they are groups that share a number of characteristics. Universities created in recent years should also be included in this classification.

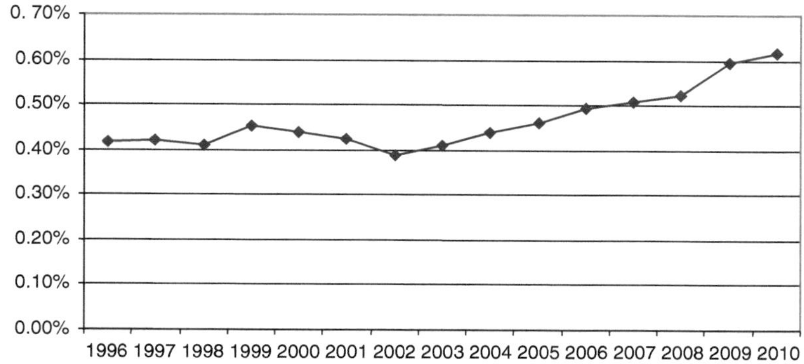

Fig. 13.1 Spending on R&D in Argentina, 1996–2010 (Source: Authors' own design based on RICYT, 2010)

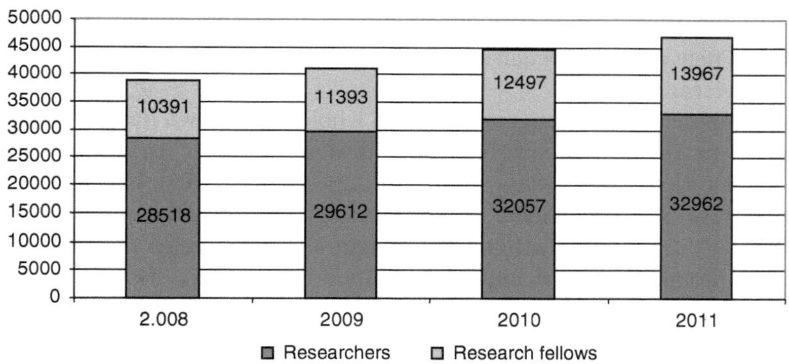

Fig. 13.2 Number of researchers and research fellows under exclusive commitment (Source: MINCYT 2012)

strategies to repatriate scientists contributed to the 98,445 employees[5] engaged in R&D activities in 2011 (Mincyt 2012). Since 2003, there has been a sharp increase in the number of researchers and research fellows, reaching 32,962 and 13,967, respectively, by 2011 (Fig. 13.2).[6]

Finally, similar growth was also recorded for the percentage of researchers within the economically active population : 2.78 in 1999, and 4.1 by 2009 (RICYT 2012).

[5] Researchers, research fellows, R&D technical staff, and R&D support staff.

[6] Exclusive commitment: implies a commitment to R&D of more than 30 h per week.

13.4.2 *New Programs for the Promotion of Networks in Argentina*

In this period (2006–2010) it is possible to identify a great variety of programs used to promote networks between national and international universities as well as relationships between research centers and production (Albornoz 2009). The Program for the Promotion of Argentinean Universities, from the Secretariat of University Politics is one such example. Since 2006, the program has engaged in annual calls for the presentation of international academic networking research projects from Argentinean universities.[7] This program aims to promote the constitution and strengthening of networks between Argentinean and foreign universities and to facilitate regional cooperation and the development and reinforcement of academic activities in an international context. Various binational programs are also highlighted, such as the Binational Program of Associated Graduate Centers of Brazil–Argentina, and the Binational Program for the Reinforcement of Argentinian–German Inter-university Networks.[8] Furthermore, there is a program for scientific–technological research in association with the Ministry of Higher Education, Science and Technology from the Republic of Slovenia, facilitating exchanges between Argentinian and foreign research groups, within the framework of a joint research project.

Furthermore, the university system, as a whole, has its own interlinked features supported by public policies that define different courses of action. One is to transfer knowledge to society and, more specifically, to link knowledge and research results to contribute to the social and productive development of Argentina. A significant particularity of regional universities is that almost all scientific production is performed by public universities. Thus, the National Agency for Promotion of Science and Technology, under the purview of MINCYT and via different instruments (e.g., the Fund for the Scientific and Technological Research and the Argentinean Technological Fund finances research projects for inter-university consortia and projects) aim to aid communication with diverse universities and enterprises.

---

[7] More than 400 projects linking Argentinian higher education institutions to 160 institutions from over 20 countries have been implemented (Pérez Lindo 2005).

[8] The Framework for Agreement between the Government of the Republic of Argentina and the Government of the Federal Republic of Germany for the constitution of an Argentinean-German University Center (CAA-DAHZ) was signed in 2010, with the following signatories: the SPU from the Ministry of Education (ME), the National Direction of International Relationships (DNRI) of the Ministry of Science, Technology and Productive Innovation (MINCYT), the German Service of Academic Exchange (DAAD), and the Argentinean-German Association of Science and Technology (ACTAA).

13.5 UNCPBA and Research Development

UNCPBA is located in the center of Buenos Aires. It is a regional institution[9] with three campuses in the cities of Tandil, Azul, and Olavarría; it was created by Law 20.753 (enacted on October 9, 1974) to consolidate the region's existing academic structures. It is one of 16 national universities established during the first half of the 1970s within the framework to decentralize the Argentinian university system. A research department at UNCPBA was established between 1977 and 1980.

The study of scientific research at UNCPBA is not possible without reference to the so-called Incentives Program (IP).[10] This central promotional instrument consists of a monetary incentive for professors–researchers[11] and is also an important mechanism for the distribution of scientific prestige and decision making (Garcia Fanelli 2011, 2005). In this regard, UNCPBA is ranked 13th (out of 40) in relation to the percentage of category I and II researchers in the IP (SPU 2009). On this point, it is important to note that the IP has five researcher categories (V, IV, III, II, and I; II and I are the highest level categories). To reach the highest categories, a distinguishable career both quantitatively and qualitatively is required. To achieve such categorization, the following factors are considered: teaching position, graduate studies, direction of research projects, outstanding teaching, scientific and transfer production, training of graduate human resources, and participation in university management bodies.

To show the true dimensions of the case study, it is important to point out that although this university can be classified as being of "medium size" in the context of public universities in Argentina, the proportion of professors–researchers categorized in the IP is larger than the national average, ranking above the largest universities of the country (e.g., the University of Buenos Aires and the National University of La Plata)

Currently, UNCPBA has a significant number of researchers categorized in the IP: 812 professors (of 1,200) have an IP categorization, with 135 (1.63 %) in categories I and II (Table 13.10). Table 13.11 shows that the IP represents 19 disciplines, with category I and II professors from the UNCPBA in 13 disciplines. The disciplines with the greatest weight in term of number of participants in categories I and II are veterinary science (21.5 %), engineering (12.6 %), physics, astronomy and geophysics (11.1 %), and anthropology, sociology, and political sciences (9.6 %); 54.8 % of all category I and II professors at UNCPBA fall within these four areas.

---

[9] According to founding Law No. 20.753, enacted by the National Congress in 1974, UNCPBA covers 19 municipalities located in the center of the province, with three categories of cities: "large towns" (from 2,000 to 20,000 inhabitants), "small cities" (20,000 to 50,000 inhabitants) and "medium-sized agglomerations" (ATI), in which Tandil, Olavarría, Azul, and Necochea are included.

[10] The Incentive Program (IP) for professors–researchers from national universities in Argentina was created by Decree No. 2427 of the National Executive Power (NEP) in November 1993 and implemented in 1994 under the ME; IP is managed by the SPU.

[11] The number of professors–researchers participating in the IP increased from 11,200 in 1994 to 30,000 in 2009 (SPU 2009).

These data highlight the importance of certain areas within the university but also within each discipline at provincial and regional levels. Within the institution, the presence of category I and II researchers in these areas marks a significant development in research. These researchers can direct projects under the IP framework; they have a solid trajectory and the necessary scientific recognition to continue promoting academic activities. In relation to their regional and provincial relevance, according to the operating mechanism of the IP, the UNCPBA academic community holds a clear position in these areas within the decision-making process; that is, these researchers are also involved in various defining aspects of the academic profession: evaluating other researchers, refereeing articles, directing projects and graduate theses, and organizing and managing research centers.

13.5.1 A Brief Description of the Position of Veterinary Sciences

Regarding the historical trajectory, it is accurate to say that the Faculty of Veterinary Sciences pioneered the scientific–technological development of UNCPBA. In fact, groups concerned with Animal Health and Preventive Medicine and Veterinary Physiology and Pharmacology of the Faculty of Veterinary Sciences were among the first six groups to be recognized as consolidated research cores in 1993,[12] complying with relevant regulatory requirements. In this process, the role of the Commission of Scientific Investigations of Buenos Aires Province (CICPBA) was fundamental via its contributions of equipment, research grants, and scholarships (Araya et al. 2006).

The UNCPBA policies regarding the structure of research cores from the 1990s aimed to avoid the dispersion and fragmentation of researchers into small groups. This was clearly applied to the Faculty of Veterinary Sciences, enjoying a similar number of active researchers at UNCPBA as Exact and Human Science faculties. This is despite the latter two having twice the number of Cores and Research Centers than Veterinary Sciences. This prevailing logic among veterinarians, beyond the particularities of the three mentioned faculties, has been a key factor that has enabled them to reach a level of development that goes beyond the internal boundaries of UNCPBA.

[12] In the early 1990s, SeCAT began registering Researchers and Research Projects based on five categories: Institutional Research Program, Consolidated Research Cores, Developing Research Cores, Small groups, and Current projects. Ordinance No. 1209/1993 (Recognition of Consolidated Research Cores) and No. 1224/1993 (Recognition of Developing Research Cores) were the fundamental documents of the period (Araya et al. 2006). From 2001, with the growth of the UNCPBA scientific system, the way in which NACT were grouped was simplified by the Secretariat of Science, Art and Technology (ex-SeCAt) into just two categories: Consolidated Cores and Developing Cores. Finally, in 2009, a new regulation differentiated NACT into Centers for Scientific–Technological Activities and Groups for Scientific–Technological Activities. Consolidated Cores become Centers, whereas those undergoing the formation process were designated as Groups.

Table 13.1 Percentage of category I and II researchers by discipline at UNCPBA

Discipline at UNCPBA	Percentage of cat. I and II researchers at the national level
Engineering	1.85 %
Anthropology, Sociology, and Political Sciences	2.33 %
Physics	4.10 %
Veterinary	17.14 %

Source: Authors' own design

Presently, Veterinary Sciences at UNCPBA comprises the Consolidated Core in Veterinary Physiology and Pharmacology (FISFARVET), the Biological Research Group (GIB), the Research Core in Veterinary Animal Production (PROANVET), and the Research Core in Animal Health and Preventive Medicine (SAMP). As of 2010, the permanent staff of the four Cores of Scientific–Technological Activities (NACT) from Veterinary Sciences totaled 204 employees, representing 21.6 % out of 944 members in the 32 NACT at UNCPBA.

Furthermore, during 2003–2010, human resources grew by 49 %. As shown in Table 13.1, this significant number of UNCPBA researchers in categories I and II placed Veterinary Sciences in a better position with respect to other disciplines. In addition, these researchers are prominent in the process of social conformation of the academic profession; their opinions regarding science, university, and academic work come into play in the evaluation process of other researchers. Table 13.1 compares the percentage of category I and II researchers by discipline at UNCPBA.

Regarding performance in comparative terms at the national level, the Faculty of Veterinary Sciences is the best positioned in the following aspect: if the National Bank of Evaluators of the SPU is considered as a parameter for Veterinary Science, it consists of 175 professors–researchers in categories I and II. Of this total, more than 50 % is concentrated in just three universities: the National University of La Plata (33 researchers), the University of Buenos Aires (32 researchers), and UNCPBA (30 researchers).

Finally, it is important to highlight some recent events and contributions to the generation and consolidation of scientific research and academic training: (1) the creation of a PhD program in Animal Science, a qualification from the top accreditation provider in Argentina, the National Council of University Evaluation and Accreditation (CONEAU) in 2008; (2) the creation of an Executor Unit[13] via a partnership between CONICET–UNCPBA, called the Center of Veterinary Research of Tandil (CIVETAN); and, (3) the opening of the School Hospital for Large Animals. The latter not only provides an unprecedented advance in the teaching of animal health but also contributes to the private sector, which offers no services of this type.

Also of relevance is the connection between the "capabilities for research" of NACT and the presence of their scientific–technological activities in the regulations, as

[13] Executor Units are centers, institutes or research units providing scientific, technological, or developmental services in various research lines.

their members sometimes contribute to the science and technology policies of their own university and those from other institutions in the sector. Consequently, Veterinary Sciences at UNCPBA is a regional and national leader in scientific and technological research.

13.5.2 Academic Networks in Veterinary Sciences at UNCPBA

This section furthers the analysis of the networks that research groups from Veterinary Sciences at UNCPBA have established with academic players at both national and international levels, focusing on the type of network in question.

13.5.2.1 International Academic Networks

In relation to academic cooperation internationally,[14] it is important to note the links with different universities and specialized institutes in areas of interest for the development of research in Veterinary Sciences (Table 13.2).[15]

The group FISFARVET has established academic cooperation networks with institutions and specialized centers, first with European countries (Italy, Slovenia, Germany, England, Ireland and Sweden) and then with North American countries (the United States and Canada). The typology of the network was oriented to scientific collaboration, the generation of joint research projects, and the training of human resources. Particularly, cooperation with Slovenia and Germany was developed within the framework of programs of scientific–technological[16] cooperation supported by MINCYT to strengthen scientific–technological capabilities in selected strategic areas in Argentina.[17] Formally, these cooperation programs fund researcher mobility, scientific exchange, and the training of human resources in the context of joint research projects (Table 13.3).[18]

---

[14] The Ministry of Science, Technology and Productive Innovation conducts a scientific and technological cooperation program with the Ministry of Higher Education, Science and Technology (MHEST) from the Republic of Slovenia, facilitating exchanges between Argentinian and Slovenian research groups within the framework of joint research projects. A joint research project refers to a program that involves two or more research groups from the two countries (Argentina and Slovenia), with clearly defined objectives to be achieved within an established timeframe, and implemented via people exchange.

[15] Cooperation between Argentina and Germany has been developed between the Ministry of Science, Technology and Productive Innovation and seven German counterparts: German Service of Academic Exchange (DAAD); Federal Ministry for German Education and Research (BMBF); Max Planck Society; Leibniz Institute; Fraunhofer Society; German Foundation for research (DFG); Argentinean-German University Center (CUAA).

[16] The SPU from the Ministry of Education, Science and Technology and the *Fundação Coordenação de Aperfeiçoamento de Pessoal de Nível Superior de Brasil*, in the framework of the Binational Program of Associated Graduate Centers of Brazil–Argentina.

[17] Regarding meat quality.

[18] Consultancy purpose: "Pre-partum food improvement and its effect on the productive and reproductive *performance* of grazing beef cows."

Table 13.2 International academic networks: FISFARVET (2006–2010)

Group	Center/university name	Type of relationship
FISFARVET	Bilateral Cooperation Project between Argentina and **Slovenia**	Joint research project and training of human resources
	Bilateral Cooperation Project between Argentina and **Germany**	Joint research project and training of human resources
	DNDi (Drugs for Neglected Diseases initiative), The Gates Foundation **USA**	Scientific collaboration
	Global Alliance for Livestock Veterinary Medicine **England**	Scientific collaboration
Grupo	Moredum Research Institute, Edinburgh **England**	Scientific collaboration
Grupo	School of Biology and Biochemistry, Queen's University of Belfast **Ireland**	Scientific collaboration
Grupo	Department of Clinical Chemistry, Faculty of Veterinary; SLU Uppsala, **Sweden**	Scientific collaboration
Grupo	Institute of Parasitology, McGill University Montreal, **Canada**	Scientific collaboration
FISFARVET	Laboratory of toxicology, Department of Animal Pathology. Università degli Studi di Torino **Italy**	Scientific Collaboration

Source: Authors' own design

PROANVET has established academic cooperation networks with institutes, centers, and universities, mostly in Latin American countries including Brazil, Mexico, Uruguay, and the Dominican Republic. Furthermore, this research group has developed a close academic relation with Spain, which involves cooperation with various specialized centers and universities[19] and diverse objectives. In this regard, the typology of the networks established with Spain includes activities such as technical support, agreements and joint research, teaching, and disclosure projects in specific areas and problems (for example, meat quality, etc.). These thematic networks enable the integration of researchers and resources for the development of specific subjects, as well as generating planned knowledge and work in solving scientific and/or technological R&D problems via the interaction of their respective contributions. In contrast, cooperation with Brazil occurred under the framework of

[19] PICT are instruments under the Funding for the Scientific and Technological Research (FONCYT) from the National Agency of Scientific and Technological Promotion, which provides grants for the funding of scientific and technological research. The presentation of projects can be performed within the following categories: I, Open topics; II, Priority areas; III, International Cooperation; IV, Start-up; and V, Internationally consolidated research groups.

Table 13.3 International academic networks: PROANVET (2006–2010)

Group	Center/university name	Type of relationship
PROANVET	Graduate Program in Zootechnics, Faculty of Agronomy, Federal University of Río Grande del Sur, Porto Alegre **Brazil**	Academic association between Graduate Studies from Argentina and Brazil
	Federal University of Río Grande del Sur, Federal University of Santa María and EMBRAPA, Bagó, **Brazil**; University of the Republic and INIA, **Uruguay**	Joint research projects
	Faculty of Veterinary, University of Zaragoza, **Spain**	Thematic network for research, teaching and disclosure
	National Agency of Research and Innovation(ANII) **Uruguay**	Consultancy
	University of Zaragoza **Spain**	Technical Support
	Andalusian Center of Ecological Apiculture. University of Córdoba **Spain**	Agreement
	CEDAF-Experimental Center for the Agricultural and Forest Development **Dominican Republic**	Agreement
	National Institute of Agricultural and Food Research and Technology (INIA) **Spain**	Agreement
	Graduate Center of the Federal University of **Mexico**	Agreement

Source: Authors' own design

the Binational Program[20] of Associated Graduate Centers of Brazil–Argentina, funded by the SPU of the Ministry of Education, Science and Technology and the *Fundação Coordenação de Aperfeiçoamento de Pessoal de Nível Superior de Brasil*, which aims at the reciprocal strengthening of training activities, encouraging the exchange of professors and graduate students (Table 13.4).

SAMP has established various academic cooperation networks with European, North American, Latin American, and Asian countries. These networks are especially oriented to scientific collaboration, the generation of joint research projects, technical support, and the generation of knowledge on specific subjects. The latter network is a prime example of a collaborative network (with the United States on *Escherichia coli*). A further example is that of Scientific and Technological Research Projects (PICT) with Canada, funded by ANPCYT (Table 13.5).

Finally, for GIB, academic cooperation networks with universities from Spain, Uruguay, and England can be observed.

[20] Collaborative network on *Escherichia coli.*

Table 13.4 International academic networks: SAMP (2006–2010)

Group	Center/university name	Type of relationship
SAMP	University of Guelph **Canada**	Joint research project (PICT)
	Institute of Biomedical Sciences University of Sao Paulo **Brazil**	NETWORK project
	University of Tennessee, Animal Department **USA**	Collaboration project
	University of Texas, Medical Branch **USA**	Thematic network for research
	Department of Clinical Chemistry, Faculty of Veterinary; SLU Uppsala, **Sweden**	Scientific collaboration
	Moredum Research Institute, Edinburgh **England**	Scientific collaboration.
	School of Biology and Biochemistry, Queen's University of Belfast **Ireland**	Scientific collaboration
	Faculty of Medicine, UNS Chile	Technical support
	Bee Research Institute of the Academy of Sciences **China**	Agreement

Source: Authors' own design

Table 13.5 International academic networks: GIB (2006–2010)

Group	Center/university name	Type of relationship
GIB	Faculty of Veterinary Sciences, Autonomous University of Barcelona **Spain**	Scientific collaboration
	Faculty of Veterinary Sciences, University of the Republic **Uruguay**	Scientific collaboration
	School of Biology and Biochemistry, The Queens University of Belfast, Northern Ireland **England**	Scientific collaboration
	Faculty of Science and Technology, University of the Basque Country **Spain**	Scientific collaboration

Source: Author's own design

Consequently, and in general terms, it can be said that Veterinary Sciences participates in numerous research joint activities among researchers, institutions, and countries, in which intellectual, economical, and physical resources for the generation and transfer of knowledge, products, and services are shared. Particularly, and considering the specificities of the areas at issue, as well as the particular trajectories of centers, the research groups analyzed have consolidated academic cooperation networks to reinforce the training of human resources via the exchange of academic to produce new knowledge from cooperation on specific subjects and to collaborate in joint research projects.

13.5.2.2 National Academic Networks

An analysis of Veterinary Sciences at UNCPBA enables the identification of network players from the national scientific–technological field, including public entities of science and technology, research institutions, universities, and other academic players.[21]

Thus, if FISFARVET is examined, the following is observed.

Table 13.6 shows that FISFARVET has developed links with players from the national scientific–technological system, including national universities, R&D institutes, and public entities involved in public health such as the Ministry of Health. Participation in the Argentinean Toxicology Network (REDARTOX) is especially of note, which was established in 1999 to: (a) improve the exchange of information; (b) contribute to the synchronization of records and regulations; (c) promote multi-center investigations, prevention and training activities and programs of analytical quality control; and (d) create virtual banks of antidote medication and laboratory standards. Currently, more than 70 specific institutions[22] or users of toxicological information participate in the network (Table 13.7).

Regarding PROANVET, academic cooperation networks with players from the national scientific–technological system are identified; these include national universities, R&D institutions, and public entities. Like FISFARVET, this group participates (together with other national institutions)[23] in the toxicology network of the Ministry of Health (Table 13.8).

SAMP has also developed a large number of cooperation networks with several players from the national scientific system, including national universities, specialized centers, and public entities. Regarding national universities, UBA and UNLP have a strong presence, followed by UNMdP. Networks were specifically guided to scientific collaboration, joint research works, technical support, and academic associations for the training of human resources. Those networks under by PICT are

Table 13.6 National academic network: FISFARVET Memories (2006–2010)

Group	Center/university name	Type of relationship
FISFARVET	National Institute of Biology (INBIAL) National University of Jujuy (UNJ)	Scientific collaboration
	Integrated Unit INTA Balcarce, Faculty of Agricultural Sciences, National University of Mar del Plata (UNMdP)	Scientific collaboration
	Ministry of Health	Laboratory participating in the network (Redartox)

Source: Authors' own design

[21] Member of the Argentinian Network of Toxicology (REDARTOX)

[22] Research Project: SECYT, UNMdP: "Utilization of natural grassland for winter feeding of beef heifers with early service."

[23] Reference Center for Lactobacilli, Tucumán, Argentina.

Table 13.7 National academic network: PROANVET Memories (2006–2010)

Group	Center/university name	Objective
PROANVET	Faculty of Agricultural Sciences UNMdP	Joint research project
	Ministry of Health	Laboratory participating in the network (Redartox)
	Integrated Unit INTA Balcarce, Faculty of Agricultural Sciences, UNMdP	Program of Scientific Collaboration
	Faculty of Agricultural Sciences, Argentinean Catholic University UCA	Participation in a project
	Faculty of Medical Sciences, National University of La Plata (UNLP)	Agreement
	National Institute of Agricultural Technology INTA	Agreement

Source: Authors' own design

Table 13.8 National academic networks: SAMP Memories (2006–2010)

Group	Center/university name	Objective
SAMP	Veterinary Diagnostic and Research Center (CIDECE-Chascomús) INTA-Balcarce INTA Mercedes-Corrientes	PICT
	Institute of Humoral Immunity Studies I(DEHU-CONICET) LELOIR Institute	PICT
	UNMdP	PICT
	Faculty of Pharmacia and Biochemistry UBA	PICT
	Microbiology Chair, Faculty of Veterinary. UBA	NETWORK Project
	Faculty of Veterinary, National University of La Plata. UNLP	Joint research project
	INTA Castelar	Joint research project
	Faculty of Veterinary UNLP	Joint research project
	INTA Balcarce	Research works and technical support
	INTA Rafaela, Faculty of Veterinary, National University of Littoral. UNL	Research work
	National University of Córdoba Faculty of Chemical Sciences. UNC	Scientific exchange
	INTA Castelar, Pathobiology Institute	Developing academic associations
	CERELA, National Council for Scientific and Technical Investigations CONICET	Developing academic associations
	Antarctic National Direction IAA	Agreement

Source: Authors' own design

Table 13.9 National academic networks: GIB Memories (2006–2010)

Group	Center/university name	Objective
GIB	Ministry of Agricultural Affairs, Buenos Aires Province	Cooperation for services and research
	Faculty of Veterinary, Equine Production. National University of Río Cuarto UNRC	Scientific collaboration
	Faculty of Chemical Sciences UNC	Scientific collaboration
	Faculty of Veterinary Sciences, Anatomy Chair UNLP	Scientific collaboration
	Faculty of Veterinary Sciences Histology Chair UBA	Scientific collaboration
	Faculty of Veterinary Sciences, Histology Chair UNRC	Scientific collaboration
	Faculty of Veterinary Sciences, Theriogenology Chair UBA	Scientific collaboration
	Faculty of Veterinary Sciences, Institute of Pathology UNLP	Scientific collaboration
	Faculty of Medicine UNC	Scientific collaboration
	Enteric Branch, Institute of Virology INTA Castelar	Scientific collaboration
	Microbiology and Parasitology Chair, Faculty of Medical Sciences UNLP	Scientific collaboration

Source: Authors' own design

funded by ANPCYT, from which recognized players of the national scientific system participate (e.g., INTA, CIDECE, Leloir Institute) (Table 13.9).

Last, GIB, has engaged in multiple academic cooperation networks with national universities, particularly, those with UBA, UNLP, and UNRC. In addition, GIB has connections with the Ministry of Agricultural Affairs at a provincial level, providing services and guidance.

Finally, the groups within Veterinary Sciences participate in various academic networks with key players within Argentina's national scientific system, especially with national and provincial entities and centers specializing in that field. Among national universities, the groups engage in academic cooperation, predominately with UBA, UNLP, and to a lesser extent UNMdP.

13.6 Extra-Academic Networks and Links with the Productive Profile of the Region

To address extra-academic networks, it is necessary to refer to the characteristics of the region in which UNCPBA exists, and more particularly, for Veterinary Sciences. In this regard, the central region of the Province of Buenos Aires shows a diversified economic structure in which agriculture and livestock remain the key areas of production. The role of the Faculty of Veterinary Sciences in its link with the producers in the sector must also be interpreted within this framework.

Many scientific–technological activities in the region relate to the provision of services and consultancy to production, mainly to the agricultural and livestock production sectors. Accordingly, research under Veterinary Sciences has existed for some time now, linked to the regional production profile.

This commitment has been made explicit in the Foundational Documents[24] of the research groups, which aim to produce technical information with regional and national interests that can be rapidly transferred to the field (FISFARVET), to conduct high-quality research on relevant topics for the industry and primary production, and to adapt technologies to solve production constraints (PROANVET).

Furthermore, the orientation of research and related activities have been conducted to respond to the changes and contextual transformations of the international field, which is characterized by, for example, an increasing global food demand, the integration of markets, development of bio-mechanization, and support and assistance for national agricultural policies. In this regard, the scientific and technological activities of the Cores of Veterinary Sciences have been mainly concentrated in topics linked to the agricultural and livestock sector: production and animal health.

These research areas address different issues facing veterinary science, and the specializations are aimed to achieve the following objectives: to optimize animal production via the search of techniques that enable the improvement of production quantity and quality (FISFARVET); to increase the efficiency and profitability of the production systems (PROANVET); to contribute to the improvement of animal and public health by means of the identification of methodologies to control and eradicate diseases (SAMP); and via the study of biological aspects of the animal medicine (GIB).

Thus, the research groups in Veterinary Sciences have established networks with extra-academic players. In this regard (and despite the restrictions at the time regarding specific economic issues in the university–enterprise relation) it is public knowledge (and is stated in academic memories) that SAMP and FISFARVET are related to enterprises largely in the pharmacy, medicine and biochemistry sectors via technological services and consultancies. More specifically, in the case of FISFARVET, technological services and consultancies have been developed in

---

[24] The Foundational Documents define the appointment of the Director and Deputy Director for each group and the mode of constitution of the Internal Council and Council of Management Control. In addition, they define topics and introduce development strategies for clustering.

Table 13.10 Percentage of professors–researchers in the IP by category

Category	Total	%
I	50	6.2
II	85	10.5
III	237	29.2
IV	212	26.1
V	212	26.1
Without commitment	16	2.0
Total	812	100.0

Source: Authors' own design

Table 13.11 Professors–researchers from UNCPBA categorized under the Incentives Program (2009)

Discipline	Researchers I and II	Total researchers in the IP
Veterinary	30	126
Engineering	17	127
Physics, astronomy and geophysics	15	38
Anthropology, sociology, and political sciences	13	90
Earth, sea and atmosphere sciences	11	24
Economy, administration and accountancy	10	68
History and geography	10	45
Education	9	98
Agronomy	7	65
Others	13	131
Total	135	812

Source: Own production

the following areas: Toxicology, Reproduction, Physiology and Pathology of Metabolism and Nutrition, Minerals, Endocrinology, and Livestock Systems.

Furthermore, it is important to note that FISFARVET, PROANVET, SAMP, and GIB are part of the Agro-Industrial Polo of UNCPBA, established in 2005, which has enabled the maintenance of the UNCPBA presence in the agricultural and livestock sectors at a local–regional level.

13.7 Conclusions

The case study in this chapter shows how Veterinary Sciences at UNCPBA has developed a scheme of diversified networks with international, national, provincial, and regional players from academic and extra-academic fields, according to the specificities of the thematic area and the particular trajectory of the relevant research centers.

Thus, the networks have had the following impacts: (a) national and international scientific presence in institutions and programs of university cooperation; and (b) presence at regional and national levels. The characteristics of this "presence" vary according to the specificities and requirements of the knowledge area at issue. Thus, for example, in some cores, the international presence is stronger than the local presence, or vice versa; for some cores, the impact is deep and specific, while the opposite may be true for others. There are groups in which the results of the bonds are easily seen because they are reflected in the daily work dynamics, while in others the greatest impacts are more obscure for those who have little awareness of work being developed. However, in all cases, and beyond these particularities, the effective presence of a research core from the Faculty of Veterinary Sciences at UNCPBA, appears linked to the generation and maintenance of an institutional network.

The claim that a set of UNCPBA research centers has achieved such a position is valid if the short timeframe is considered: most of the networks were only established (created from national, public policies) a little more than 10 years ago, certainly a too short a timeframe in international terms.

Additionally, the case at issue refers to a medium-sized public university, located in the interior of the country and away from the centers of knowledge generation, centers in which most researchers were trained to return to Argentina and take the first steps in research (maintaining the personal connections achieved abroad). Thus, it is possible to have a glimpse at the true achievement that, out of context, may seem inherent and typical to any research center.

Nevertheless, at the time of writing this article and as observed in the methodology section, this research is continuing via interviews with researchers, and from here, new research hypotheses are being generated. Thus, the issue of networking in higher education will be further developed in future research.

References

Ahuja, G. (2000). Collaboration networks, structural holes, and innovation: A longitudinal study. *Administrative Science Quarterly, 45*, 425–455.

Albornoz, M. (2009). Desarrollo y Políticas Públicas en Ciencia y Tecnología en América Latina RIPS. *Revista de Investigaciones Políticas y Sociológicas, 8*(1), 65–75.

Albornoz, M., & Estebánez, M. (1998). What do we mean by networking? Selected Latin American experiences in cooperation. In *New approaches to science and technology cooperation and capacity building* (pp. 75–112). Ginebra: UNCTAD.

Albornoz, M., & Sebastián, J. (1993). Jorge Sábato revisitado: del triángulo a las redes. *Arbor, 575*, 117–128.

Almuiñas Rivero, J.L., & Galarza López, J. (2010). Las redes académicas: una vía para fortalecer la gestión del conocimiento y la cooperación internacional universitaria. Conferencia Universidad La Habana.

Araya, J., Taborga, A., & Pérez, C. (2006). Análisis de las transformaciones en las políticas científico-tecnológicas de la Universidad Nacional del Centro (UNCPBA), Argentina 1974–2005. In *En: VI Jornadas Latinoamericanas de Estudios Sociales de la Ciencia y la Tecnología*. Bogotá: ESOCITE.

Bianco, M., & Sutz, J. (2005). Las formas colectivas de la investigación universitaria. *Revista Iberoamericana de Ciencia, Tecnología y Sociedad, 2*(6), 25–44.

Bowen, H. P., & Wiersema, M. F. (1999). Matching method to paradigm in strategy research: Limitations of cross-sectional analysis and some methodological alternatives. *Strategic Management, 20*, 625–636.

Callon, M., Laredo, P., Rabeharisoa, V., Gonard, T., & Leray, T. (1992). The management and evaluation of technological programs and the dynamics of techno-economics network. *Research Policy, 21*, 215–236.

Casas, R. (1991). Las capacidades de investigación biotecnológica en México en el área agroalimentaria: Tendencias durante la década de 1980. *Revista Sociológica, 6*(16), 54–65.

Casas, R. (2001). El enfoque de redes y flujos de conocimiento en el análisis de las relaciones entre ciencia, tecnología y sociedad. *Kairos, 5*(8), 1–17.

Cassaigne, R. (1997). La transferencia de tecnología como caso exitoso de la vinculación Universidad-Empresa. In C. Pallan & G. Ávila (Eds.), *Estrategias para el impulso de la vinculación Universidad-Empresa* (pp. 98–115). México: Asociación Nacional de Universidades e Instituciones de Educación Superior.

Chetty, S. (1996). The case study method for research in small- and medium-sized firms. *International Small Business Journal, 15*(1), 173–185.

Di Meglio, F., & Avendaño, R. (2012). Ciencia, tecnología y producción en la Argentina reciente: alcances, límites y potencialidades para la transformación posneoliberal. In F. Piñero & J. M. Araya (Eds.), *Ciencia y Tecnología en la Argentina contemporánea. Dimensiones para su análisis* (pp. 85–112). Tandil: CEIPIL-UNCPBA.

Eisenhardt, K. M. (1989). Building theories from case study research. *Academy of Management Review, 14*(4), 532–550.

Emiliozzi, S. (2011). Políticas en ciencia y tecnología, definición de áreas prioritarias y universidad en Argentina. *Revista Sociedad, 29*(30), 120–128.

Etzkowitz, H., & Leydesdorff, L. (1997). *Universities in the global knowledge economy: A triple helix of university-industry-government relations*. London: Cassell.

Garcia Fanelli, A. (2005). *Universidad, organización e incentivos: desafíos de la política de financiamientos frente a la complejidad institucional*. Buenos Aires: Miño y Dávila.

Garcia Fanelli, A. (2011). *Financiación de la educación superior argentina, Revista Educación Superior y Sociedad*. Venezuela: IESALC- UNESCO.

García Hernández, A. (2013). Las redes de colaboración científica y su efecto en la productividad. Un análisis bibliométrico. *Investigación Bibliotecológica, 27*(59), 159–175.

Gibbons, M., Limoges, C., Nowotny, H., Schwartzmann, S., Scott, P., & Trow, M. (1994). *The new production of knowledge*. London: Sage.

Láredo, P. (2003). Six major challenges facing public intervention in higher education, science, technology and innovation. *Science and Public Policy, 30*(1), 4–12.

MINCYT (2012) Indicators of Science and Technology Argentina, 2012 Edition. Secretary of Planning and Policies in Science, Technology and Innovation. www.mincyt.gob.ar/publicaciones. Accessed 27 September 2015.

Muro, P. (2009). El sentido de las redes. In S. Didou Aupetit & E. Gérard (Eds.), *Fuga de cerebros, movilidad académica, redes científicas. Perspectivas latinoamericanas*. Mexico: IESALC-CINVESTAV-IRD.

Pelkonen, A., Teravainen, T., & Waltari, S. (2008). Assessing policy coordination capacity: Higher education, science, and technology policies in Finland. *Science and Public Policy, 35*(4), 241–252.

Pérez Lindo, A. (2005). Políticas de investigación en las universidades argentinas, IESALC-UNESCO. http://eco.mdp.edu.ar/cendocu/repositorio/00105.pdf. Accessed 25 May 2014.

Pirróe Longo, W., Rocha, I., & Costa Telles, M. (2000). Reengineering' engineering research and education in Brazil: Cooperative networks and coalitions. *Science and Public Policy, 27*(1), 37–44.

RICYT (2010) Indicators of Science and Technology Ibero-American Annual Report 1996–2010. www.ricyt.org/indicadores. Accessed 27 September 2015.

RICYT (2012) Indicators of Science and Technology Ibero-American. Annual Report 1996-2012. www.ricyt.org/indicadores. Accessed 27 September 2015.

Royero, J. (2005). Las redes de investigación y desarrollo (I+D) en América Latina. *Departamento de Tecnología Administrativa, 1*(2), 1–19.

Sábato, J., & Botana, N. (1968). La ciencia y la tecnología en el desarrollo futuro de América Latina. *Revista de la Integración, 1*(3), 15–36.

Sebastián, J. (2000). Las redes de cooperación como modelo organizativo y funcional para la I+D. *Redes, 7*(15), 97–111.

Sebastián, J. (2003). *Estrategias de cooperación universitaria para la formación de investigadores en Iberoamérica.* España: Organización de Estados Iberoamericanos para la Educación, la Ciencia y la Cultura (OEI).

Stoeker, R. (1991). Evaluating and rethinking the case study. *The Sociological Review, 39*(1), 88–112.

SPU (2009) University Statistics Report, Secretary of Research and University, Statistics. www.me.gov.ar/spu/Publicaciones/publicaciones.html. Accessed 27 September 2015.

Vacarezza, L. (2000). Las estrategias de desempeño de la profesión académica. Ciencia periférica y sustentabilidad del rol de investigador universitario. *Redes, 7*(015), 15–43.

Venkatraman, N., & Grant, J. H. (1986). Construct measurement in organizational strategy research: A critique and proposal. *Academy of Management Review, 11*(1), 71–87.

Yin, R. K. (1989). *Case study research: Design and methods, applied social research methods series.* Newbury Park: Sage.

Zarur, M. (2008). Integración regional e internacionalización de la educación de la educación superior en América Latina y el Caribe. In A. Gazzola & A. Didriksson (Eds.), *Tendencias de la Educación Superior en América Latina y el Caribe* (pp. 179–240). Caracas: IESALC-UNESCO.

Zirene, J., & Mejias, A. (2011). Las redes científicas en los grupos de investigación de la Universidad Pontificia Bolivariana Sede Medellín-Colombia. *Revista Ciencias Estratégicas, 19*(26), 25–37.

Part IV
New Horizons for Scientific and Higher Education Development in Latin America

Chapter 14
Looking Forward: Building Capacity in Latin America

Manuel Heitor, Hugo Horta, and Jamil Salmi

14.1 Introduction: Research Framework and Hypothesis

In this concluding chapter, we explore ideas surrounding modern systems of technical change in Latin America, with a particular focus on new developments in emerging societies and developing regions. The basic premise of this chapter is that the central locus of innovation is becoming broader and is increasingly dependent upon linkages among many types of institutions and sources of knowledge worldwide. First, the increasingly transnational nature of business, technology, and science requires a shift from nationalistic approaches to new collaborative policy frameworks. Among these, large international collaborative arrangements play an important role. Second, although science and technology (S&T) performance sectors, namely government, industry, and academia, remain key players, the connectivity, links, and associations with other institutional players and agencies are no less significant. In particular, the increasingly relevant role played by new technology-based firms on a global scale needs to be recognized. This requires the further development of science and innovation policies, the promotion of

M. Heitor (✉)
Center for Innovation, Technology and Policy Research, IN+, Instituto Superior Técnico, University of Lisbon, Lisbon, Portugal
e-mail: mheitor@ist.utl.pt

H. Horta
Division of Policy, Administration and Social Sciences Education, Faculty of Education, The University of Hong Kong, Hong Kong, Hong Kong

Center for Innovation, Technology and Policy Research, IN+, Instituto Superior Técnico, University of Lisbon, Lisbon, Portugal

J. Salmi
University Diego Portales, Santiago, Chile

H. Horta et al. (eds.), *Trends and Challenges in Science and Higher Education*, Knowledge Studies in Higher Education 3,
DOI 10.1007/978-3-319-20964-7_14

investments in research and development (R&D), the inclusion of multiple public and private agents, and the stimulation of global research networks towards socioeconomic resilience and active learning mechanisms worldwide.

These issues are gaining greater relevance as much of the political debate worldwide is centered on long-term economic competiveness, largely under a rather "nationalistic" approach to innovation for growth (McKinsey Global Institute 2012; Mazzucato 2013). One question that does arise is whether the acceleration of knowledge investments in China and the impact of the international context in the United States and the European Union (with the notable exception of Germany) should be countered by aggressive "technonationalism" elsewhere? Any new narrative on global research networks must be based on a careful analysis of the events in recent decades. Furthermore, we should consider the seminal work of Sylvia Ostry and Dick Nelson (1995), who have directed our attention to the tension between the globalization of firms and the nationalism of governments and to the related interplay of cooperation and competition that characterizes high-tech and knowledge-based environments.

It should be noted that the Brookings Institution project described by Ostry and Nelson (1995) promoted this debate in the early 1990s, although in a different international context, clearly showing that tensions concerning deeper integration arise from three broad sources: cross-border spillovers, diminished national autonomy, and challenges to political sovereignty. As a result, the technoglobalism of the 1980s gave rise to national policies designed to help high-tech industries become more innovative and, consequently, to the emergence of technonationalism.

It is in this context that the concept of "national systems of innovation" emerged in academia, mainly via economists and related schools of thought, to explain and explore how systems have evolved differently in the major industrial nations, mainly the United States, Japan, the United Kingdom, Germany, and France. It was clear by then that the increasing international tensions and economic instability (e.g., Galbraith 2012 and Easterly 2013) were largely the result of government attempts to implement national technology and innovation policies in a world in which business and technology are increasingly transnational.

It is in this context that the chapters in this book contribute to the analysis of the challenges and opportunities to accelerate technical change in Latin America in coming years. They have been written mindful of the unique opportunities that many regions worldwide face to develop new and modern universities (Mazzucato 2013). The key role for policymakers and governments, in those regions where new investments are being made, is to select priority actions and make the correct decisions: where and how to start the process?

For the purpose of this chapter, we use international comparisons but also draw from fieldwork conducted in recent years in Latin America, as well as our own experience as researchers and policymakers in the field of technology management and innovation policy. This chapter also considers various presentations from a major event organized by some of the authors in Porto, Portugal, in October 2013, bringing together over 800 experts in science, technology, and innovation (ST&I) policies in Latin America (ALTEC 2013, see http://www.altec2013.org).

It should be noted that this book is not intended to provide any particular formula. Rather, it aims to launch a new agenda for research in technology management and innovation, based on lessons learned. We attempt to explore the dynamic relationship between economy and knowledge production, and consider the social construction of technological systems, as seminally described by Bijker et al. (1987). Following observations by Conceição and Heitor (2002) and Nowotny et al. (2003), where "science is contextualized," we foster the idea that knowledge diffusion processes, and therefore innovation, are "context sensitive" and should be pursued towards "inclusive learning." In other words, any region in the world has to learn its own way and build its own development path. However, it is also relevant to continuously adapt and improve on lessons learned from others.

From this book it is possible to draw some key implications for future S&T policies in Latin America in times of increasing uncertainty and increased globalization. We start by discussing in Sect. 14.2, the context for technical change in the 2000s and the relative positioning of Latin America from an international perspective. Section 14.3 briefly presents our conceptual framework, discussing the need to move away from the "old" paradigm of "national systems of innovation," which is unable to appropriately address emerging patterns of openness and international cooperation. We examine, in particular, the need to strengthen the main pillars of research and education, together with an industrial base for socioeconomic resilience closely articulated with all relevant stakeholders. Then, in Sect. 14.4, we focus our analysis on the main implications for science and innovation policies in Latin America, including two critical issues: the internationalization of the knowledge base and the increasingly important role of new technology-based firms. Section 14.5 looks at cultural dimensions, and addresses the critical role of assessment and evaluation practices beyond quantitative methodologies. The final section briefly discusses lessons learned and provides a brief summary.

14.2 Latin America and the Context for Technical Change and Innovation in the 2000s

Most Latin American countries are becoming aware of the opportunities for investment in their knowledge base, but they still face serious obstacles in actually being able to do so, as revealed by the low level of investment in R&D in relation to most industrialized countries. Figure 14.1 extends data published by UNESCO (2010; for 2002–2007) to the 2002–2012 period and compares world shares of GDP and GERD (gross expenditure in R&D) for the G20. It is important to note that the more dynamic economies (including the United States, Germany, and China) keep increasing their gross expenditure in R&D and, above all, are characterized by a higher share of world GERD than their share of world GDP. The most notable is China, increasing its world share of GERD from 5 % to 15 % between 2002 and 2012, surpassing its share of world GDP. In contrast, South American levels of

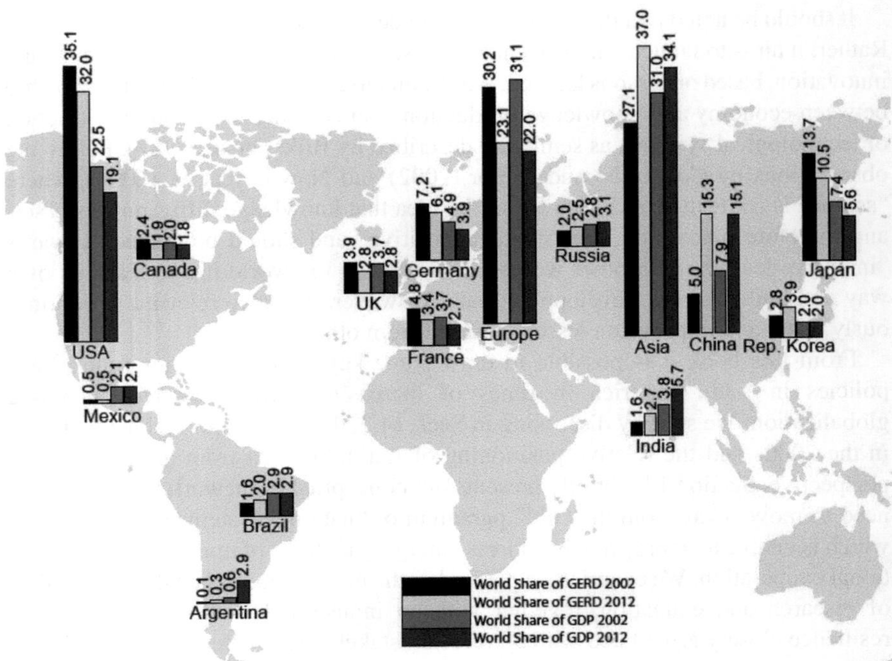

Fig. 14.1 World share of GDP and GERD for the G20 (2002–2012); values in %

R&D investment remain relatively small (Battelle 2014), albeit with some noteworthy national initiatives. For example, gross expenditure in R&D in Brazil has not surpassed 1.3 % of GDP, and in Argentina it is as low as 0.6 % (Fig. 14.2). Overall, the region lags in R&D capacity, with Brazil appearing to underperform expectations, with 1.6 fewer publications (in the Science Citation Index) by million inhabitants than Chile, and 5.5 fewer than Germany (Fig. 14.3).

Although Brazil, Argentina, Mexico, and many other Latin America countries have more than doubled their scientific capacity over the last decade, when measured in terms of publications by million inhabitants, a systematic gap in relation to industrialized countries clearly exists. This has been the case for a number of decades now and has been attributed to continually low investment levels in R&D. In terms of the cumulative gross domestic expenditure on R&D for the first decade of this century (2000–2010), Latin America shows very low levels when compared internationally. For example, the accumulation of investment in R&D in Brazil between 2000 and 2010 was half that of South Korea's. Furthermore, investment rates in Germany are three times higher than in Brazil, and China's are approximately 3.6 times higher. As suggested in the introductory chapter by Heitor and Horta in this book, these very low levels of investment in R&D do not help ensure adequate training levels for skilled employees, nor do they instill modern values to foster the desire to create, explore, and meet emerging challenges.

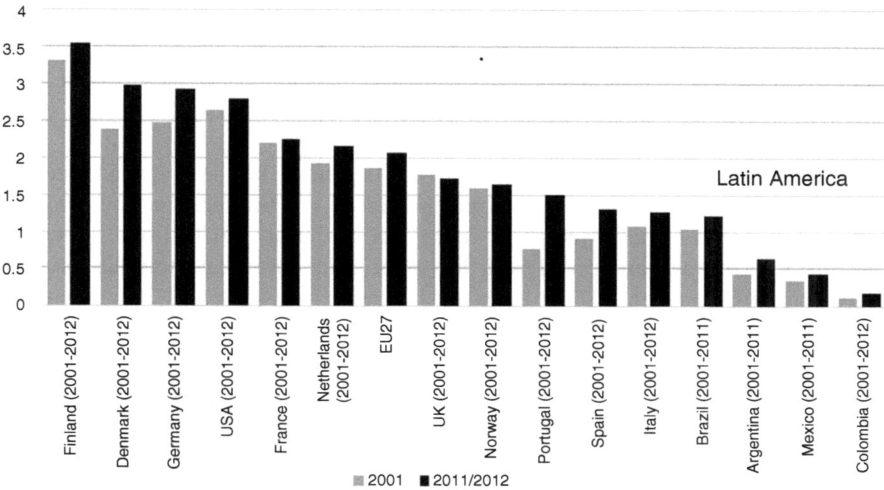

Fig. 14.2 GERD/GDP (2000–2012) for a sample of selected countries (Source: UNESCO Institute for Statistics, EuroStat)

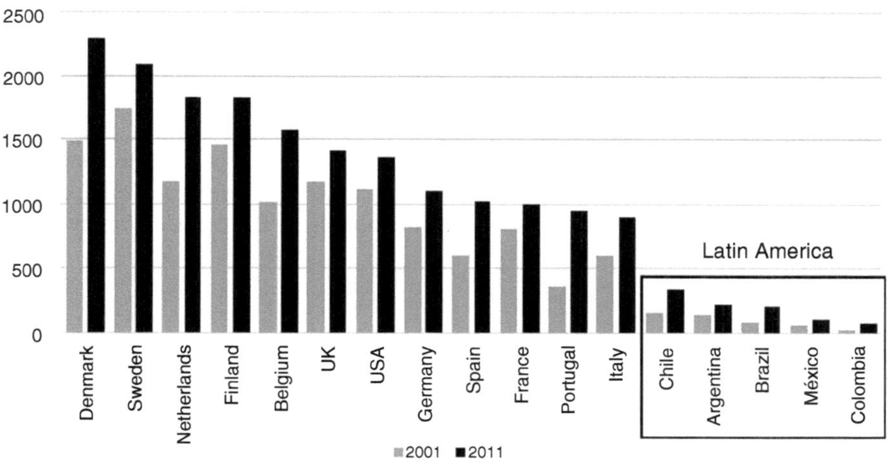

Fig. 14.3 Number of publications (in the Science Citation Index) by million inhabitants (2000–2012) for a sample of selected countries (Source: DGEEC, RICYT)

Although the broad picture taken from international comparative assessments suggests a "lost decade" for technical change in Latin America, analysis has shown that Brazil and most Latin American countries are becoming more conscious of the need and opportunity for large investments in S&T and higher education. These countries are now responding to the explosive social demand for higher education and to the vast social and political transformations already induced by new waves of educated youth (e.g., UNESCO, 2010). These investments not only seek new skills but also the certification of quality that may be expected from working with well-established academic and scientific institutions from developed countries. In addition, new research on the design of higher education at an international level and in very different socioeconomic and cultural contexts is expected to help guide the modernization of Latin America societies.

Analysis has also shown that Latin America, in general, has been characterized by low levels of integration among scientific institutions, and between them and productive and social demands (UNESCO 2010; Navarro et al. 2010). Low innovation dynamics have been associated with low-performing firms and this has implied important shifts in recent S&T policies to "market-driven" strategies. They represent efforts by many national governments to leverage traditional structures and stimulate new forms of promoting cooperation among industry, academy, and government.

This situation occurs in challenging times, considering the growth of China and other emerging economies (Castro and Castro 2012), and the current international financial situation. Furthermore, people and corporations increasingly believe in S&T, accompanied by unprecedented ways to access knowledge and diffuse consumer goods worldwide. These are also times of emerging risks, most of them associated with the concentration of populations in cities, involving complex technological systems, in many cases without the proper understanding of how to speed up and improve those processes that enable investments in R&D and human capital for economic growth and productivity gains. However, these are also times of unprecedented opportunities for S&T and for the modernization of higher education to promote new learning systems across disciplines to help build human capital and stimulate wealth creation across regions worldwide.

To assist our understanding of these and other related issues, while bearing in mind the influencing factors of emerging globalized economies and the current international situation, Sect. 14.3 presents the necessary conceptual framework to guide future policies.

14.3 A Conceptual Approach: Beyond National Systems of Innovation

The concept of "national systems of innovation" has evolved during the last three decades, first in association with the need to eliminate "market failures" and then against "system failures" (e.g., Lundvall 1992; Nelson 1993). It fueled new nationalistic policies worldwide, paradoxically at a time when business and science were becoming increasingly transnational. The end result has been a relative failure of national policies on the one hand, and a further move toward the multi-nationalization of business on the other.

This observation enables the furthering of the debate in relation to the current economic and social situation in the United States and Europe, compared with that in newly industrialized regions and in particular, Latin America and Asia. First, the myth of "national" high-tech industries and related policies to protect them require a better understanding, if analyzed in terms of increasing unemployment rates. Second, the debate on "national innovation policies" is in any case naïve. No country, even in nondemocratic regimes, has had a broad, well-coordinated innovation policy, mainly because of the complex structures associated with any "innovation ecosystem." Furthermore, national policies do not imply closing boundaries and restricting international trade, foreign investment, and international cooperation. Rather, they concentrate on improving the spillover effects of such international interactions.

Looking at the last two decades, the picture that emerges at a global level is not very different from that discussed by Ostry and Nelson (1995) in the early 1990s. In other words, it is one of the growing internationalization of private business strategies, while government innovation polices and science funding agencies remain overwhelmingly national. This presents new dilemmas for policymaking and new sources of international friction, although with new boundaries and new players. The key issue to answer concerns the implication of increasing technoglobalism for national and international innovation policies, in the present case innovation policies in Latin America countries. Furthermore, what new approaches are required to reduce international frictions and where do public policies require wider integration?

In the case of the United States, the key message that emerges from the analysis of long-term patterns of investments in S&T is the diversity of policies that led over time to increased opportunities for citizens, as well as to increased institutional specialization based on a clear separation of the roles of private and public incentives to support S&T (Conceição et al. 2004). Regarding Europe, a recent study (Aghion et al. 2011) also argues that debates on climate change, the recent financial crisis, and the new Chinese dominance of the world market represent the necessity to revisit the role and design of industrial policy. This has been used to justify the need for renewed targeted sector-based intervention of governments, namely to redirect production and innovation towards clean technologies, as well as to make industrial policy more competition-friendly and innovation-enhancing.

The literature also clearly shows that while China's capacity to innovate is evolving, it is still limited in comparison with the United States (e.g., McKinsey Global Institute, 2012). A similar comment could be made about Brazil and Latin America, suggesting that there is much scope to better develop innovation policies in a broader international context, well beyond national borders. In addition, a new paradigm of international academic, scientific, and technological cooperation has emerged, as will be discussed below.

14.3.1 Strengthening Pillars

At a time of increasing financial difficulties because of public budget constraints, there is the expectation that the links between research activity and its application in society will be reflected in more direct and immediate financial flows. However, this perception has led to a process of institutional convergence between the role of universities (and supposed role) and that of firms and other agents (Heitor 2008).

Almost two decades after Burton Clark launched the idea of "entrepreneurial universities" (Clark 1998), there is still much to learn about their impact. Some researchers have considered this convergence a potential threat to the institutional integrity of universities and the future of scientific research because of the commoditization of knowledge (Nelson 2004).

The issue is not to "save the university," but to understand who will play the fundamental and unique role that universities previously played in the cumulative system of knowledge generation and diffusion. It is clear that many elites worldwide (including in the United States and the European Union) are not willing to allow this integrity to be jeopardized. By misunderstanding national policies towards university-based research, there is a grave danger that university policy in Latin America (and elsewhere) will destroy these basic functions. This would be detrimental to the global production of knowledge, and would also certainly harm the development prospects of many countries in Latin America.

Overall, making teaching and learning more student-centered and interactive, and strengthening the role of research and university–science relationships are the ultimate goals of many leading institutions. Following such practices, skills, attitudes, and values, education at all levels should take into account that learning requires moving through discovery, invention, and production not once, but many times, in different contexts and combinations.

To achieve these objectives, we must learn from new research and foster evidence-based projects and experimental research, as well as focus our attention on the necessary transferable skills that students should acquire in the twenty-first century. We also need to reduce dropout rates in tertiary education and involve students in research activities from an early stage. In summary, we need to go beyond the present structure of tertiary education and gradually concentrate our efforts on measuring and taking stock of the diversity and evolution of specific student-centered parameters.

We also argue for a deeper examination of the complexity of stakeholder engagement and the politics of trust-building in S&T worldwide. This is essential because, beyond any single measure, it is the public understanding of science and the related level of trust in academic and scientific institutions that determine the success of science and innovation policies. It is under this condition that the systematic development and promotion of activities to foster science awareness, science education, and the role of science in the daily life of citizens has been implemented in many regions and countries with a high level of priority assigned to the innovation policy agenda.

14.3.2 Innovation and Socioeconomic Resilience

It is clear that technoglobalism and the globalization of trade and supply chains have led to the emergence of increasingly competitive global markets that facilitate access to new suppliers, independently of their geographic location (Berger 2013; Mazzucato 2013; Locke and Wellhausen 2014). This has enabled countries and regions with strong technological and industrial bases to profit from the lowering of trade barriers to access new markets, while the majority of firms located in other regions remain confined to local markets.

In addition, the analysis of the overall trend towards knowledge intensive services and their impact on job creation and economic growth requires significant pragmatism. This is because in parallel to technoglobalism came post-industrialism, reflecting the growth of services—not manufacturing industries—as the main driver of growth. Captivated by the prospects of accelerated and cost-effective economic growth, many countries, the United States included, shifted their focus from manufacturing industries to knowledge-intensive services (Hepburn 2011; Ghani and Kharas 2010).

Looking at the United States and other highly developed economies (including Germany) for comparative purposes, we can identify not only common factors but also opportunities that need to be understood in international comparative terms: strong industrial base, diversified economy, and the complexity of supply chain and knowledge networks (Amsden 2001; Hidalgo and Hausmann 2009).

Approaching this question for Latin America would require establishing a large task force for the "observation" of industrialization, to cover various aspects including:

- The geography and dynamics of economic development and specialization—how do scientific, technological, and industrial bases evolve and impact socio-economic development?
- The structure, geography and dynamics of supply chains and knowledge networks in different sectors and markets.
- Policy tools to foster local industrialization processes (e.g., public procurement, local production agreements, public expenditure in R&D and training).
- Deindustrialization processes, characterizing them, and identifying, analyzing, and governing related risks.

It should be clear that a new generation of industries will drive economic recovery over the next decade, fuelled by long-term changes in technology, society, and geopolitics. The international recession has not only been a point of change—it has also acted as a catalyst for growth. As the business landscape alters, we will see the emergence of new ways of conducting business in an increasingly interconnected world.

14.4 Stimulating Research, Economic Development, and University–Industry Relationships

It is well known that university–industry relationships are strongly linked to cultural traditions and are particularly influenced by the active role that firms play in the innovation process. This may explain why those Latin America countries characterized by recent attempts to foster knowledge production have not been able to actively turn those investments into commercial and economic results. Furthermore, most small and medium-sized enterprises (SMEs) have been excluded from this process because of a lack of qualified people and the inability to identify sophisticated technical problems and value-based propositions. In this context, our analysis focuses on two main topics: (i) managing R&D on a global scale, together with international flows of human capital, the increasing internationalization of knowledge institutions and industry–science relationships towards productivity gains and new markets (as described in this section); and (ii) the social construction of technological systems, involving the public understanding of S&T (see Sect. 14.5).

In this section we focus again on the issue of reinforcing innovation in Latin America. It has become commonplace to look at the US system as a world reference, although analysis has shown that it is of utmost importance to understand its policy diversity and mix of public and private incentives (Conceição et al. 2004). Moreover, its long history of past investments and current division of labor or specialization cannot be replicated in systems of a lower scale and complexity (Mazzucato 2013). The key elements of US academic history are those of diverse policies and increasing "institutional specialization," and of the clarification of the unique roles played by private and public incentives to support S&T (Conceição et al. 2006).

Just as the US innovation system is taken as a worldwide reference, the US university system is also used as a role model for its fast rate of response to economic change and contribution to the creation of wealth via close links to firms (NAE 2003). The view that universities are considered important engines of economic growth and development instead of mere institutions of higher education learning has been evident for many years (e.g., Saxenian 1986). There is increasing evidence of their importance as pillars of regional industrial and technological development (Cooke and Huggins 1996); this is a role that US universities, especially research universities, have assumed throughout the second half of the twentieth century (Rosenberg 2002).

Here, too, as with the entire US system, there is the perception that private funding associated with a high level of industry-science relationships is very significant and stimulates a very dynamic academia. This is considered to contribute in a much more direct way and with bigger impacts on socioeconomic development at both regional and national levels. The possibility of obtaining funding from private sources and the availability of private incentives (e.g., intellectual property rights) are very appealing for universities that strive to respond to increasing demands for change and the expectation of being more closely engaged with society. However, research suggests this to be potentially dangerous for the development of universities (Conceição et al. 2006). At a time when universities face increasing financial difficulties, derived from public budget constraints, there is the expectation that these closer links between research and application and usefulness in society will be translated into more direct and immediate financial flows (Neave 1995). This perception has led to an institutional convergence between what universities do (and are supposed to do) and that of firms and other agents. Above all, we follow Charles Vest (2007), the former president of MIT, who stated: "what is best about American higher education—we create opportunity. That is our mission. That is our business. That is first and foremost what society expects of us."

To address this issue, international academic and scientific cooperation may help create, monitor, or coach and steer collaborative research programs with industry in developing countries, their early inclusion in international networks, and the affiliation of private companies to academic and research programs (Heitor 2014).

14.4.1 Promoting Internationalization

Following our observations in previous sections on the role of international networks in the production and dissemination of knowledge, we now develop the issue of how to reinforce higher education systems by focusing on the increasingly relevant theme of internationalizing higher education and industry–science relationships. Our main hypothesis is that a new paradigm of international academic and scientific cooperation is emerging as a major influencing factor for development at an unprecedented level. It is well known that universities from developed countries are now operating internationally, not only addressing potential students individually (this was the traditional paradigm) but also increasingly addressing foreign universities and firms, local authorities and governments to develop new types of institutional arrangements (Knight 2011). These include helping to create or evaluate emerging institutions in other countries, transfer organizational skills, commercialize technology, operate training programs for teachers and researchers, contribute to higher education and research capacity abroad, and to the marketing of its benefits for economic and social progress in other societies (Altbach and Knight 2007). Such new arrangements may also include the coaching and steering of research programs in developing countries, their early inclusion in international networks, and the affiliation of private companies to academic and research programs (Heitor and Bravo 2010).

However, this new paradigm in international academic cooperation does not appear to match the usual model for exporting services. Franchising, for instance, may seem attractive in the short term but its glamour fades away under increasing academic and political criticism (Kim and Zhu 2010). It seems that a new reality is emerging, in which the export of services is intimately associated with the development of national institutional capacities, deriving their strengths from the much needed accumulation of qualified human resources and from institutional participation in, and recognition from, international academic and research networks.

Although it is well known that this internationalization process depends on the level of involvement of foreign faculty and students, analysis has also shown that it is critical that the internationalization process is aligned with the characteristics and institutional missions of the university to preserve its institutional integrity. In this context, Horta (2009) argued that research-oriented universities must focus the internationalization of the student body for postgraduate students. He also commented that in research-oriented universities, the internationalization of the faculty is strongly related to the internationalization of postgraduate students and thus, in alignment with the focus of research activities.

It is in this context that our focus is to broaden the ultimate goal of internationalizing universities in emerging economies. However, we do so in terms of university networks to foster attractive and competitive research and learning environments and to attract and train highly qualified human resources (Horta 2010). The key issue is the creation of international partnerships that can strengthen institutions and achieve the necessary critical masses to compete at an international level and, at the same time, guarantee an adequate level of institutional integrity of universities in emerging and developing regions (Marginson 2004). These networks may have an important impact on doctoral education, helping to attract students, and to train their future teaching staff in times when higher education systems in those regions are becoming increasingly relevant.

A greater knowledge of the operational advantages and shortcomings of large international research consortia and organizations will assist in better understanding the new paradigm of international partnerships in higher education will gain from. This also requires understanding the local characteristics of the technical change process, including the specific regulatory and institutional constraints that affect our knowledge of the social construction of technological systems. This new model of academic cooperation, that includes but does not seem to be hostage to the traditional forms of services and international commerce, may derive its uniqueness from the very nature of academic communities and the strong meritocratic and universalistic ideals that prevail in science on an international scale. It also draws on the flow of students and researchers, and the sense of being part of a "mission" for scientific and social development that motivates some of the best professionals in academic institutions worldwide.

Under what conditions is such a model sustainable? To answer this question, Table 14.1 summarizes the main lessons learned from the Portuguese experience in

Table 14.1 Potential guidelines to foster international research networks

Major objectives and policy instruments	Justification
People Train, attract, and co-hire researchers, fostering their exchange and the training of a teaching body	Sustain excellence and internationalization in doctoral programs Foster and systematize the hiring of researchers with PhDs
Institutions Reinforce and promote the role of scientific institutions in society, and their links with the private sector (promoting R&D in business enterprises)	Reinforce institutional evaluation mechanisms to improve systemic and organizational efficiencies Adopt policies that foster the creation of critical mass including policies oriented towards fostering R&D consortia; promote the training of a new generation of technicians and other human resources to support R&D activities
Test beds and thematic R&D networks Facilitate the integration of researchers and scientific institutions in international networks focused on "test beds" as living laboratories for the production and dissemination of knowledge with local relevance and facilitating ideas for markets worldwide	Reinforce international partnerships and foster participation in international knowledge-based networks to improve scientific quality and the employability of researchers Foster S&T thematic networks in terms of test beds and living laboratories that can boost companies' capacity to export and access emerging markets

Source: Authors' own design

establishing international research networks (Heitor and Bravo 2010). It considers three main focuses: (i) training via co-hiring young researchers and exchange programs for faculty; (ii) institutional building by promoting the role of scientific institutions in society and their links with the private sector, and adopting policies that foster the creation of critical mass, including those oriented towards promoting R&D consortia; and (iii) test beds and thematic R&D networks, facilitating the integration of researchers and scientific institutions into international thematic networks with local relevance, as living laboratories for the production and dissemination of knowledge and facilitating ideas for markets worldwide. Test beds should be assembled and integrated in international collaborative programs in such a way as to boost local companies' capacity to export and access emerging markets.

The above discussion leads to the idea that a new paradigm of technology commercialization via international academic and scientific cooperation is emerging at an unprecedented level. We refer to the capacity to turn science-based inventions into commercially viable innovations and related new potential factors of progress on a global scale, in association with growing evidence of the potential benefits resulting from economic appropriation of the results and methods of science by society. Sustained growth in emerging and developing regions can only occur via the continuous introduction of truly new goods and services, namely in the form of radical technological innovations that disrupt markets and create new industries.

At this stage it should be remembered that the accumulation of knowledge by skilled people and institutions in the area of technology-based entrepreneurship requires a specific learning process that occurs with the creation of the necessary critical mass in the research community, and must be oriented to external and emerging markets worldwide. Creating local knowledge-intensive communities—associated with local and specific institutional and university contexts—that are able to operate in global and sophisticated markets requires organized networks fostering new competencies in international technology commercialization and diffusion. In other words, we underline the challenges associated with implementing country- and regional-wide "university technology enterprise networks" to stimulate competencies in a way that fosters access to emerging markets worldwide for technology-based start-ups.

14.4.2 Engaging New Technology-Based Firms

Technology-based entrepreneurship is increasingly seen as a key element of regional competitiveness, taken as "the model" for many other regions and countries worldwide. Silicon Valley in California and Route 128 in the Boston–Cambridge area, the most dynamic regions in the world today in terms of growth and innovation, were largely propelled by new technology and the creation of startups, Apple, HP, Google, and Intel, to name a few. At the same time, start-up companies are also becoming global enterprises, engaging in services, manufacturing, and research throughout the world, with strong links to universities and research groups. Others are going beyond their borders to procure products and services at lower prices, often from new companies or subsidiaries in emerging economies like China, India, and Brazil. Well-trained engineers and computer scientists from Bangalore and Shanghai are successfully competing for jobs that traditionally went to their counterparts in Europe and the United States.

At the same time, universities in Latin America (and worldwide) are attempting to "emulate" their US counterparts and foster a range of technology transfer offices and commercialization activities together with industrial liaison programs, mostly intended to foster entrepreneurial environments and the launch of technology-based start-ups. The large number of technology parks established throughout Latin America over the last 15 years is clear evidence that the concept has become widespread; their key aim is to bring ideas to the market.

Beyond the concentration of people and skills in a number of regions, a key issue that has differentiated North America from many other countries and regions is the availability of a mix of public and private funding sources available to SMEs. In this context, some countries have tried to emulate the Small Business Innovation Research program (SBIR), which remains unique in many of its characteristics. Although SBIR has faced many funding difficulties as a result of the financial crisis, its enormous success and impact should be further acknowledged. This is a program of the upmost importance and relevance that has helped innovative firms in the United States grow. In addition, many other schemes to fund and support new technology-based firms have been used in the United States in original ways, namely via public procurement by the defense and energy departments.

14.5 Promoting a Science and Technology Culture

Our second level of analysis concerns the need to strengthen external societal links as a critical step in fostering the role of ST&I and meeting the needs of global competition and the knowledge economy. This issue has been discussed in the United States and Europe over the last five decades, either in terms of renewing science education, or creating a science culture; here we reinforce this argument with a specific focus on Latin America.

As mentioned in Chap. 1 of this book by Heitor and Horta, there is a need to foster the public's understanding of science, and to better explain to society the role of scientific and technical development. The continued implementation of actions fostering "science for all," as well as that of "thinkering" (e.g., Michalko 2011), are essential. Furthermore, the concept of "knowledge-integrated communities" appears particularly suitable to facilitate the joint participation of researchers, universities, and basic and secondary schools in specific projects serving society at large. It is clear that this requires new knowledge about social behaviors, as well as new methodological developments to help transform emerging regions worldwide into knowledge societies. The objective is to integrate systems of knowledge and ways of applying S&T, where schools interact cooperatively with universities in systematic ways.

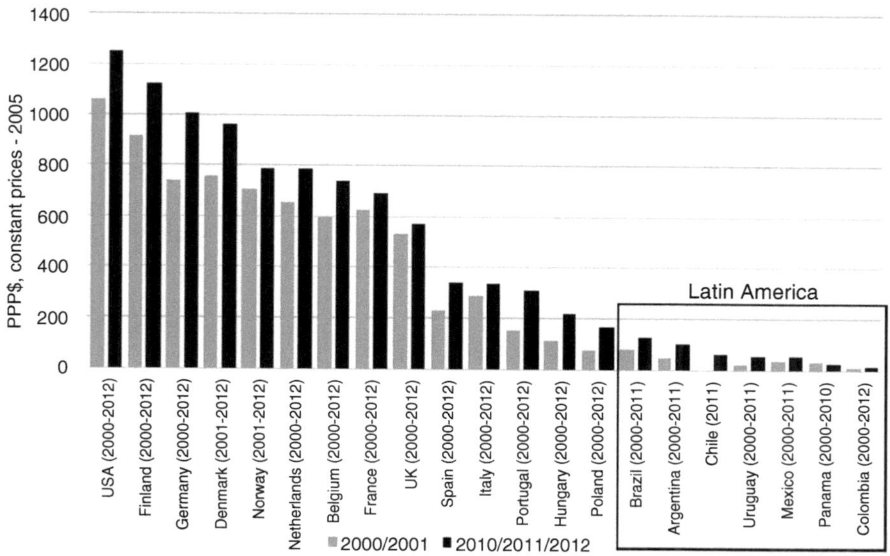

Fig. 14.4 GERD per capita (in PPP$, constant 2005 prices) (2000–2012) (Source: UNESCO Institute for Statistics)

Figure 14.4 complements the analysis of the previous section and shows that the gross expenditure on R&D per capita in Latin America countries is still well below acceptable international levels. In addition, the increase in R&D expenditure has not matched that of the industrial world over the last decade. For example, after correcting for parity and at constant 2005 prices, the annual per capita expenditure in R&D in Brazil is approximately eight times smaller than that in Germany and one-tenth of that in North America. Although Brazil saw a 50 % increase during 2000–2011, the corresponding increase in Germany was 67 % and approximately 37 % in North America. These figures clearly show a significant gap concerning Brazil's (and, in general, Latin America) attitude towards prioritizing investment in R&D.

Analysis has also shown that the traditional public policy instruments in many Latin America regions have been very ineffective to promote innovation in the productive sectors for the following reasons. (1) The policies do not consider the firm as the focal point of the innovation process. (2) There is low continuity and high oscillation over time. (3) The policies are extremely bureaucratic to access and use. (4) They are not appealing enough to attract firms and modify their behavior regarding innovation and industry–science relationships.

Addressing societal links as a critical step in fostering the role of ST&I, and meeting the needs of the knowledge economy would help strengthen the process and establish a solid framework to foster "beliefs in science." Some 45 years after John Ziman launched the discussion on "public knowledge" (Ziman 1968) and 35 years after his work on "reliable knowledge" (Ziman, 1978), few countries appreciate the significance of scientific knowledge in understanding the nature of science as a complex whole. In *Real Science* (Ziman 2000), we are reminded that, "science is social," referring to "the whole network of social and epistemic practices where scientific beliefs actually emerge and are sustained."

Our goals are to seek the overhaul and expansion of the social basis for S&T development in Latin America on the basis of effective assessment and evaluation practices. This requires a strong conviction not only from S&T professions and public and private research organizations, but also from students and the general public. Thus, the growing appropriation of a S&T culture by society is a central aspect of the argument presented in this chapter.

14.5.1 Contribution of Evaluation Practices Beyond Quantitative Methodologies

In identifying lessons learned over the last decades regarding the development of ST&I capacity, we must also refer to assessment practices at two distinct but related levels: first, research evaluation practices set as a system independent from education and teaching; second, the implementation of an independent institutional accreditation and assessment system of teaching programs and higher education institutions based on best international practices.

Restructuring and strengthening the network of research centers in Latin America, including those in universities and related nonprofit institutions, requires the implementation of a systematic research assessment exercise of a periodic nature (e.g., every 3 years) and with direct impact on funding levels. This has been widely established in industrialized countries, and we argue that there is no other way to foster research capacities in emerging regions worldwide. Furthermore, it should involve external experts and/or internationally relevant and independent institutions to guarantee complete objectivity. In addition, accreditation and assessment systems of teaching programs already in place in Latin America should be able to operate in a flexible and independent manner. They should be responsible for the assessment and accreditation of universities and their study programs, and foster the integration of local quality assurance into international systems of higher education.

In this process, it is important to note that the assessment of knowledge production results should not be solely based on quantitative methodologies. This is because although quantitative evaluative methods may complement peer-review practices, overcoming the recognized criticisms and pointing out the faults of "pure" peer-review assessments (Relman 1990), the "extreme" quantification of the academic activity may bring forth a culture of "bureaucratization of knowledge," which is to be avoided. Furthermore, although research evaluation is a well-consolidated assessment, it is still difficult to measure and evaluate teaching performance (Dixit 1998).

While certain authors have defended the robustness of teaching evaluation processes such as student ratings and student engagement surveys (Marsh 2007), faculty and higher education administrators have expressed doubts about their meaningfulness and suitability (Gilliot 2001). The same problem arises when evaluating faculty work as a whole, or universities as a whole, where institutional evaluation assessments are often flawed as they rely too heavily on either subjective aspects or an excess of quantitative features (Schloegl et al. 2003). In other words, the robustness of performance indicators for research, if complemented by more qualitative procedures, provides a sound basis for making judgments.

14.6 Discussion and Summary

The rapid development of S&T at a global level, particularly in many "transition economies" and developing regions, and the growth in higher education worldwide demand a better understanding and definition of inclusive ST&I policy actions in Latin America. However, developing knowledge-based practices in Latin America requires a greater focus on the myriad of public policies and institutional strategies. This will ultimately open the "Black Box" associated with the creation of a new set of relationships with society at large and introducing an "intelligent accountability" associated with a renewed structure of incentives.

Two main emerging issues have driven the rationale for this book. First is the recent explosion in demand for higher education by millions of young people around the world (Altbach et al. 2009), related to the growing perception of the potential benefits of the economic appropriation of scientific results by society. This is associated with the changing perception of the "academic–scientific divide." Many developing regions and countries, including Latin America, now recognize the importance of large investments in S&T and higher education (public and private), aiming to respond to the growing social demand for higher education and the vast social and political transformations already induced by new waves of educated youths (Roberts and Hite 2007). This is illustrated by the recent investments in hundreds of new public and private campuses in Northern Brazil, Ecuador, and Colombia (among many others in, for example, China, Russia, South Africa, Rwanda, and Turkey). These investments not only seek to foster new skills and knowledge, but also the certification of quality that may be expected from working in partnership with well-established academic and scientific institutions from developed countries. Such institutional arrangements provide new forms of expansion, as they tend to help secure new financial and human resources, and challenge their own traditional competences and agendas. The question that arises is whether these investments are enough to foster economic development in the short term.

Figure 14.5 quantifies the cumulative gross domestic expenditure in R&D for the first decade of this century (2000–2010) for a sample of countries; it shows very low levels for Latin America, when compared internationally. The figure considers a logarithmic scale to facilitate potential comparisons and reveals, for example, that the accumulation of investment in R&D in Brazil is similar to that of the Netherlands and Italy, three times smaller than that of Germany, and approximately 20 times smaller than that of the United States. We argue that these relatively low levels of

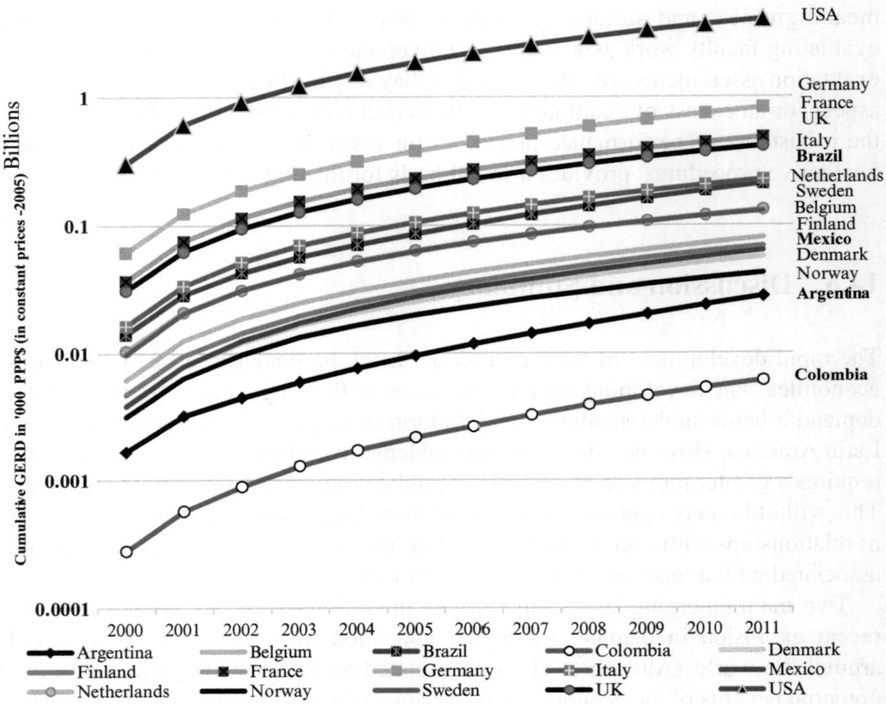

Fig. 14.5 Cumulative GERD (2000–2011), in logarithmic scale (base 10) and in '000' PPP$ (in constant 2005 prices) (Source: UNESCO Institute for Statistics)

investment in R&D are not sufficient to adequately train the skilled people that are essential to Brazil's development.

Second, the relatively low level of sophistication of firms in Latin America and the low per capita income that still characterizes the region are associated with very low technology intensity and R&D investments. Figure 14.6 compares the share of gross expenditures in R&D, GERD, by sector of performance for a sample of countries in 2011 and identifies the critically low level of business R&D in most Latin American countries and regions.

This is important because technoglobalism and the globalization of trade and supply chains have led to the emergence of increasingly competitive global markets with easier access to new suppliers, independent of their geographic location (Ostry and Nelson 1995; Mazzucato 2013). This has enabled countries and regions with strong technological and industrial bases to benefit from the lowering of trade barriers to access new markets, while the majority of firms located in other regions remain confined to local markets.

The analysis of the two effects described above aims to foster a broader research framework to approach industrial innovation and technology management in Latin America. That research should look at how industry is distributed, how production is conducted across countries and regions, and how capabilities are used in different

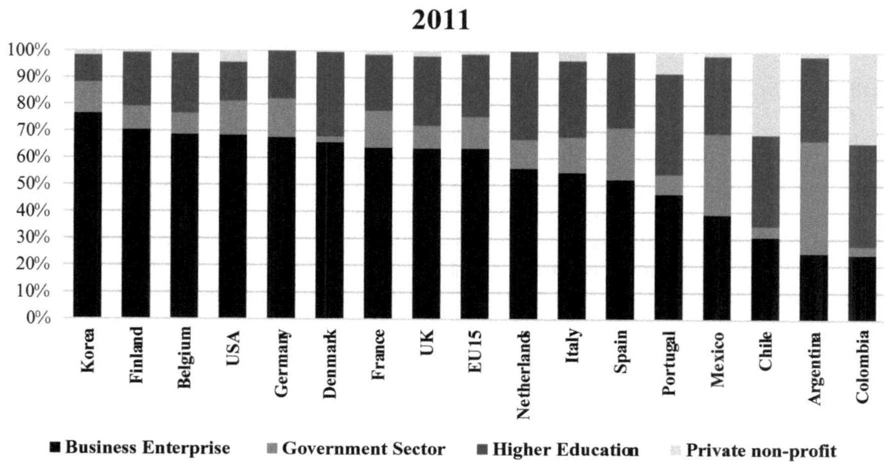

Fig. 14.6 GERD by sector of performance in 2011 (Source: RICYT, OECD)

countries and regions. In fact, this framework is being used in several research programs in centers and universities around the world, including the "Production in the Innovation Economy" project at the MIT (Berger 2013; Locke and Wellhausen 2014), and has led to a significant body of research on industrialization in Brazil, looking namely at the role of regulation and local content policies (Apolinário e Silva 2011; Cassiolato et al. 2008; Salles-Filho et al. 2010).

We should note that by attempting to analyze both the supply side of technical change (via the identification of human capacity and skills) and the demand side (via the examination of firm competencies and knowledge intensity), we aim to promote a dialogue oriented towards industry–science relationships and the building of a modern labor force in Latin America. This requires strengthening experimentation in social networks, which necessarily involves flows of people. It is the organized cooperation among networks of knowledge workers, together with different arrays of users that will help to diffuse innovation and the design of new products and services (Ernst and Kim 2002). However, establishing these innovation communities implies the systematic development of new patterns of collaboration on the basis of formal education programs, sophisticated research projects, and a diversified and unstructured array of informal processes of networking (e.g., Saxenian 2006; Tung 2008). We argue that this requires a new paradigm of public policies to foster international cooperation in S&T and the economy, bringing together scientific institutions and firms worldwide.

To cope with such a variety of demands and with a continuously changing environment, higher education systems need to be diversified. The challenge of establishing modern higher education systems requires effective international networks and a platform of research institutions. Thus, these will stimulate political debate among various stakeholders and assist in the networking of national constituencies to promote the positioning of Latin American and Caribbean institutions in the emerging pathways of "brain circulation" worldwide.

References

Aghion, P., Boulanger, J., & Cohen, E. (2011, June). Rethinking industrial policy. *Bruegel Policy Brief.*

Altbach, P. G., & Knight, J. (2007). The internationalization of higher education: Motivations and realities. *Journal of Studies in International Education, 11*(3–4), 290–305.

Altbach, P., Reisberg, L., & Rumbley, L. E. (2009). *Trends in global higher education: Tracking an academic revolution*. Paris: UNESCO.

Amsden, A. H. (2001). *The rise of "the rest"—challenges to the west from late-industrializing economies*. Oxford University Press: Oxford.

Apolinário, V., & Silva, M. L. (Eds.). (2011). *Impacto dos grandes projectos federais sobre os estados do nordeste*. Natal: EDUFRN.

Battelle. (2014). Global R&D funding forecast, Dec 2013. http:www.battelle.org/docs/tpp/2014_global_rd_funding_forecast.pdf

Berger, S. (2013). *Production in the innovation economy*. Cambridge: MIT Press.

Bijker, W. E., Hughes, T. P., & Pinch, T. (1987). *The social construction of technological systems*. Cambridge: MIT Press.

Cassiolato, J. E., Matos, M. P., & Lastres, H. M. M. (Eds.). (2008). *Arranjos produtivos locais—uma alternativa para o desenvolvimento. Experiências de política* (Vol. 2). Rio de Janeiro: E-chapters.

Castro, C., & Castro, L. B. (2012). *Do pensamento renegado ao desafio sinocêntrico—reflexões de antónio barros de castro sobre o Brasil*. Rio de Janeiro: Elsevier.

Clark, B. R. (1998). *Creating entrepreneurial university: Organizational pathways of transformation*. Oxford: Pergamon Press.

Conceição, P., & Heitor, M. V. (2002). Knowledge interaction towards inclusive learning: Promoting systems of innovation and competence building. *Technological Forecasting and Social Change, 69*(7), 641–651.

Conceição, P., Heitor, M. V., Sirilli, G., & Wilson, R. (2004). The swing of the pendulum from public to market support for science and technology: Is the US leading the way? *Technological Forecasting and Social Change, 71*(5), 553–578.

Conceição, P., Heitor, M. V., & Horta, H. (2006). R&D funding in US universities: From public to private support or public policies strengthening diversification? In J. Enders & B. Jongbloed (Eds.), *Public-private dynamics in higher education: Expectations, developments and outcomes* (p. 328). Piscataway: Distributed by Transaction Publishers.

Cooke, P., & Huggins, R. (1996). *University–industry relations in Wales. Working chapter* (Center for advanced studies in the social sciences). Wales: UWCC.

Dixit, A. K. (1998). *The making of economic policy: A transition-cost politics perspective. Munich lectures in economics*. Cambridge: MIT Press.

Easterly, W. (2013). *The tyranny of experts: Economists, dictators and the forgotten rights of the poor*. New York: Basic Books.

Ernst, D., & Kim, L. (2002). Global production networks, knowledge diffusion, and local capability formation. *Research Policy, 31*(8–9), 1417–1429.

Galbraith, J. (2012). *Inequality and instability: A study of the world economy just before the great crisis*. Oxford: Oxford University Press.

Ghani, E., & Kharas, H. (2010, May). *The service revolution*. Economic Premise, The World Bank.

Gilliot, D. (2001). Incentives in academia. In M. Dewatripont, F. Thys-Clemenat, & L. Wilkin (Eds.), *The strategic analysis of universities: Microeconomic and management perspectives* (pp. 57–71). Brussels: Editions de L' Université de Bruxelles.

Heitor, M. V. (2008). A system approach to tertiary education institutions: Towards knowledge networks and enhanced societal trust. *Science and Public Policy, 35*(8), 607–617.

Heitor, M. V. (2014). How far university global partnerships may facilitate a new era of international affairs and foster political and economic relations? *Technological Forecasting and Social Change*, accepted for publication.

Heitor, M. V., & Bravo, M. (2010). Portugal on the crosstalk of change, facing the shock of the new: People, knowledge and ideas fostering the social fabric to facilitate the concentration of knowledge integrated communities. *Technological Forecasting and Social Change, 77*, 218–247.

Hepburn, D. (2011, July). *Mapping the world's changing industrial landscape* (The world's industrial transformation series, IE WIT BP 2011/01). Chatham House.

Hidalgo, C. A., & Hausmann, R. (2009). The building blocks of economic complexity. *Proceedings of the National Academy of Sciences of the United States of America, 106*(26), 10570–10575.

Horta, H. (2009). Global and national prominent universities: Internationalization, competitiveness and the role of the state. *Higher Education, 58*(3), 387–405.

Horta, H. (2010). The role of the state in the internationalization of universities in catching-up countries: An analysis of the Portuguese higher education system. *Higher Education Policy, 23*, 63–81.

Kim, H. E., & Zhu, M. (2010). Universities as firms: The case of US overseas programs. In C. T. Clotfelter (Ed.), *American universities in a global market* (pp. 163–201). Chicago: Chicago University Press.

Knight, J. (2011). Education hubs: A fad, a brand, an innovation? *Journal of Studies in International Education, 15*(3), 221–240.

Locke, R. M., & Wellhausen, R. L. (2014). *Production in the innovation economy*. Cambridge: MIT Press.

Lundvall, B.-Å. (1992). *National systems of innovation: Towards a theory of innovation and interactive learning*. London: Pinter Publishers.

Marginson, S. (2004). Competition and markets in higher education: A 'glonacal' analysis. *Policy Futures in Education, 2*(2), 175–244.

Marsh, H. W. (2007). Students' evaluations of university teaching: A multidimensional perspective. In R. P. Perry & J. C. Smart (Eds.), *The scholarship of teaching and learning in higher education: An evidence based perspective* (pp. 319–384). New York: Springer.

Mazzucato, M. (2013). *The entrepreneurial state—debunking public vs. Private sector myths*. London: Anthem Press.

McKinsey Global Institute. (2012). *Manufacturing the future: The next era of global growth and innovation*.

Michalko, M. (2011). *Creative thinkering—Putting your imagination to work*. New York: New York Library.

NAE. (2003). *The impact of academic research on industrial performance*. Washington: NAE Press.

Navarro, J. C., Llisterri, J. J., & Zuniga, P. (2010). The importance of ideas: Innovation and productivity in Latin America. In C. Pages-Serra (Ed.), *The age of productivity: Transforming economies from the bottom up*. Washington: Development in the Americas/Inter-American Development Bank/Palgrave-McMillan.

Neave, G. (1995). The stirring of the prince and the silence of the lambs: The changing assumptions beneath higher education policy, reform and society. In D. D. Dill & B. Sporn (Eds.), *Emerging patterns of social demand and university reform: Through a glass darkly* (pp. 54–71). Oxford: Pergamon.

Nelson, R. R. (Ed.). (1993). *National innovation systems: A comparative analysis*. New York/Oxford: Oxford University Press.

Nelson, R. R. (2004). The market economy, and the scientific commons. *Research Policy, 33*, 455–471.

Nowotny, H., Scott, P., & Gibbons, M. (2003). Mode 2' revisited: The new production of knowledge. *Minerva, 41*, 179–194.

Ostry, S., & Nelson, R. (1995). *Techno-nationalism and techno-globalism: Conflict and cooperation*. Washington: The Brookings Institution.

Relman, A. S. (1990). Peer review in scientific journals: What good is it? *Western Journal of Medicine, 153*, 520–522.

Roberts, J. T., & Hite, A. B. (2007). *The globalization and development reader: Perspectives on development and global change*. Malden: Blackwell.

Rosenberg, N. (2002). Knowledge and innovation for economic development: Should universities be economic institutions? In P. Conceição, D. V. Gibson, M. V. Heitor, G. Sirilli, & F. M. Veloso (Eds.), *Knowledge for inclusive development* (pp. 34–47). Westport: Quorum.

Salles-Filho, S., Avila, F. D., Sepulveda, J., & Colugnati, F. A. B. (2010). Multidimensional assessment of technology and innovation programs: The impact evaluation of INCAGRO-Peru. *Research Evaluation, 19*, 1.

Saxenian, A. (1986). *Regional advantage: Culture and competition in Silicon Valley and Route 128*. Cambridge: Harvard University Press.

Schloegl, C., Gorraiz, J., Bart, C., & Bargmann, M. (2003). Evaluating two Australian university departments: Lessons learned. *Scientometrics, 56*(3), 287–299.

Tung, R. L. (2008). Brain circulation, Diaspora, and international competitiveness. *European Management Journal, 26*, 298–304.

UNESCO. (2010). *Science report 2010*. Paris: UNESCO.

Vest, C. M. (2007). *The American research university—from world war II to worldwide web: Governments, the private sector and the emerging meta-university*. Berkeley: University of California Press.

Ziman, J. (1968). *Public knowledge: The social dimension of science*. Cambridge: Cambridge University Press.

Ziman, J. (1978). *Reliable knowledge: An exploration of the grounds for belief in science*. Cambridge: Cambridge University Press.

Ziman, J. (2000). *Real science: What it is, and what it means*. Cambridge: Cambridge University Press.

MIX
Papier aus verantwortungsvollen Quellen
Paper from responsible sources
FSC® C105338

If you have any concerns about our products,
you can contact us on
ProductSafety@springernature.com

In case Publisher is established outside the EU,
the EU authorized representative is:
Springer Nature Customer Service Center GmbH
Europaplatz 3, 69115 Heidelberg, Germany

Printed by Libri Plureos GmbH
in Hamburg, Germany